Association for Women in Mathematics Series

Volume 21

Series Editor
Kristin Lauter
Microsoft Research
Redmond, Washington, USA

Association for Women in Mathematics Series

Focusing on the groundbreaking work of women in mathematics past, present, and future, Springer's Association for Women in Mathematics Series presents the latest research and proceedings of conferences worldwide organized by the Association for Women in Mathematics (AWM). All works are peer-reviewed to meet the highest standards of scientific literature, while presenting topics at the cutting edge of pure and applied mathematics, as well as in the areas of mathematical education and history. Since its inception in 1971, The Association for Women in Mathematics has been a non-profit organization designed to help encourage women and girls to study and pursue active careers in mathematics and the mathematical sciences and to promote equal opportunity and equal treatment of women and girls in the mathematical sciences. Currently, the organization represents more than 3000 members and 200 institutions constituting a broad spectrum of the mathematical community, in the United States and around the world.

More information about this series at http://www.springer.com/series/13764

Bahar Acu • Donatella Danielli • Marta Lewicka
Arati Pati • Saraswathy RV
Miranda Teboh-Ewungkem
Editors

Advances in Mathematical Sciences

AWM Research Symposium, Houston, TX,
April 2019

 Springer

Editors

Bahar Acu
Department of Mathematics
Northwestern University
Evanston, IL, USA

Donatella Danielli
Department of Mathematics
Purdue University
West Lafayette, IN, USA

Marta Lewicka
Department of Mathematics
University of Pittsburgh
Pittsburgh, PA, USA

Arati Pati
Department of Mathematics
St. Thomas University
Houston, TX, USA

Saraswathy RV
Security and Privacy
Bosch Research and Technology Center
Pittsburgh, PA, USA

Miranda Teboh-Ewungkem
Department of Mathematics
Lehigh University
Bethlehem, PA, USA

ISSN 2364-5733   ISSN 2364-5741   (electronic)
Association for Women in Mathematics Series
ISBN 978-3-030-42689-7   ISBN 978-3-030-42687-3   (eBook)
https://doi.org/10.1007/978-3-030-42687-3

Mathematics Subject Classification: 00B25

This Springer imprint is published by the registered company Springer Nature Switzerland AG.
The registered company address is: Gewerbestrasse 11, 6330 Cham, Switzerland

Preface

This volume highlights the mathematical research presented at the 2019 Association for Women in Mathematics (AWM) Research Symposium. This event, fifth in the biennial series launched in 2011, was held at Rice University on April 6–7, 2019. The objective of the AWM Research Symposia Series is to showcase research from women across the mathematical sciences, working in academia, government, and industry. Additionally, these symposia facilitate the creation of new research collaboration networks and support the ones already existing in many areas of mathematics. They feature women across the whole career spectrum: undergraduates, graduate students, postdocs, and professionals. The symposia also include career panels and social events to enable networking among women in different paths or career stages while promoting the discussion of prospects, visibility, and recognition.

About the 2019 AWM Research Symposium

The 2019 AWM Research Symposium was organized by Ruth Haas, Shelly Harvey, Raegan Higgins, Magnhild Lien, Omayra Ortega, Karoline Pershell, Ami Radunskaya, and Beatrice Rivière. The event was attended by over 340 participants, gathering from 42 states in the USA, Canada, Colombia, Germany, Mexico, and Turkey. The Symposium featured:

- 3 plenary talks by distinguished women mathematicians (see Table 1);
- 16 special sessions addressing a broad range of research in pure mathematics, applied mathematics, and mathematics education (Table 2);

Table 1 Plenary talks

Speaker	Title
Susanne C. Brenner	Higher Order Elliptic Problems
Kristin Lauter	How to Keep Your Secrets in a Post-Quantum World
Chelsea Walton	Quantum Symmetry

Table 2 Special sessions

Title	Organizers
Analysis and Numerical Methods for Kinetic Transport and Related Models	Liu Liu
Applied and Computational Harmonic Analysis	Julia Dobrosotskaya and Xuemei Chen
Braid Groups and Quantum Computing	Colleen Delaney, Jennifer Vasquez, and Helen Wong
Combinatorial Algebra	Christine Berkesch and Laura Felicia Matusevich
Combinatorial Commutative Algebra	Sara Faridi and Susan Morey
Current Challenges in Mathematical Biology	Renee Dale
Education Partnerships: University Mathematics Faculty and K-12 Mathematics Teachers	Evan Rushton
Graph Theory	Carolyn Reinhart and Kate Lorenzo
Math on the EDGE	Sarah Chehade
Multiphysics and Multiscale problems	Yue Yu and Xingjie Li
New Advances in Symplectic and Contact Topology	Jo Nelson and Morgan Weiler
New Developments in Algebraic Biology	Anne Shiu and Brandilyn Stigler
Origami, Belyi Maps, and Dessins D'Enfants	Rachel Davis and Edray Goins
Recent Developments in the Analysis of Obstacle problems	Donatella Danielli and Camelia Pop
On Advances and New Techniques of Fluid Dynamics and Dispersive Equations	Betul Orcan Ekmekci
Topology of 3- and 4-Manifolds	Allison N. Miller and Arunima Ray
WDS: Women in Data Science	Jing Qin and Yifei Lou

Table 3 Invited sessions by research networks supported by the AWM ADVANCE Grant

Title	Organizers
ACxx: Women in Algebraic Combinatorixx: Enumerative and Algebraic Combinatorics	Elizabeth Niese and Elizabeth Drellich
WIC: Women in Control: Control in Infinite Dimensional Systems	Lorena Bociu and Irena Lasiecka
WICA: Women in Commutative Algebra	Sandra Spiroff and Adela Vraciu
WIG: Women in Geometry	Liz Stanhope and Chikako Mese
WIMB: Women in Math Biology: Advances in Mathematical Biology	Angela Peace and Wenjing Zhang
WIMM: Women in Math Materials	Malena Espanol and Hala AH Shehadeh
WIN: Women in Numbers	Michelle Manes, Ila Varma
WINART: Women in Noncommutative Algebra and Representation Theory: Homological Methods in Noncommutative Algebra and Representation Theory	Van C. Nguyen, Julia Plavnik, and Sarah Witherspoon
WINASC: Women in Numerical Analysis and Scientific Computing: Recent advances in numerical methods and its applications	Bo Dong and Adrianna Gillman
WinCompTop: Women in Computational Topology: Trends in Computational Topology	Erin Chambers, Brittany Terese Fasy, and Elizabeth Munch
WiSDM: Women in the Science of Data and Mathematics: Data Science Theory and Practice	Linda Ness and Carlotta Domeniconi
WiSh: Women in Shape Modeling	Kathryn Leonard and Terry Knight
WIT: Women in Topology: Topics in Homotopy Theory	Sarah Yeakel and Martina Rovelli

- 14 invited sessions organized by research networks supported by the AWM ADVANCE grant (Table 3);
- a poster session for graduate students and recent Ph.D.s;
- a Wikipedia Edit-a-Thon;
- a professional development panel on mathematics in government and industry;
- a presentation of funding opportunities by the NSF Program Director Yuliya Gorb, followed by a Q&A session;
- an informational meeting about establishing and maintaining research networks, hosted by AWM ADVANCE Director Magnhild Lien and co-PI Kristin Lauter;
- a reception and a banquet.

The keynote speakers at the banquet were: Provost Marie Lynn Miranda (Rice University) and PhD candidate and Science Master Teacher Mariam Manuel (teachHOUSTON). Provost Miranda shared insights about her efforts towards diversifying faculty at Rice University and asked the audience to be role models for the next generation. Manuel talked about cultivating creative confidence in girls and women. As a part of the banquet program, local initiatives that make a difference in encouraging girls and women to pursue mathematics were recognized. The honored individuals were: Anne Papakonstantinou (Rice University Mathematics School Project Director), Paula Myrick Short (University of Houston Provost, Director of the Center for ADVANCING UH Faculty Success), Kelsey Friedemann (Houston Museum of Natural Science, Girls Exploring Math and Science Program), Tricia Berry (UT Austin, Director, Texas Girls Collaborative Project), and Joanna Papakonstantinou (Episcopal High School, Math Educator).

The Symposium also featured the *Remembering Maryam Mirzakhani Exhibit*, which includes photos of, and artwork based on the life of Maryam Mirzakhani, the first woman to be awarded the Fields Medal. The exhibition was created by the International Mathematical Union's Committee for Women in Mathematics (CWM) with Curator Thais Jordao and Designer Rafael Meireles Barroso. It was first shown at the 2018 World Meeting for Women in Mathematics in Rio and during the International Congress of Mathematicians (ICM2018).

For the full symposium schedule with a list of all talks, poster sessions, presenters, and other activities, please follow the program link on the conference's website: https://awm-math.org/meetings/awm-research-symposium/.

About This Volume

This volume opens with a part entitled *From the Plenary Talks*, featuring a survey of the finite element methods for an elliptic optimal control problem with pointwise state constraints, written by Susanne Brenner. After that, the papers are grouped together in several parts based on subject areas. Part II, *Algebraic Combinatorics and Graph Theory*, contains both original research and survey papers written by presenters in the session "ACxx: Women in Algebraic Combinatorics" and "Graph Theory." Part III, *Algebraic Biology*, consists of four papers based on talks in the session "New Developments in Algebraic Biology." Part IV, *Commutative Algebra*, features original research presented in the session "Combinatorial Commutative Algebra," and a survey by a speaker in the session "WICA: Women in Commutative Algebra." Part V, *Analysis, Probability, and PDEs*, collects research featured in the sessions "WDS: Women in Data Science" and "On Advances and New Techniques of Fluid Dynamics and Dispersive Equations," together with a survey paper written by a speaker in the session "Recent Developments in the Analysis of Obstacle problems." Part VI, *Topology*, consists of an original contribution by a presenter in the session "WIT: Women in Topology" and a survey by a speaker in the session "New Advances in Symplectic and Contact Topology." This is followed by Part VII, "Applied Mathematics," which collects original research presented in the session "WIC: Women in Control," together with survey papers authored by speakers in the sessions "WIMM: Women in Math Materials," "WISDM: Women in the Science of Data and Mathematics," and "WiSh: Women in Shape Modeling." The volume concludes with Part VIII, *Math Education*, containing survey articles by presenters in the session "Education Partnerships: University Mathematics Faculty and K-12 Mathematics Teachers."

Acknowledgments

Firstly, we express gratitude to the authors who submitted their manuscripts, and to the referees who carefully reviewed them and provided valuable feedback. All articles in this volume have been peer-reviewed.

An acknowledgement extends to the organizers of the Symposium for their time, effort, and dedication, and to the host institution for providing the necessary resources and infrastructures. Further acknowledgement is due to the sponsors of the 2019 AWM Research Symposium, which are: Rice University, the American Mathematical Society, the Association of Members of the Institute for Advanced Studies, the Mathematical Association of America, Microsoft Research, the National Security Agency, the National Science Foundation, Overleaf, and Springer. An appreciation is, likewise, extended to all Symposium Exhibitors.

Finally, the Editors would like to thank Kristin Lauter, Editor of the AWM Springer Series, for the opportunity of publishing this volume.

Evanston, IL, USA   Bahar Acu
West Lafayette, IN, USA   Donatella Danielli
Pittsburgh, PA, USA   Marta Lewicka
Houston, TX, USA   Arati Pati
Pittsburgh, PA, USA   Saraswathy RV
Bethlehem, PA, USA   Miranda Teboh-Ewungkem

December 2019

Contents

Part I
From the Plenary Talks

Finite Element Methods for Elliptic Distributed Optimal Control Problems with Pointwise State Constraints (Survey)

Susanne C. Brenner

1 Model Problem

Let Ω be a convex bounded polygonal/polyhedral domain in $\mathbb{R}^2/\mathbb{R}^3$, $y_d \in L_2(\Omega)$, β be a positive constant, $\psi \in H^3(\Omega) \cap W^{2,\infty}(\Omega)$ and $\psi > 0$ on $\partial\Omega$. The model problem [1] is to find

$$(\bar{y}, \bar{u}) = \operatorname*{argmin}_{(y,u)\in\mathbb{K}} \frac{1}{2}\Big[\|y - y_d\|_{L_2(\Omega)}^2 + \beta\|u\|_{L_2(\Omega)}^2\Big], \tag{1}$$

where $(y, u) \in H_0^1(\Omega) \times L_2(\Omega)$ belongs to $\mathbb{K}$ if and only if

$$\int_\Omega \nabla y \cdot \nabla z \, dx = \int_\Omega uz \, dx \qquad \forall\, z \in H_0^1(\Omega), \tag{2}$$

$$y \leq \psi \qquad \text{a.e. on } \Omega. \tag{3}$$

Throughout this paper we will follow the standard notation for operators, function spaces and norms that can be found for example in [2, 3].

In this model problem y (resp., u) is the state (resp., control) variable, y_d is the desired state and β is a regularization parameter. Similar linear-quadratic optimization problems also appear as subproblems when general PDE constrained optimization problems are solved by sequential quadratic programming (cf. [4, 5]).

In view of the convexity of Ω, the constraint (2) implies $y \in H^2(\Omega)$ (cf. [6–8]). Therefore we can reformulate (1)–(3) as follows:

S. C. Brenner (✉)
Department of Mathematics and Center for Computation & Technology, Louisiana State University, Baton Rouge, LA, USA
e-mail: brenner@math.lsu.edu

© The Author(s) and the Association for Women in Mathematics 2020
B. Acu et al. (eds.), *Advances in Mathematical Sciences*, Association for Women in Mathematics Series 21, https://doi.org/10.1007/978-3-030-42687-3_1

$$\text{Find} \quad \bar{y} = \underset{y \in K}{\operatorname{argmin}} \frac{1}{2}\left[\|y - y_d\|^2_{L_2(\Omega)} + \beta\|\Delta y\|^2_{L_2(\Omega)}\right], \tag{4}$$

where

$$K = \{y \in H^2(\Omega) \cap H_0^1(\Omega) : y \le \psi \text{ on } \Omega\}. \tag{5}$$

Note that K is nonempty because $\psi > 0$ on $\partial\Omega$. It follows from the classical theory of calculus of variations [9] that (4)–(5) has a unique solution $\bar{y} \in K$ characterized by the fourth order variational inequality

$$a(\bar{y}, y - \bar{y}) \ge \int_\Omega y_d(y - \bar{y})dx \qquad \forall\, y \in K, \tag{6}$$

where

$$a(y, z) = \beta \int_\Omega (\Delta y)(\Delta z)dx + \int_\Omega yz\, dx. \tag{7}$$

Furthermore, by the Riesz-Schwartz Theorem for nonnegative linear functionals [10, 11], we can rewrite (6) as

$$a(\bar{y}, z) = \int_\Omega y_d z\, dx + \int_\Omega z\, d\mu \qquad \forall\, z \in H^2(\Omega) \cap H_0^1(\Omega), \tag{8}$$

where

$$\mu \text{ is a nonpositive finite Borel measure} \tag{9}$$

that satisfies the complementarity condition

$$\int_\Omega (\bar{y} - \psi)d\mu = 0. \tag{10}$$

Note that (10) is equivalent to the statement that

$$\mu \text{ is supported on } \mathcal{A}, \tag{11}$$

where the active set $\mathcal{A} = \{x \in \Omega : \bar{y}(x) = \psi(x)\}$ satisfies

$$\mathcal{A} \subset\subset \Omega \tag{12}$$

because $\psi > 0$ on $\partial\Omega$ and $\bar{y} = 0$ on $\partial\Omega$.

According to the elliptic regularity theory in [6–8, 12, 13], we have

$$\bar{y} \in H^3_{loc}(\Omega) \cap W^{2,\infty}_{loc}(\Omega) \cap H^{2+\alpha}(\Omega), \tag{13}$$

where $\alpha \in (0, 1]$ is determined by the geometry of Ω. It then follows from (8), (11)–(13) and integration by parts that

$$\mu \in H^{-1}(\Omega). \tag{14}$$

Details for (13) and (14) can be found in [14].

Remark 1 Note that (cf. [6, 15])

$$\int_\Omega (\Delta y)(\Delta z) dx = \int_\Omega D^2 y : D^2 z \, dx \qquad \forall \, y, z \in H^2(\Omega) \cap H^1_0(\Omega),$$

where $D^2 y : D^2 z$ denotes the Frobenius inner product between the Hessian matrices of y and z. Therefore we can rewrite the bilinear form $a(\cdot, \cdot)$ in (7) as

$$a(y, z) = \beta \int_\Omega D^2 y : D^2 z \, dx + \int_\Omega yz \, dx. \tag{15}$$

2   Finite Element Methods

In the absence of the state constraint (3), we have $K = H^2(\Omega) \cap H^1_0(\Omega)$ and (6) becomes the boundary value problem

$$a(\bar{y}, z) = \int_\Omega y_d z \, dx \qquad \forall \, z \in H^2(\Omega) \cap H^1_0(\Omega). \tag{16}$$

Since (16) is essentially a bending problem for simply supported plates, it can be solved by many finite element methods such as (1) conforming methods, (2) classical nonconforming methods, (3) discontinuous Galerkin methods, and (4) mixed methods. For the sake of brevity, below we will consider these methods for $\Omega \subset \mathbb{R}^2$. But all the results can be extended to three dimensions.

Let V_h be a finite element space associated with a triangulation $\mathcal{T}_h$ of Ω. The approximate solution $\bar{y}_h \in V_h$ is determined by

$$a_h(\bar{y}_h, z) = \int_\Omega y_d z \, dx \qquad \forall \, z \in V_h, \tag{17}$$

where the choice of the bilinear form $a_h(\cdot, \cdot)$ depends on the type of finite element method being used.

2.1 Conforming Methods

In this case $V_h \subset H^2(\Omega) \cap H_0^1(\Omega)$ is a C^1 finite element space and we can take $a_h(\cdot, \cdot)$ to be $a(\cdot, \cdot)$. This class of methods includes the Bogner-Fox-Schmit element [16], the Argyris elements [17], the macro elements [18–20], and generalized finite elements [21–23].

2.2 Classical Nonconforming Methods

In this case $V_h \subset L_2(\Omega)$ consists of finite element functions that are weakly continuous up to first order derivatives across element boundaries, and the bilinear form $a_h(\cdot, \cdot)$ is given by

$$a_h(y, z) = \beta \sum_{T \in \mathcal{T}_h} \int_\Omega D^2 y : D^2 z \, dx + \int_\Omega yz \, dx. \tag{18}$$

Here we are using the piecewise version of (15), which provides better local control of the nonconforming energy norm $\| \cdot \|_{a_h} = \sqrt{a_h(\cdot, \cdot)}$.

This class of methods includes the Adini element [24], the Zienkiewicz element [25], the Morley element [26], the Fraeijs de Veubeke element [27], and the incomplete biquadratic element [28].

2.3 Discontinuous Galerkin Methods

In this case V_h consists of functions that are totally discontinuous or only discontinuous in the normal derivatives across element boundaries, and stabilization terms are included in the bilinear form $a_h(\cdot, \cdot)$. The simplest choice is a Lagrange finite element space $V_h \subset H_0^1(\Omega)$, resulting in the C^0 interior penalty methods [29–31], where the bilinear form $a_h(\cdot, \cdot)$ is given by

$$a_h(y, z) = \beta \left[\sum_{T \in \mathcal{T}_h} \int_T D^2 y : D^2 z \, dx + \sum_{e \in \mathcal{E}_h^i} \int_e \{\!\{\partial^2 y / \partial n^2\}\!\} [\![\partial z / \partial n]\!] \, ds \right.$$

$$+ \sum_{e \in \mathcal{E}_h^i} \int_e \{\!\{\partial^2 z / \partial n^2\}\!\} [\![\partial y / \partial n]\!] \, ds \tag{19}$$

$$\left. + \sigma \sum_{e \in \mathcal{E}_h^i} |e|^{-1} \int_e [\![\partial y / \partial n]\!] [\![\partial z / \partial n]\!] \, ds \right] + \int_\Omega yz \, dx.$$

Here $\mathcal{E}_h^i$ is the set of the interior edges of $\mathcal{T}_h$, $\{\!\{\partial^2 y/\partial n^2\}\!\}$ (resp., $[\![\partial y/\partial n]\!]$) is the average (resp., jump) of the second (resp., first) normal derivative of y across the edge e, $|e|$ is the length of the edge e, and σ is a (sufficiently large) penalty parameter.

Other discontinuous Galerkin methods for fourth order problems can be found in [32–34].

2.4   Mixed Methods

In this case $V_h \subset H_0^1(\Omega)$ is a Lagrange finite element space. The approximate solution $\bar{y}_h$ is determined by

$$\int_\Omega \bar{y}_h z \, dx + \beta \int_\Omega \nabla \bar{u}_h \cdot \nabla z \, dx = \int_\Omega y_d z \, dx \qquad \forall z \in V_h, \tag{20}$$

$$\int_\Omega \nabla \bar{y}_h \cdot \nabla v \, dx - \int_\Omega \bar{u}_h v \, dx = 0 \qquad \forall v \in V_h. \tag{21}$$

By eliminating $\bar{u}_h$ from (20)–(21), we can recast $\bar{y}_h$ as the solution of (17) where

$$a_h(y, z) = \beta \int_\Omega (\Delta_h y)(\Delta_h z) \, dx + \int_\Omega yz \, dx, \tag{22}$$

and the discrete Laplace operator $\Delta_h : V_h \longrightarrow V_h$ is defined by

$$\int_\Omega (\Delta_h y)z \, dx = -\int_\Omega \nabla y \cdot \nabla z \, dx \qquad \forall y, z \in V_h. \tag{23}$$

2.5   Finite Element Methods for the Optimal Control Problem

With the finite element methods for (16) in hand, we can now simply discretize the variational inequality (6) as follows: Find $\bar{y}_h \in V_h$ such that

$$a_h(\bar{y}_h, y - \bar{y}_h) \geq \int_\Omega y_d(y - \bar{y}_h)dx \qquad \forall y \in K_h, \tag{24}$$

where

$$K_h = \{y \in V_h : I_h y \leq I_h \psi \text{ on } \Omega\}, \tag{25}$$

and I_h is the nodal interpolation operator for the conforming P_1 finite element space associated with $\mathcal{T}_h$. In other words, the constraint (3) is only imposed at the vertices of $\mathcal{T}_h$.

Remark 2 Conforming, nonconforming, C^0 interior penalty and mixed methods for (6) were investigated in [14, 35–41].

3   Convergence Analysis

For simplicity, we will only provide details for the case of conforming finite element methods and briefly describe the extensions to other methods at the end of the section.

For conforming finite element methods, we have $a_h(\cdot, \cdot) = a(\cdot, \cdot)$ and the energy norm $\| \cdot \|_a = \sqrt{a(\cdot, \cdot)}$ satisfies, by a Poincaré-Friedrichs inequality [42],

$$\|v\|_a \approx \|v\|_{H^2(\Omega)} \qquad \forall\, v \in H^2(\Omega). \tag{26}$$

Our goal is to show that

$$\|\bar{y} - \bar{y}_h\|_a \leq Ch^\alpha, \tag{27}$$

where α is the index of elliptic regularity that appears in (13).

We assume (cf. [43]) that there exists an operator $\Pi_h : H^2(\Omega) \cap H_0^1(\Omega) \longrightarrow V_h$ such that

$$\Pi_h \zeta = \zeta \quad \text{at the vertices of } \mathcal{T}_h \tag{28}$$

and

$$\|\zeta - \Pi_h \zeta\|_{L_2(\Omega)} + h|\zeta - \Pi_h \zeta|_{H^1(\Omega)} + h^2|\zeta - \Pi_h \zeta|_{H^2(\Omega)} \leq Ch^{2+\alpha}|\zeta|_{H^{2+\alpha}(\Omega)} \tag{29}$$

for all $\zeta \in H^{2+\alpha}(\Omega) \cap H_0^1(\Omega)$, where $h = \max_{T \in \mathcal{T}_h} \operatorname{diam} T$ is the mesh size of the triangulation $\mathcal{T}_h$. Here and below we use C to denote a generic positive constant independent of h.

In particular (5), (25) and (28) imply

$$\Pi_h \text{ maps } K \text{ into } K_h. \tag{30}$$

Therefore K_h is nonempty and the discrete problem defined by (24)–(25) has a unique solution.

We will also use the following standard properties of the interpolation operator I_h (cf. [2, 3]):

$$\|\zeta - I_h \zeta\|_{L_\infty(T)} \leq Ch_T^2|\zeta|_{W^{2,\infty}(T)} \qquad \forall\, \zeta \in W^{2,\infty}(T),\ T \in \mathcal{T}_h, \tag{31}$$

$$|\zeta - I_h \zeta|_{H^1(T)} \leq Ch_T|\zeta|_{H^2(T)} \qquad \forall\, \zeta \in H^2(T),\ T \in \mathcal{T}_h, \tag{32}$$

where h_T is the diameter of T.

We begin with the estimate

$$
\begin{aligned}
\|\bar{y} - \bar{y}_h\|_a^2 &= a(\bar{y} - \bar{y}_h, \bar{y} - \bar{y}_h) \\
&= a(\bar{y} - \bar{y}_h, \bar{y} - \Pi_h \bar{y}) + a(\bar{y}, \Pi_h \bar{y} - \bar{y}_h) - a(\bar{y}_h, \Pi_h \bar{y} - \bar{y}_h) \quad (33) \\
&\leq C_1 \|\bar{y} - \bar{y}_h\|_a h^\alpha + \left[a(\bar{y}, \Pi_h \bar{y} - \bar{y}_h) - \int_\Omega y_d (\Pi_h \bar{y} - \bar{y}_h) dx \right]
\end{aligned}
$$

that follows from (13), (24), (26), (29), (30) and the Cauchy-Schwarz inequality.

Remark 3 Note that an estimate analogous to (33) also appears in the error analysis for the boundary value problem (16). Indeed the second term on the right-hand side of (33) vanishes in the case of (16) and we would have arrived at the desired estimate $\|\bar{y} - \bar{y}_h\|_a \leq Ch^\alpha$.

The idea now is to show that

$$
a(\bar{y}, \Pi_h \bar{y} - \bar{y}_h) - \int_\Omega y_d (\Pi_h \bar{y} - \bar{y}_h) dx \leq C_2 \left[h^{2\alpha} + h^\alpha \|\bar{y} - \bar{y}_h\|_a \right], \quad (34)
$$

which together with (33) implies

$$
\|\bar{y} - \bar{y}_h\|_a^2 \leq C_3 h^\alpha \|\bar{y} - \bar{y}_h\|_a + C_2 h^{2\alpha}. \quad (35)
$$

The estimate (27) then follows from (35) and the inequality

$$
ab \leq \frac{\epsilon}{2} a^2 + \frac{1}{2\epsilon} b^2
$$

that holds for any positive ϵ.

Let us turn to the derivation of (34). Since $K_h \subset V_h \subset H^2(\Omega) \cap H_0^1(\Omega)$, we have, according to (8),

$$
\begin{aligned}
a(\bar{y}, \Pi_h \bar{y} - \bar{y}_h) - \int_\Omega y_d (\Pi_h \bar{y} - \bar{y}_h) dx &= \int_\Omega (\Pi_h \bar{y} - \bar{y}_h) d\mu \\
&= \int_\Omega (\Pi_h \bar{y} - \bar{y}) d\mu + \int_\Omega (\bar{y} - \psi) d\mu + \int_\Omega (\psi - I_h \psi) d\mu \\
&\quad + \int_\Omega (I_h \psi - I_h \bar{y}_h) d\mu + \int_\Omega (I_h \bar{y}_h - \bar{y}_h) d\mu,
\end{aligned} \quad (36)
$$

and, in view of (9), (10) and (25),

$$
\int_\Omega (\bar{y} - \psi) d\mu = 0 \quad \text{and} \quad \int_\Omega (I_h \psi - I_h \bar{y}_h) d\mu \leq 0. \quad (37)
$$

We can estimate the other three integrals on the right-hand side of (36) as follows:

$$\int_\Omega (\Pi_h \bar{y} - \bar{y}) d\mu \le \|\mu\|_{H^{-1}(\Omega)} \|\Pi_h \bar{y} - \bar{y}\|_{H^1(\Omega)} \le Ch^{1+\alpha} \tag{38}$$

by (13), (14) and (29);

$$\int_\Omega (\psi - I_h \psi) d\mu \le |\mu(\Omega)| \|\psi - I_h \psi\|_{L_\infty(\Omega)} \le Ch^2 \tag{39}$$

by (9) and (31);

$$\int_\Omega (I_h \bar{y}_h - \bar{y}_h) d\mu = \int_\Omega \big[I_h(\bar{y}_h - \bar{y}) - (\bar{y}_h - \bar{y})\big] d\mu + \int_\Omega (I_h \bar{y} - \bar{y}) d\mu$$

$$\le \|\mu\|_{H^{-1}(\Omega)} |I_h(\bar{y}_h - \bar{y}) - (\bar{y}_h - \bar{y})|_{H^1(\Omega)} + |\mu(\Omega)| \|I_h \bar{y} - \bar{y}\|_{L_\infty(\mathcal{A})} \tag{40}$$

$$\le C\big[h|\bar{y}_h - \bar{y}|_{H^2(\Omega)} + h^2\big]$$

$$\le C\big(h\|\bar{y} - \bar{y}_h\|_a + h^2\big)$$

by (11)–(13), (26), (31) and (32).

The estimate (34) follows from (36)–(40) and the fact that $\alpha \le 1$.

The estimate (27) can be extended to the other finite element methods in Sect. 2 provided $\| \cdot \|_a$ is replaced by $\| \cdot \|_{a_h} = \sqrt{a_h(\cdot, \cdot)}$.

For classical nonconforming finite element methods and discontinuous Galerkin methods, the key ingredient for the convergence analysis, in addition to an operator $\Pi_H : H^2(\Omega) \cap H_0^1(\Omega) \longrightarrow V_h$ that satisfies (28) and (29), is the existence of an *enriching* operator $E_h :\longrightarrow H^2(\Omega) \cap H_0^1(\Omega)$ with the following properties:

$$(E_h v)(p) = v(p) \quad \text{for all vertices } p \text{ of } \mathcal{T}_h, \tag{41}$$

$$\|v - E_h v\|_{L_2(\Omega)} + h\Big(\sum_{T \in \mathcal{T}_h} |v - E_h v|^2_{H^1(T)} \Big)^{\frac{1}{2}} + h^2 |E_h v|_{H^2(\Omega)}$$

$$\le Ch^2 \|v\|_h \quad \forall v \in V_h, \tag{42}$$

$$\|\zeta - E_h \Pi_h \zeta\|_{H^1(\Omega)} \le Ch^{1+\alpha} \|\zeta\|_{H^{2+\alpha}(\Omega)} \quad \forall \zeta \in H^{2+\alpha}(\Omega) \cap H_0^1(\Omega), \tag{43}$$

$$|a_h(\Pi_h \zeta, v) - a(\zeta, E_h v)| \le Ch^\alpha \|\zeta\|_{H^{2+\alpha}(\Omega)} \|v\|_h \tag{44}$$

for all $\zeta \in H^{2+\alpha}(\Omega) \cap H_0^1(\Omega)$ and $v \in V_h$.

Property (41) is related to the fact that the discrete constraints are imposed at the vertices of $\mathcal{T}_h$; property (42) indicates that in some sense $\|v - E_h v\|_h$ measures the distance between V_h and $H^2(\Omega) \cap H_0^1(\Omega)$; property (43) means that $E_h \Pi_h$ behaves like a quasi-local interpolation operator; property (44) states that E_h is essentially the adjoint of Π_h with respect to the continuous and discrete bilinear forms. The idea is to use (42) and (44) to reduce the error estimate to the continuous level, and then the error analysis can proceed as in the case of conforming finite element method by using (41) and (43). Details can be found in [44].

Remark 4 The operator E_h maps V_h to a conforming finite element space and its construction is based on averaging. The history of using such enriching operators to handle nonconforming finite element methods is discussed in [45].

In the case of the mixed method where $V_h \subset H_0^1(\Omega)$ is a Lagrange finite element space, the operator $E_h : V_h \longrightarrow H^2(\Omega) \cap H_0^1(\Omega)$ is defined by

$$\int_\Omega \nabla E_h v \cdot \nabla w \, dx = \int_\Omega \nabla v \cdot \nabla w \, dx \qquad \forall v \in V_h, \; w \in H_0^1(\Omega). \tag{45}$$

The properties (42)–(44) remain valid provided Π_h is replaced by the Ritz projection operator $R_h : H_0^1(\Omega) \longrightarrow V_h$ defined by

$$\int_\Omega \nabla R_h \zeta \cdot \nabla v \, dx = \int_\Omega \nabla \zeta \cdot \nabla v \, dx \qquad \forall v \in V_h. \tag{46}$$

In fact (45) and (46) imply $\zeta - E_h R_h \zeta = 0$ and property (43) becomes trivial. However the properties (28) and (41) no longer hold, which necessitates the use of the more sophisticated interior error estimates (cf. [46]) in the convergence analysis. Details can be found in [14].

Remark 5 Since the elliptic regularity index α in (13) is determined by the singularity of the Laplace equation near the boundary of Ω, various finite element techniques [47, 48] can be employed to improve the estimate (27) to

$$\|\bar{y} - \bar{y}_h\|_{a_h} \leq Ch. \tag{47}$$

One can also compute an approximation $\bar{u}_h$ for the optimal control $\bar{u}$ from the approximate optimal state $\bar{y}_h$ through post-processing processes [49].

Remark 6 The discrete problems generated by the finite element methods in Sect. 2, which only involve simple box constraints, can be solved efficiently by a primal-dual active set algorithm [50–52].

4 Concluding Remarks

In this paper finite element methods for elliptic distributed optimal control problems with pointwise state constraints are treated from the perspective of finite element methods for the boundary value problem of simply supported plates.

The discussion in Sect. 2 shows that one can solve elliptic distributed optimal control problems with pointwise state constraints by a straightforward adaptation of many finite element methods for simply supported plates. The convergence analysis in Sect. 3 demonstrates that the gap between the finite element analysis for boundary value problems and the finite element analysis for elliptic optimal control problems is in fact quite narrow. Thus the vast arsenal of finite element techniques developed for elliptic boundary value problems over several decades can be applied to elliptic optimal control problems with only minor modifications.

Note that in the traditional approach to elliptic optimal control problems, the optimal control $\bar{u}$ is treated as the primary unknown and the resulting finite element methods in [35, 39] are equivalent to the method defined by (24), where the bilinear form is given by (22). Therefore the approach based on the reformulation (4)–(5) expands the scope of finite element methods for elliptic optimal control problems from a special class of methods (i.e., mixed methods) to all classes of methods. In addition to the finite element mentioned in Sect. 2, one can also consider recently developed finite element methods for fourth order problems on polytopal meshes [53–60].

The new approach has been extended to problems with the Neumann boundary condition [61, 62] and to problems with pointwise constraints on both control and state [63]. It has also been extended to problems on nonconvex domains [14, 62, 64].

Below are some open problems related to the finite element methods presented in Sect. 2.

1. It follows from the error estimates (27) and (47) that

$$\|\bar{y} - \bar{y}_h\|_{H^1(\Omega)} + \|\bar{y} - \bar{y}_h\|_{L_\infty(\Omega)} \leq Ch^\gamma, \tag{48}$$

where $\gamma = \alpha$ (without special treatment) or 1 (with special treatments). For conforming or mixed finite element methods, the estimate (48) is a direct consequence of the fact that the energy norm is equivalent to the $H^2(\Omega)$ norm and that we have the Sobolev inequality

$$\|\zeta\|_{L_\infty(\Omega)} \leq C\|\zeta\|_{H^2(\Omega)}.$$

For classical nonconforming and discontinuous Galerkin methods, the estimate (48) follows from the Poincaré-Friedrichs inequality and Sobolev inequality for piecewise H^2 functions in [65, 66].

Comparing to $\|\cdot\|_{H^2(\Omega)}$, the norms $\|\cdot\|_{H^1(\Omega)}$ and $\|\cdot\|_{L_\infty(\Omega)}$ are lower order norms and, based on experience with finite element methods for the boundary value problem (16), the convergence in $\|\cdot\|_{H^1(\Omega)}$ and $\|\cdot\|_{L_\infty(\Omega)}$ should be of higher order, and this is observed in numerical experiments. But the theoretical justifications for the observed higher order convergence is missing. In the case of the boundary value problem (16), one can show higher order convergence for lower order norms through a duality argument. However duality arguments do not work for variational inequalities even in one dimension [67]. New ideas are needed.

2. An interesting phenomenon concerning fourth order variational inequalities is that a posteriori error estimators originally designed for fourth order boundary value problems can be directly applied to fourth order variational inequalities [61, 68]. This is different from the second order case where a posteriori error estimators for boundary value problems are not directly applicable to variational inequalities. This difference is essentially due to the fact that Dirac point measures belong to $H^{-2}(\Omega)$ but not $H^{-1}(\Omega)$.

   Optimal convergence of these adaptive finite element methods have been observed in numerical experiments. However the proofs of convergence and optimality are missing.

3. Fast solvers for fourth order variational inequalities is an almost completely open area. Some recent work on additive Schwarz preconditioners for the subsystems that appear in the primal-dual active set algorithm can be found in [69, 70]. Much remains to be done.

Acknowledgements This paper is based on research supported by the National Science Foundation under Grant Nos. DMS-13-19172, DMS-16-20273 and DMS-19-13035.

References

1. Casas, E.: Control of an elliptic problem with pointwise state constraints. SIAM J. Control Optim. **24**, 1309–1318 (1986)
2. Ciarlet, P.G.: The Finite Element Method for Elliptic Problems. North-Holland, Amsterdam (1978)
3. Brenner, S.C., Scott, L.R.: The Mathematical Theory of Finite Element Methods (Third Edition). Springer-Verlag, New York (2008)
4. Hinze, M. and Pinnau, R. and Ulbrich, M. and Ulbrich, S.: Optimization with PDE Constraints. Springer, New York (2009)
5. Tröltzsch, F.: Optimal Control of Partial Differential Equations. American Mathematical Society, Providence (2010)
6. Grisvard, P.: Elliptic Problems in Non Smooth Domains. Pitman, Boston (1985)
7. Dauge, M.: Elliptic Boundary Value Problems on Corner Domains. Springer-Verlag, Berlin-Heidelberg (1988)
8. Maz'ya, V., Rossmann, J.: Elliptic Equations in Polyhedral Domains. American Mathematical Society, Providence (2010)
9. Kinderlehrer, D., Stampacchia, G.: An Introduction to Variational Inequalities and Their Applications. Society for Industrial and Applied Mathematics, Philadelphia (2000)
10. Rudin, W.: Real and Complex Analysis. McGraw-Hill, New York (1966)
11. Schwartz, L.: Théorie des Distributions. Hermann, Paris (1966)
12. Frehse, J.: Zum Differenzierbarkeitsproblem bei Variationsungleichungen höherer Ordnung. Abh. Math. Sem. Univ. Hamburg **36**, 140–149 (1971)
13. Frehse, J.: On the regularity of the solution of the biharmonic variational inequality. Manuscripta Math. **9**, 91–103 (1973)
14. Brenner, S.C., Gedicke, J., Sung, L.-Y.: P_1 finite element methods for an elliptic optimal control problem with pointwise state constraints. IMA J. Numer. Anal. (2018). https://doi.org/10.1093/imanum/dry071

15. Ladyženskaya, O.A.: On integral estimates, convergence, approximate methods, and solution in functionals for elliptic operators. Vestnik Leningrad. Univ. **13**, 60–69 (1958)

16. Bogner, F.K., Fox, R.L., Schmit, L.A.: The generation of interelement compatible stiffness and mass matrices by the use of interpolation formulas. In: Proceedings Conference on Matrix Methods in Structural Mechanics, pp. 397–444. Wright Patterson A.F.B., Dayton, Ohio (1965)

17. Argyris, J.H., Fried, I., Scharpf, D.W.: The TUBA family of plate elements for the matrix displacement method. Aero. J. Roy. Aero. Soc. **72**, 701–709 (1968)

18. Clough, R.W., Tocher, J.L.: Finite element stiffbess matrices for analysis of plate bending. In: Proceedings Conference on Matrix Methods in Structural Mechanics, pp. 515–545. Wright Patterson A.F.B., Dayton, Ohio (1965)

19. Ciarlet, P.G.: Sur l'élément de Clough et Tocher. RAIRO Anal. Numér. **8**, 19–27 (1974)

20. Douglas J.Jr., Dupont, T., Percell, P., Scott, L.R.: A family of C^1 finite elements with optimal approximation properties for various Galerkin methods for 2nd and 4th order problems. R.A.I.R.O. Modél. Math. Anal. Numér. **13**, 227–255 (1979)

21. Melenk, J.M., Babuška, I.: The partition of unity finite element method: basic theory and applications Comput. Methods Appl. Mech. Engrg. **139**, 289–314 (1996)

22. Babuška, I. and Banerjee, U. and Osborn, J.E.: Survey of meshless and generalized finite element methods: a unified approach. Acta Numer. **12**, 1–125 (2003)

23. Oh, H.S., Davis, C.B., Jeong, J.W.: Meshfree particle methods for thin plates. Comput. Methods Appl. Mech. Engrg. **209**, 156–171 (2012)

24. Adini, A., Clough, R.W.: Analysis of plate bending by the finite element method. NSF Report G. 7337 (1961)

25. Bazeley, G.P., Cheung, Y.K., Irons, B.M., Zienkiewicz, O.C.: Triangular elements in bending - conforming and nonconforming solutions. In: Proceedings Conference on Matrix Methods in Structural Mechanics, pp. 547–576. Wright Patterson A.F.B., Dayton, Ohio (1965)

26. Morley, L.S.D.: The triangular equilibrium problem in the solution of plate bending problems. Aero. Quart. **19**, 149–169 (1968)

27. de Veubeke, B.F.: Variational principles and the patch test. Internat. J. Numer. Methods Engrg. **8**, 783–801 (1974)

28. Shi, Z.-C.: On the convergence of the incomplete biquadratic nonconforming plate element. Math. Numer. Sinica. **8**, 53–62 (1986)

29. Engel, G., Garikipati, K., Hughes, T.J.R., Larson, M.G., Mazzei, L., Taylor, R.L.: Continuous/discontinuous finite element approximations of fourth order elliptic problems in structural and continuum mechanics with applications to thin beams and plates, and strain gradient elasticity. Comput. Methods Appl. Mech. Engrg. **191**, 3669–3750 (2002)

30. Brenner, S.C., Sung, L.-Y.: C^0 interior penalty methods for fourth order elliptic boundary value problems on polygonal domains. J. Sci. Comput. **22/23**, 83–118 (2005)

31. Brenner, S.C.: C^0 Interior Penalty Methods. In Blowey, J., Jensen, M. (eds.) Frontiers in Numerical Analysis-Durham 2010, pp. 79–147. Springer-Verlag, Berlin-Heidelberg (2012)

32. Süli, E., Mozolevski, I.: hp-version interior penalty DGFEMs for the biharmonic equation. Comput. Methods Appl. Mech. Engrg. **196**, 1851–1863 (2007)

33. Huang, J., Huang, X., Han, W.: A new C^0 discontinuous Galerkin method for Kirchhoff plates. Comput. Methods Appl. Mech. Engrg. **199**, 1446–1454 (2010)

34. Huang, X. and Huang, J.: A superconvergent C^0 discontinuous Galerkin method for Kirchhoff plates: error estimates, hybridization and postprocessing. J. Sci. Comput. **69**, 1251–1278 (2016)

35. Meyer, C.: Error estimates for the finite-element approximation of an elliptic control problem with pointwise state and control constraints. Control Cybernet. **37**, 51–83 (2008)

36. Liu, W., Gong, W., Yan, N.: A new finite element approximation of a state-constrained optimal control problem. J. Comput. Math. **27**, 97–114 (2009)

37. Gong, W., Yan, N.: A mixed finite element scheme for optimal control problems with pointwise state constraints. J. Sci. Comput. **46**, 82–203 (2011)

38. Brenner, S.C., Sung, L.-Y., Zhang, Y.: A quadratic C^0 interior penalty method for an elliptic optimal control problem with state constraints. The IMA Volumes in Mathematics and its Applications. **157**, 97–132 (2013)

39. Casas, E., Mateos, M., Vexler, B.: New regularity results and improved error estimates for optimal control problems with state constraints. ESAIM Control Optim. Calc. Var. **20**, 803–822 (2014)
40. Brenner, S.C., Davis, C.B., Sung, L.-Y.: A partition of unity method for a class of fourth order elliptic variational inequalities. Comp. Methods Appl. Mech. Engrg. **276**, 612–626 (2014)
41. Brenner, S.C., Oh, M., Pollock, S., Porwal, K., Schedensack, M., Sharma, N.: A C^0 interior penalty method for elliptic distributed optimal control problems in three dimensions with pointwise state constraints. The IMA Volumes in Mathematics and its Applications. **160**, 1–22 (2016)
42. Nečas, J.: Direct Methods in the Theory of Elliptic Equations, Springer, Heidelberg (2012)
43. Girault, V., Scott, L.R.: Hermite interpolation of nonsmooth functions preserving boundary conditions. Math. Comp. **71**, 1043–1074 (2002)
44. Brenner, S.C., Sung, L.-Y.: A new convergence analysis of finite element methods for elliptic distributed optimal control problems with pointwise state constraints. SIAM J. Control Optim. **55**, 2289–2304 (2017)
45. Brenner, S.C.: Forty years of the Crouzeix-Raviart element. Numer. Methods Partial Differential Equations. **31**, 367–396 (2015)
46. Wahlbin, L.B. Local Behavior in Finite Element Methods. In: Ciarlet, P.G., Lions, J.L. (eds.) Handbook of Numerical Analysis, II, pp. 353–522. North-Holland, Amsterdam (1991)
47. Fix, G.J., Gulati, S., Wakoff, G.I.: On the use of singular functions with finite element approximations. J. Computational Phys. **13**, 209–228 (1973)
48. Babuška, I., Kellogg, R.B., Pitkäranta, J.: Direct and inverse error estimates for finite elements with mesh refinements. Numer. Math. **33**, 447–471 (1979)
49. Brenner, S.C., Sung, L.-Y., Zhang, Y.: Post-processing procedures for a quadratic C^0 interior penalty method for elliptic distributed optimal control problems with pointwise state constraints. Appl. Numer. Math. **95**, 99–117 (2015)
50. Bergounioux, M., Kunisch, K.: Primal-dual strategy for state-constrained optimal control problems. Comput. Optim. Appl. **22**, 193–224 (2002)
51. Hintermüller, M., Ito, K., Kunisch, K.: The primal-dual active set strategy as a semismooth Newton method. SIAM J. Optim. **13**, 865–888 (2003)
52. Ito, K. and Kunisch, K.: Lagrange Multiplier Approach to Variational Problems and Applications. Society for Industrial and Applied Mathematics, Philadelphia (2008)
53. Brezzi, F., Marini, L.D.: Virtual element methods for plate bending problems. Comput. Methods Appl. Mech. Engrg. **253**, 455–462 (2013)
54. Mu, L. and Wang, J. and Ye, X.: Weak Galerkin finite element methods for the biharmonic equation on polytopal meshes. Numer. Methods Partial Differential Equations. **30**, 1003–1029 (2014)
55. Wang, C. and Wang, J.: An efficient numerical scheme for the biharmonic equation by weak Galerkin finite element methods on polygonal or polyhedral meshes. Comput. Math. Appl. **68**, 2314–2330 (2014)
56. Chinosi, C., Marini, L.D.: Virtual element method for fourth order problems: L^2-estimates. Comput. Math. Appl. **72**, 1959–1967 (2016)
57. Antonietti, P.F. and Manzini, G. and Verani, M.: The fully nonconforming virtual element method for biharmonic problems. Math. Models Methods Appl. Sci. **28**, 387–407 (2018)
58. Zhao, J. and Zhang, B. and Chen, S. and Mao, S.: The Morley-type virtual element for plate bending problems. J. Sci. Comput. **76**, 610–629 (2018)
59. Bonaldi, F., Di Pietro, D.A., Geymonat, G., Krasucki, F.: A hybrid high-order method for Kirchhoff-Love plate bending problems. ESAIM Math. Model. Numer. Anal. **52**, 393–421 (2018)
60. Beirão da Veiga, L., Dassi, F., Russo, A.: A C^1 virtual element method on polyhedral meshes. arXiv:1808.01105v2 [math.NA] (2019)
61. Brenner, S.C., Sung, L-Y., Zhang, Y.: C^0 interior penalty methods for an elliptic state-constrained optimal control problem with Neumann boundary condition. J. Comput. Appl. Math. **350**, 212–232 (2019)

62. Brenner, S.C., Oh, M., Sung, L.-Y.: P_1 finite element methods for an elliptic state-constrained distributed optimal control problem with Neumann boundary conditions. Preprint (2019)

63. Brenner, S.C., Gudi, T. and Porwal, K. and Sung, L.-Y.: A Morley finite element method for an elliptic distributed optimal control problem with pointwise state and control constraints. ESAIM:COCV. **24**, 1181–1206 (2018)

64. Brenner, S.C., Gedicke, J., Sung, L.-Y.: C^0 interior penalty methods for an elliptic distributed optimal control problem on nonconvex polygonal domains with pointwise state constraints. SIAM J. Numer. Anal. **56**, 1758–1785 (2018)

65. Brenner, S.C., Wang, K., Zhao, J.: Poincaré-Friedrichs inequalities for piecewise H^2 functions. Numer. Funct. Anal. Optim. **25**, 463–478 (2004)

66. Brenner, S.C., Neilan, M., Reiser, A., Sung, L.-Y.: A C^0 interior penalty method for a von Kármán plate. Numer. Math. **135**, 803–832 (2017)

67. Christof, C. and Meyer, C.: A note on a priori L^p-error estimates for the obstacle problem. Numer. Math. **139**, 27–45 (2018)

68. Brenner, S.C., Gedicke, J., Sung, L.-Y., Zhang, Y.: An a posteriori analysis of C^0 interior penalty methods for the obstacle problem of clamped Kirchhoff plates. SIAM J. Numer. Anal. **55**, 87–108 (2017)

69. Brenner, S.C., Davis, C.B., Sung, L.-Y.: Additive Schwarz preconditioners for the obstacle problem of clamped Kirchhoff plates. Electron. Trans. Numer. Anal. 49, 274–290 (2018)

70. Brenner, S.C., Davis, C.B., Sung, L.-Y.: Additive Schwarz preconditioners for a state constrained elliptic distributed optimal control problem discretized by a partition of unity method. arXiv:1811.07809v1 [math.NA] (2018)

Part II
Algebraic Combinatorics
and Graph Theory

Some q-Exponential Formulas Involving the Double Lowering Operator ψ for a Tridiagonal Pair (Research)

Sarah Bockting-Conrad

1 Introduction

Throughout this paper, $\mathbb{K}$ denotes an algebraically closed field. We begin by recalling the notion of a tridiagonal pair. We will use the following terms. Let V denote a vector space over $\mathbb{K}$ with finite positive dimension. For a linear transformation $A : V \to V$ and a subspace $W \subseteq V$, we say that W is an *eigenspace* of A whenever $W \neq 0$ and there exists $\theta \in \mathbb{K}$ such that $W = \{v \in V | Av = \theta v\}$. In this case, θ is called the *eigenvalue* of A associated with W. We say that A is *diagonalizable* whenever V is spanned by the eigenspaces of A.

Definition 1 ([9, Definition 1.1]) Let V denote a vector space over $\mathbb{K}$ with finite positive dimension. By a *tridiagonal pair* (or *TD pair*) on V we mean an ordered pair of linear transformations $A : V \to V$ and $A^* : V \to V$ that satisfy the following four conditions.

(i) Each of A, A^* is diagonalizable.
(ii) There exists an ordering $\{V_i\}_{i=0}^{d}$ of the eigenspaces of A such that

$$A^* V_i \subseteq V_{i-1} + V_i + V_{i+1} \qquad (0 \leq i \leq d), \qquad (1)$$

where $V_{-1} = 0$ and $V_{d+1} = 0$.
(iii) There exists an ordering $\{V_i^*\}_{i=0}^{\delta}$ of the eigenspaces of A^* such that

$$A V_i^* \subseteq V_{i-1}^* + V_i^* + V_{i+1}^* \qquad (0 \leq i \leq \delta), \qquad (2)$$

S. Bockting-Conrad (✉)
DePaul University, Chicago, IL, USA
e-mail: sarah.bockting@depaul.edu

© The Author(s) and the Association for Women in Mathematics 2020
B. Acu et al. (eds.), *Advances in Mathematical Sciences*, Association for
Women in Mathematics Series 21, https://doi.org/10.1007/978-3-030-42687-3_2

where $V_{-1}^* = 0$ and $V_{\delta+1}^* = 0$.

(iv) There does not exist a subspace W of V such that $AW \subseteq W$, $A^*W \subseteq W$, $W \neq 0$, $W \neq V$.

We say the pair A, A^* is *over* $\mathbb{K}$.

Note 1 According to a common notational convention A^* denotes the conjugate-transpose of A. We are not using this convention. In a TD pair A, A^* the linear transformations A and A^* are arbitrary subject to (i)–(iv) above.

Referring to the TD pair in Definition 1, by [9, Lemma 4.5] the scalars d and δ are equal. We call this common value the *diameter* of A, A^*. To avoid trivialities, throughout this paper we assume that the diameter is at least one.

TD pairs first arose in the study of Q-polynomial distance-regular graphs and provided a way to study the irreducible modules of the Terwilliger algebra associated with such a graph. Since their introduction, TD pairs have been found to appear naturally in a variety of other contexts including representation theory [1, 7, 10–12, 14, 15, 25], orthogonal polynomials [23, 24], partially ordered sets [22], statistical mechanical models [3, 6, 19], and other areas of physics [16, 18]. As a result, TD pairs have become an area of interest in their own right. Among the above papers on representation theory, there are several works that connect TD pairs to quantum groups [1, 5, 7, 11, 12]. These papers consider certain special classes of TD pairs. We call particular attention to [5], in which the present author describes a new relationship between TD pairs in the q-Racah class and quantum groups. The present paper builds off of this work.

In the present paper, we give a new relationship between the maps $\Delta, \psi :$ $V \rightarrow V$ introduced in [4], as well as describe a new decomposition of the underlying vector space that, in some sense, lies between the first and second split decompositions associated with a TD pair. In order to motivate our results, we now recall some basic facts concerning TD pairs. For the rest of this section, let A, A^* denote a TD pair on V, as in Definition 1. Fix an ordering $\{V_i\}_{i=0}^d$ (resp. $\{V_i^*\}_{i=0}^d$) of the eigenspaces of A (resp. A^*) which satisfies (1) (resp. (2)). For $0 \leq i \leq d$ let θ_i (resp. θ_i^*) denote the eigenvalue of A (resp. A^*) corresponding to V_i (resp. V_i^*). By [9, Theorem 11.1] the ratios

$$\frac{\theta_{i-2} - \theta_{i+1}}{\theta_{i-1} - \theta_i}, \qquad \frac{\theta_{i-2}^* - \theta_{i+1}^*}{\theta_{i-1}^* - \theta_i^*}$$

are equal and independent of i for $2 \leq i \leq d-1$. This gives two recurrence relations, whose solutions can be written in closed form. There are several cases [9, Theorem 11.2]. The most general case is called the q-Racah case [12, Section 1]. We will discuss this case shortly.

We now recall the split decompositions of V [9]. For $0 \leq i \leq d$ define

$$U_i = (V_0^* + V_1^* + \cdots + V_i^*) \cap (V_i + V_{i+1} + \cdots + V_d),$$

$$U_i^{\Downarrow} = (V_0^* + V_1^* + \cdots + V_i^*) \cap (V_0 + V_1 + \cdots + V_{d-i}).$$

By [9, Theorem 4.6], both the sums $V = \sum_{i=0}^{d} U_i$ and $V = \sum_{i=0}^{d} U_i^{\Downarrow}$ are direct. We call $\{U_i\}_{i=0}^{d}$ (resp. $\{U_i^{\Downarrow}\}_{i=0}^{d}$) the first split decomposition (resp. second split decomposition) of V. In [9], the authors showed that A, A^* act on the first and second split decomposition in a particularly attractive way. This will be described in more detail in Sect. 3.

We now describe the q-Racah case. We say that the TD pair A, A^* has q-Racah type whenever there exist nonzero scalars $q, a, b \in \mathbb{K}$ such that $q^4 \neq 1$ and

$$\theta_i = aq^{d-2i} + a^{-1}q^{2i-d}, \qquad \theta_i^* = bq^{d-2i} + b^{-1}q^{2i-d}$$

for $0 \leq i \leq d$. For the rest of this section assume that A, A^* has q-Racah type.

We recall the maps K and B [13, Section 1.1]. Let $K : V \to V$ denote the linear transformation such that for $0 \leq i \leq d$, U_i is an eigenspace of K with eigenvalue q^{d-2i}. Let $B : V \to V$ denote the linear transformation such that for $0 \leq i \leq d$, $U_i^{\Downarrow}$ is an eigenspace of B with eigenvalue q^{d-2i}. The relationship between K and B is discussed in considerable detail in [5].

We now bring in the linear transformation $\Psi : V \to V$ [4, Lemma 11.1]. As in [5], we work with the normalization $\psi = (q - q^{-1})(q^d - q^{-d})\Psi$. A key feature of ψ is that by [4, Lemma 11.2, Corollary 15.3],

$$\psi U_i \subseteq U_{i-1}, \qquad \psi U_i^{\Downarrow} \subseteq U_{i-1}^{\Downarrow}$$

for $1 \leq i \leq d$ and both $\psi U_0 = 0$ and $\psi U_0^{\Downarrow} = 0$. In [5], it is shown how ψ is related to several maps, including the maps K, B, as well as the map Δ which we now recall. By [4, Lemma 9.5], there exists a unique linear transformation $\Delta : V \to V$ such that

$$\Delta U_i \subseteq U_i^{\Downarrow} \qquad\qquad (0 \leq i \leq d),$$
$$(\Delta - I)U_i \subseteq U_0 + U_1 + \cdots + U_{i-1} \quad (0 \leq i \leq d).$$

In [4, Theorem 17.1], the present author showed that both

$$\Delta = \sum_{i=0}^{d} \left(\prod_{j=1}^{i} \frac{aq^{j-1} - a^{-1}q^{1-j}}{q^j - q^{-j}} \right) \psi^i, \quad \Delta^{-1} = \sum_{i=0}^{d} \left(\prod_{j=1}^{i} \frac{a^{-1}q^{j-1} - aq^{1-j}}{q^j - q^{-j}} \right) \psi^i.$$

The primary goal of this paper is to provide factorizations of these power series in ψ and to investigate the consequences of these factorizations. We accomplish this goal using a linear transformation $\mathcal{M} : V \to V$ given by

$$M = \frac{aK - a^{-1}B}{a - a^{-1}}.$$

By construction, $M^{\Downarrow} = M$. One can quickly check that M is invertible. We show that the map M is equal to each of

$$(I - a^{-1}q\psi)^{-1}K, \quad K(I - a^{-1}q^{-1}\psi)^{-1}, \quad (I - aq\psi)^{-1}B, \quad B(I - aq^{-1}\psi)^{-1}.$$

We give a number of different relations involving the maps M, K, B, ψ, the most significant of which are the following:

$$K \exp_q\left(\frac{a^{-1}}{q-q^{-1}}\psi\right) = \exp_q\left(\frac{a^{-1}}{q-q^{-1}}\psi\right)M,$$

$$B \exp_q\left(\frac{a}{q-q^{-1}}\psi\right) = \exp_q\left(\frac{a}{q-q^{-1}}\psi\right)M.$$

Using these equations, we obtain our main result which is that both

$$\Delta = \exp_q\left(\frac{a}{q-q^{-1}}\psi\right)\exp_{q^{-1}}\left(-\frac{a^{-1}}{q-q^{-1}}\psi\right),$$

$$\Delta^{-1} = \exp_q\left(\frac{a^{-1}}{q-q^{-1}}\psi\right)\exp_{q^{-1}}\left(-\frac{a}{q-q^{-1}}\psi\right).$$

Due to its important role in the factorization of Δ, we explore the map M further. We show that M is diagonalizable with eigenvalues $q^d, q^{d-2}, q^{d-4}, \ldots, q^{-d}$. For $0 \leq i \leq d$, let W_i denote the eigenspace of M corresponding to the eigenvalue q^{d-2i}. We show that for $0 \leq i \leq d$,

$$U_i = \exp_q\left(\frac{a^{-1}}{q-q^{-1}}\psi\right)W_i, \qquad U_i^{\Downarrow} = \exp_q\left(\frac{a}{q-q^{-1}}\psi\right)W_i,$$

$$W_i = \exp_{q^{-1}}\left(-\frac{a^{-1}}{q-q^{-1}}\psi\right)U_i, \qquad W_i = \exp_{q^{-1}}\left(-\frac{a}{q-q^{-1}}\psi\right)U_i^{\Downarrow}.$$

In light of this result, we interpret the decomposition $\{W_i\}_{i=0}^d$ as a sort of halfway point between the first and second split decompositions. We explore this decomposition further and give the actions of $\psi, K, B, \Delta, A, A^*$ on $\{W_i\}_{i=0}^d$. We then give the actions of $M^{\pm 1}$ on $\{U_i\}_{i=0}^d, \{U_i^{\Downarrow}\}_{i=0}^d, \{V_i\}_{i=0}^d, \{V_i^*\}_{i=0}^d$. We conclude the paper with a discussion of the special case when A, A^* is a Leonard pair.

The present paper is organized as follows. In Sect. 2 we discuss some preliminary facts concerning TD pairs and TD systems. In Sect. 3 we discuss the split decompositions of V as well as the maps K and B. In Sect. 4 we discuss the map ψ. In Sect. 5 we recall the map Δ and give Δ as a power series in ψ. In Sect. 6 we introduce the map M and describe its relationship with A, K, B, ψ. In Sect. 7 we express Δ as a product of two linear transformations; one is a q-exponential in ψ and the other is a q^{-1}-exponential in ψ. In Sect. 8 we describe the eigenvalues and eigenspaces of $\mathcal{M}$ and discuss how the eigenspace decomposition of $\mathcal{M}$ is related to the first and second split decompositions. In Sect. 9 we discuss the actions of $\psi, K, B, \Delta, A, A^*$ on the eigenspace decomposition of $\mathcal{M}$. In Sect. 10 we describe the action of $\mathcal{M}$ on the first and second split decompositions of V, as well as on the eigenspace decompositions of A, A^*. In Sect. 11 we consider the case when A, A^* is a Leonard pair.

2 Preliminaries

When working with a tridiagonal pair, it is useful to consider a closely related object called a tridiagonal system. In order to define this object, we first recall some facts from elementary linear algebra [9, Section 2].

We use the following conventions. When we discuss an algebra, we mean a unital associative algebra. When we discuss a subalgebra, we assume that it has the same unit as the parent algebra.

Let V denote a vector space over $\mathbb{K}$ with finite positive dimension. By a *decomposition* of V, we mean a sequence of nonzero subspaces whose direct sum is V. Let $\text{End}(V)$ denote the $\mathbb{K}$-algebra consisting of all linear transformations from V to V. Let A denote a diagonalizable element in $\text{End}(V)$. Let $\{V_i\}_{i=0}^d$ denote an ordering of the eigenspaces of A. For $0 \le i \le d$ let θ_i be the eigenvalue of A corresponding to V_i. Define $E_i \in \text{End}(V)$ by $(E_i - I)V_i = 0$ and $E_i V_j = 0$ if $j \ne i$ $(0 \le j \le d)$. In other words, E_i is the projection map from V onto V_i. We refer to E_i as the *primitive idempotent* of A associated with θ_i. By elementary linear algebra, (i) $AE_i = E_i A = \theta_i E_i$ $(0 \le i \le d)$; (ii) $E_i E_j = \delta_{ij} E_i$ $(0 \le i, j \le d)$; (iii) $V_i = E_i V$ $(0 \le i \le d)$; (iv) $I = \sum_{i=0}^d E_i$. Moreover

$$E_i = \prod_{\substack{0 \le j \le d \\ j \ne i}} \frac{A - \theta_j I}{\theta_i - \theta_j} \qquad (0 \le i \le d).$$

Let $\mathcal{M}$ denote the subalgebra of $\text{End}(V)$ generated by A. Note that each of $\{A^i\}_{i=0}^d$, $\{E_i\}_{i=0}^d$ is a basis for the $\mathbb{K}$-vector space $\mathcal{M}$.

Let A, A^* denote a TD pair on V. An ordering of the eigenspaces of A (resp. A^*) is said to be *standard* whenever it satisfies (1) (resp. (2)). Let $\{V_i\}_{i=0}^d$ denote a standard ordering of the eigenspaces of A. By [9, Lemma 2.4], the ordering $\{V_{d-i}\}_{i=0}^d$ is standard and no further ordering of the eigenspaces of A is standard. A similar result holds for the eigenspaces of A^*. An ordering of the primitive idempotents of A (resp. A^*) is said to be *standard* whenever the corresponding ordering of the eigenspaces of A (resp. A^*) is standard.

Definition 2 ([17, Definition 2.1]) Let V denote a vector space over $\mathbb{K}$ with finite positive dimension. By a *tridiagonal system* (or *TD system*) on V, we mean a sequence

$$\Phi = (A; \{E_i\}_{i=0}^d; A^*; \{E_i^*\}_{i=0}^d)$$

that satisfies (i)–(iii) below.

(i) A, A^* is a tridiagonal pair on V.
(ii) $\{E_i\}_{i=0}^d$ is a standard ordering of the primitive idempotents of A.
(iii) $\{E_i^*\}_{i=0}^d$ is a standard ordering of the primitive idempotents of A^*.

We call d the *diameter* of Φ, and say Φ is *over* $\mathbb{K}$. For notational convenience, set $E_{-1} = 0, E_{d+1} = 0, E_{-1}^* = 0, E_{d+1}^* = 0$.

In Definition 2 we do not assume that the primitive idempotents $\{E_i\}_{i=0}^d, \{E_i^*\}_{i=0}^d$ all have rank 1. A TD system for which each of these primitive idempotents has rank 1 is called a Leonard system [20]. The Leonard systems are classified up to isomorphism [20, Theorem 1.9].

For the rest of this paper, fix a TD system Φ on V as in Definition 2. Our TD system Φ can be modified in a number of ways to get a new TD system [9, Section 3]. For example, the sequence

$$\Phi^{\Downarrow} = (A; \{E_{d-i}\}_{i=0}^d; A^*; \{E_i^*\}_{i=0}^d)$$

is a TD system on V. Following [9, Section 3], we call $\Phi^{\Downarrow}$ the *second inversion* of Φ. When discussing $\Phi^{\Downarrow}$, we use the following notational convention. For any object f associated with Φ, let $f^{\Downarrow}$ denote the corresponding object associated with $\Phi^{\Downarrow}$.

Definition 3 For $0 \le i \le d$ let θ_i (resp. θ_i^*) denote the eigenvalue of A (resp. A^*) associated with E_i (resp. E_i^*). We refer to $\{\theta_i\}_{i=0}^d$ (resp. $\{\theta_i^*\}_{i=0}^d$) as the *eigenvalue sequence* (resp. *dual eigenvalue sequence*) of Φ.

By construction $\{\theta_i\}_{i=0}^d$ are mutually distinct and $\{\theta_i^*\}_{i=0}^d$ are mutually distinct. By [9, Theorem 11.1], the scalars

$$\frac{\theta_{i-2} - \theta_{i+1}}{\theta_{i-1} - \theta_i}, \qquad \frac{\theta_{i-2}^* - \theta_{i+1}^*}{\theta_{i-1}^* - \theta_i^*}$$

are equal and independent of i for $2 \le i \le d - 1$. For this restriction, the solutions have been found in closed form [9, Theorem 11.2]. The most general solution is called q-Racah [12, Section 1]. This solution is described as follows.

Definition 4 Let Φ denote a TD system on V as in Definition 2. We say that Φ has *q-Racah type* whenever there exist nonzero scalars $q, a, b \in \mathbb{K}$ such that such that $q^4 \ne 1$ and

$$\theta_i = aq^{d-2i} + a^{-1}q^{2i-d}, \qquad \theta_i^* = bq^{d-2i} + b^{-1}q^{2i-d} \qquad (3)$$

for $0 \le i \le d$.

Note 2 Referring to Definition 4, the scalars q, a, b are not uniquely defined by Φ. If q, a, b is one solution, then their inverses give another solution.

For the rest of the paper, we make the following assumption.

Assumption 1 *We assume that our TD system Φ has q-Racah type. We fix q, a, b as in Definition 4.*

Lemma 1 ([5, Lemma 2.4]) *With reference to Assumption 1, the following hold.*

(i) *Neither of a^2, b^2 is among $q^{2d-2}, q^{2d-4}, \ldots, q^{2-2d}$.*
(ii) *$q^{2i} \neq 1$ for $1 \leq i \leq d$.*

Proof The result follows from the comment below Definition 3. $\qquad\square$

3   The First and Second Split Decomposition of V

Recall the TD system Φ from Assumption 1. In this section we consider two decompositions of V associated with Φ, called the first and second split decomposition.
For $0 \leq i \leq d$ define

$$U_i = (E_0^* V + E_1^* V + \cdots + E_i^* V) \cap (E_i V + E_{i+1} V + \cdots + E_d V).$$

For notational convenience, define $U_{-1} = 0$ and $U_{d+1} = 0$. Note that for $0 \leq i \leq d$,

$$U_i^{\Downarrow} = (E_0^* V + E_1^* V + \cdots + E_i^* V) \cap (E_0 V + E_1 V + \cdots + E_{d-i} V).$$

By [9, Theorem 4.6], the sequence $\{U_i\}_{i=0}^d$ (resp. $\{U_i^{\Downarrow}\}_{i=0}^d$) is a decomposition of V. Following [9], we refer to $\{U_i\}_{i=0}^d$ (resp. $\{U_i^{\Downarrow}\}_{i=0}^d$) as the *first split decomposition* (resp. *second split decomposition*) of V with respect to Φ. By [9, Corollary 5.7], for $0 \leq i \leq d$ the dimensions of $E_i V, E_i^* V, U_i, U_i^{\Downarrow}$ coincide; we denote the common dimension by ρ_i. By [9, Theorem 4.6],

$$E_i V + E_{i+1} V + \cdots + E_d V = U_i + U_{i+1} + \cdots + U_d, \quad (4)$$

$$E_0 V + E_1 V + \cdots + E_i V = U_{d-i}^{\Downarrow} + U_{d-i+1}^{\Downarrow} + \cdots + U_d^{\Downarrow}, \quad (5)$$

$$E_0^* V + E_1^* V + \cdots E_i^* V = U_0 + U_1 + \cdots + U_i = U_0^{\Downarrow} + U_1^{\Downarrow} + \cdots + U_i^{\Downarrow}. \quad (6)$$

By [9, Theorem 4.6], A and A^* act on the first split decomposition in the following way:

$$(A - \theta_i I)U_i \subseteq U_{i+1} \quad (0 \leq i \leq d-1), \qquad (A - \theta_d I)U_d = 0,$$

$$(A^* - \theta_i^* I)U_i \subseteq U_{i-1} \quad (1 \leq i \leq d), \qquad (A^* - \theta_0^* I)U_0 = 0.$$

By [9, Theorem 4.6], A and A^* act on the second split decomposition in the following way:

$$(A - \theta_{d-i}I)U_i^{\Downarrow} \subseteq U_{i+1}^{\Downarrow} \qquad (0 \le i \le d-1), \qquad (A - \theta_0 I)U_d^{\Downarrow} = 0,$$

$$(A^* - \theta_i^* I)U_i^{\Downarrow} \subseteq U_{i-1}^{\Downarrow} \qquad (1 \le i \le d), \qquad (A^* - \theta_0^* I)U_0^{\Downarrow} = 0.$$

Definition 5 ([5, Definitions 3.1 and 3.2]) Define $K, B \in \text{End}(V)$ such that for $0 \le i \le d$, U_i (resp. $U_i^{\Downarrow}$) is the eigenspace of K (resp. B) with eigenvalue q^{d-2i}. In other words,

$$(K - q^{d-2i}I)U_i = 0, \qquad (B - q^{d-2i}I)U_i^{\Downarrow} = 0 \qquad (0 \le i \le d). \quad (7)$$

Observe that $B = K^{\Downarrow}$.

By construction each of K, B is invertible and diagonalizable on V.

We now describe how K and B act on the eigenspaces of the other one.

Lemma 2 ([5, Lemma 3.3]) *For* $0 \le i \le d$,

$$(B - q^{d-2i}I)U_i \subseteq U_0 + U_1 + \cdots + U_{i-1}, \tag{8}$$

$$(K - q^{d-2i}I)U_i^{\Downarrow} \subseteq U_0^{\Downarrow} + U_1^{\Downarrow} + \cdots + U_{i-1}^{\Downarrow}. \tag{9}$$

Next we describe how A, K, B are related.

Lemma 3 ([13, Section 1.1]) *Both*

$$\frac{qKA - q^{-1}AK}{q - q^{-1}} = aK^2 + a^{-1}I, \qquad \frac{qBA - q^{-1}AB}{q - q^{-1}} = a^{-1}B^2 + aI. \tag{10}$$

Lemma 4 ([5, Theorem 9.9]) *We have*

$$aK^2 - \frac{a^{-1}q - aq^{-1}}{q - q^{-1}} KB - \frac{aq - a^{-1}q^{-1}}{q - q^{-1}} BK + a^{-1}B^2 = 0. \tag{11}$$

4 The Linear Transformation ψ

We continue to discuss the situation of Assumption 1. In [4, Section 11] we introduced an element $\Psi \in \text{End}(V)$. In [5] we used the normalization $\psi = (q - q^{-1})(q^d - q^{-d})\Psi$. In [5, Theorem 9.8], we showed that ψ is equal to some rational expressions involving K, B. We now recall this result. We start with a comment.

Lemma 5 ([5, Lemma 9.7]) *Each of the following is invertible:*

$$aI - a^{-1}BK^{-1}, \qquad a^{-1}I - aKB^{-1}, \qquad (12)$$

$$aI - a^{-1}K^{-1}B, \qquad a^{-1}I - aB^{-1}K. \qquad (13)$$

Lemma 6 ([5, Theorem 9.8]) *The following four expressions coincide:*

$$\frac{I - BK^{-1}}{q(aI - a^{-1}BK^{-1})}, \qquad \frac{I - KB^{-1}}{q(a^{-1}I - aKB^{-1})}, \qquad (14)$$

$$\frac{q(I - K^{-1}B)}{aI - a^{-1}K^{-1}B}, \qquad \frac{q(I - B^{-1}K)}{a^{-1}I - aB^{-1}K}. \qquad (15)$$

In (14), (15) *the denominators are invertible by Lemma 5.*

Definition 6 Define $\psi \in \text{End}(V)$ to be the common value of the four expressions in Lemma 6.

We now recall some facts concerning ψ.

Lemma 7 ([5, Lemma 5.4]) *Both*

$$K\psi = q^2\psi K, \qquad B\psi = q^2\psi B. \qquad (16)$$

Lemma 8 ([4, Lemma 11.2, Corollary 15.3]) *We have*

$$\psi U_i \subseteq U_{i-1}, \qquad \psi U_i^{\Downarrow} \subseteq U_{i-1}^{\Downarrow} \qquad (1 \leq i \leq d) \qquad (17)$$

and also $\psi U_0 = 0$ and $\psi U_0^{\Downarrow} = 0$. Moreover $\psi^{d+1} = 0$.

In Lemma 6 we obtained ψ as a rational expression in BK^{-1} or $K^{-1}B$. Next we solve for BK^{-1} and $K^{-1}B$ as a rational function in ψ. In order to state the answer, we will need the following result.

Lemma 9 ([5, Lemma 9.2]) *Each of the following is invertible:*

$$I - aq\psi, \qquad I - a^{-1}q\psi, \qquad I - aq^{-1}\psi, \qquad I - a^{-1}q^{-1}\psi. \qquad (18)$$

Their inverses are as follows:

$$(I - aq\psi)^{-1} = \sum_{i=0}^{d} a^i q^i \psi^i, \qquad (I - a^{-1}q\psi)^{-1} = \sum_{i=0}^{d} a^{-i} q^i \psi^i, \qquad (19)$$

$$(I - aq^{-1}\psi)^{-1} = \sum_{i=0}^{d} a^i q^{-i} \psi^i, \qquad (I - a^{-1}q^{-1}\psi)^{-1} = \sum_{i=0}^{d} a^{-i} q^{-i} \psi^i. \qquad (20)$$

The next result is an immediate consequence of Lemma 6, Definition 6, and Lemma 9.

Theorem 1 ([5, Theorem 9.4]) *The following hold:*

$$BK^{-1} = \frac{I - aq\psi}{I - a^{-1}q\psi}, \qquad\qquad KB^{-1} = \frac{I - a^{-1}q\psi}{I - aq\psi}, \qquad (21)$$

$$K^{-1}B = \frac{I - aq^{-1}\psi}{I - a^{-1}q^{-1}\psi}, \qquad\qquad B^{-1}K = \frac{I - a^{-1}q^{-1}\psi}{I - aq^{-1}\psi}. \qquad (22)$$

In (21), (22) *the denominators are invertible by Lemma 9.*

Lemma 10 ([5, Equation (22)]) *We have*

$$\frac{\psi A - A\psi}{q - q^{-1}} = (I - aq\psi) K - \left(I - a^{-1}q^{-1}\psi\right) K^{-1}. \qquad (23)$$

Proof This result is a reformulation of [5, Equation (22)] using [5, Equation (14)].
□

5   The Linear Transformation Δ

We continue to discuss the situation of Assumption 1. In [4, Section 9] we introduced an invertible element $\Delta \in \mathrm{End}(V)$. In [4] we showed that Δ, ψ commute and in fact both Δ, Δ^{-1} are power series in ψ. These power series will be the central focus of this paper. We will show that each of those power series factors as a product of two power series, each of which is a quantum exponential in ψ.

Lemma 11 ([4, Lemma 9.5]) *There exists a unique* $\Delta \in \mathrm{End}(V)$ *such that*

$$\Delta U_i \subseteq U_i^{\Downarrow} \qquad\qquad (0 \le i \le d), \qquad (24)$$

$$(\Delta - I)U_i \subseteq U_0 + U_1 + \cdots + U_{i-1} \qquad (0 \le i \le d). \qquad (25)$$

Lemma 12 ([4, Lemmas 9.3 and 9.6]) *The map* Δ *is invertible. Moreover* $\Delta^{-1} = \Delta^{\Downarrow}$ *and*

$$(\Delta^{-1} - I)U_i \subseteq U_0 + U_1 + \cdots + U_{i-1} \qquad (0 \le i \le d). \qquad (26)$$

Lemma 13 *The map* $\Delta - I$ *is nilpotent. Moreover* $\Delta K = B\Delta$.

Proof The first assertion follows from (25). The last assertion follows from (24) and Definition 5.
□

The map Δ is characterized as follows.

Lemma 14 ([4, Lemma 9.8]) *The map* Δ *is the unique element of* $\mathrm{End}(V)$ *such that*

$$(\Delta - I)E_i^* V \subseteq E_0^* V + E_1^* V + \cdots + E_{i-1}^* V \qquad (0 \le i \le d),$$

$$(27)$$

$$\Delta(E_i V + E_{i+1} V + \cdots + E_d V) = E_0 V + E_1 V + \cdots + E_{d-i} V \qquad (0 \le i \le d).$$

$$(28)$$

Theorem 2 ([4, Theorem 17.1]) *Both*

$$\Delta = \sum_{i=0}^{d} \left(\prod_{j=1}^{i} \frac{aq^{j-1} - a^{-1}q^{1-j}}{q^j - q^{-j}} \right) \psi^i, \tag{29}$$

$$\Delta^{-1} = \sum_{i=0}^{d} \left(\prod_{j=1}^{i} \frac{a^{-1}q^{j-1} - aq^{1-j}}{q^j - q^{-j}} \right) \psi^i. \tag{30}$$

In (29) and (30), the elements Δ, Δ^{-1} are expressed as a power series in ψ. In the present paper, we factor these power series and interpret the results. This interpretation will involve a linear transformation $\mathcal{M}$. We introduce $\mathcal{M}$ in the next section.

6   The Linear Transformation $\mathcal{M}$

We continue to discuss the situation of Assumption 1. In this section we introduce an element $\mathcal{M} \in \mathrm{End}(V)$. We explain how $\mathcal{M}$ is related to K, B, ψ, A.

Definition 7 Define $\mathcal{M} \in \mathrm{End}(V)$ by

$$\mathcal{M} = \frac{aK - a^{-1}B}{a - a^{-1}}. \tag{31}$$

By construction, $\mathcal{M}^{\Downarrow} = \mathcal{M}$. Evaluating (31) using Lemma 5, we see that $\mathcal{M}$ is invertible.

Lemma 15 *The map $\mathcal{M}$ is equal to each of:*

$$(I - a^{-1}q\psi)^{-1}K, \qquad K(I - a^{-1}q^{-1}\psi)^{-1}, \qquad (I - aq\psi)^{-1}B, \qquad B(I - aq^{-1}\psi)^{-1}.$$

Proof We first show that $\mathcal{M} = (I - a^{-1}q\psi)^{-1}K$. By Definition 7,

$$(a - a^{-1})\mathcal{M}K^{-1} = aI - a^{-1}BK^{-1}.$$

The result follows from this fact along with the equation on the left in (21).

The remaining assertions follow from Theorem 1.   □

Lemma 15 can be reformulated as follows.

Lemma 16 *We have*

$$K = \left(I - a^{-1}q\psi\right)M, \qquad\qquad K = M\left(I - a^{-1}q^{-1}\psi\right), \qquad (32)$$

$$B = (I - aq\psi)M, \qquad\qquad B = M\left(I - aq^{-1}\psi\right). \qquad (33)$$

For later use, we give several descriptions of $M^{\pm 1}$.

Lemma 17 *The map M^{-1} is equal to each of:*

$$K^{-1}(I - a^{-1}q\psi), \quad (I - a^{-1}q^{-1}\psi)K^{-1}, \quad B^{-1}(I - aq\psi), \quad (I - aq^{-1}\psi)B^{-1}.$$

Proof Immediate from Lemma 15.   □

Lemma 18 *The map M is equal to each of:*

$$K\sum_{n=0}^{d} a^{-n}q^{-n}\psi^{n}, \qquad \sum_{n=0}^{d} a^{-n}q^{n}\psi^{n}K, \qquad B\sum_{n=0}^{d} a^{n}q^{-n}\psi^{n}, \qquad \sum_{n=0}^{d} a^{n}q^{n}\psi^{n}B$$

$$(34)$$

Proof Use Lemmas 9 and 15.   □

We now give some attractive equations that show how M is related to ψ, K, B, A.

Lemma 19 *We have*

$$M\psi = q^{2}\psi M. \qquad (35)$$

Proof Use Lemma 7 and Definition 7.   □

Lemma 20 *We have*

$$\frac{qM^{-1}K - q^{-1}KM^{-1}}{q - q^{-1}} = I, \qquad\qquad \frac{qM^{-1}B - q^{-1}BM^{-1}}{q - q^{-1}} = I. \qquad (36)$$

Proof Use Lemma 17.   □

Lemma 21 *We have*

$$\frac{qAM^{-1} - q^{-1}M^{-1}A}{q - q^{-1}} = (a + a^{-1})I - (q + q^{-1})\psi. \qquad (37)$$

Proof Use Lemmas 3, 7, 10, and 17.   □

Lemma 22 *We have*

$$M^{-2}A - (q^2+q^{-2})M^{-1}AM^{-1} + AM^{-2} = -(q-q^{-1})^2(a+a^{-1})M^{-1}. \quad (38)$$

Proof Use Lemmas 19 and 21. □

7 A Factorization of Δ

We continue to discuss the situation of Assumption 1. We now bring in the q-exponential function [8]. In [4, Theorem 17.1] we expressed Δ as a power series in ψ. In this section we strengthen this result in the following way. We express Δ as a product of two linear transformations; one is a q-exponential in ψ and the other is a q^{-1}-exponential in ψ.

For an integer n, define

$$[n]_q = \frac{q^n - q^{-n}}{q - q^{-1}} \quad (39)$$

and for $n \geq 0$, define

$$[n]_q^! = [n]_q[n-1]_q \cdots [1]_q. \quad (40)$$

We interpret $[0]_q^! = 1$.

We now recall the q-exponential function [8]. For a nilpotent $T \in \text{End}(V)$,

$$\exp_q(T) = \sum_{n=0}^{\infty} \frac{q^{\binom{n}{2}}}{[n]_q^!} T^n. \quad (41)$$

The map $\exp_q(T)$ is invertible. Its inverse is given by

$$\exp_{q^{-1}}(-T) = \sum_{n=0}^{\infty} \frac{(-1)^n q^{-\binom{n}{2}}}{[n]_q^!} T^n. \quad (42)$$

Using (41) we obtain

$$(I - (q^2 - 1)T)\exp_q(q^2 T) = \exp_q(T). \quad (43)$$

For $S \in \text{End}(V)$ such that $ST = q^2 TS$, we have

$$S\exp_q(T)S^{-1} = \exp_q(STS^{-1}) = \exp_q(q^2 T).$$

Consequently

$$S \exp_q(T) = \exp_q(q^2 T) S. \tag{44}$$

Combining (43) and (44),

$$(I - (q^2 - 1)T)S \exp_q(T) = \exp_q(T)S. \tag{45}$$

We return our attention to K, B, ψ, M.

Proposition 1 *Both*

$$K \exp_q\left(\frac{a^{-1}}{q - q^{-1}}\psi\right) = \exp_q\left(\frac{a^{-1}}{q - q^{-1}}\psi\right) M, \tag{46}$$

$$B \exp_q\left(\frac{a}{q - q^{-1}}\psi\right) = \exp_q\left(\frac{a}{q - q^{-1}}\psi\right) M. \tag{47}$$

Proof Recall from Lemma 19 that $M\psi = q^2\psi M$. We first obtain (46). To do this, in (45) take $S = M$ and $T = \frac{a^{-1}}{q-q^{-1}}\psi$. Evaluate the result using the equation $M = (I - a^{-1}q\psi)^{-1}K$ from Lemma 15.

Next we obtain (47). To do this, in (45) take $S = M$ and $T = \frac{a}{q-q^{-1}}\psi$. Evaluate the result using the equation $M = (I - aq\psi)^{-1}B$ from Lemma 15. $\square$

The following is our main result.

Theorem 3 *Both*

$$\Delta = \exp_q\left(\frac{a}{q - q^{-1}}\psi\right) \exp_{q^{-1}}\left(-\frac{a^{-1}}{q - q^{-1}}\psi\right), \tag{48}$$

$$\Delta^{-1} = \exp_q\left(\frac{a^{-1}}{q - q^{-1}}\psi\right) \exp_{q^{-1}}\left(-\frac{a}{q - q^{-1}}\psi\right). \tag{49}$$

Proof We first show (48). Let $\tilde{\Delta}$ denote the expression on the right in (48). Combining (46) and (47), we see that $\tilde{\Delta}K = B\tilde{\Delta}$. Therefore $\tilde{\Delta}U_i = U_i^{\Downarrow}$ for $0 \le i \le d$. Observe that $\tilde{\Delta} - I$ is a polynomial in ψ with zero constant term. By Lemma 8, $(\tilde{\Delta} - I)U_i \subseteq U_0 + U_1 + \cdots + U_{i-1}$ for $0 \le i \le d$. By Lemma 11, $\tilde{\Delta} = \Delta$.

To obtain (49) from (48), use (42). $\square$

Corollary 1 *We have*

$$\exp_q\left(\frac{a}{q - q^{-1}}\psi\right) \exp_{q^{-1}}\left(-\frac{a^{-1}}{q - q^{-1}}\psi\right) = \sum_{i=0}^{d}\left(\prod_{j=1}^{i}\frac{aq^{j-1} - a^{-1}q^{1-j}}{q^j - q^{-j}}\right)\psi^i,$$

$$\exp_q \left(\frac{a^{-1}}{q-q^{-1}} \psi \right) \exp_{q^{-1}} \left(-\frac{a}{q-q^{-1}} \psi \right) = \sum_{i=0}^{d} \left(\prod_{j=1}^{i} \frac{a^{-1}q^{j-1} - aq^{1-j}}{q^j - q^{-j}} \right) \psi^i.$$

Proof Combine Theorems 2 and 3. The equations can also be obtained directly by expanding their left-hand sides using (41) and (42), and evaluating the results using the q-binomial theorem [2, Theorem 10.2.1]. □

8 The Eigenvalues and Eigenspaces of $\mathcal{M}$

We continue to discuss the situation of Assumption 1. In Sect. 6 we introduced the linear transformation $\mathcal{M}$. Proposition 1 indicates the role of $\mathcal{M}$ in the factorization of Δ in Theorem 3. In this section we show that $\mathcal{M}$ is diagonalizable. We describe the eigenvalues and eigenspaces of $\mathcal{M}$. We also explain how the eigenspace decomposition for $\mathcal{M}$ is related to the first and second split decompositions.

Lemma 23 *The map $\mathcal{M}$ is diagonalizable with eigenvalues* $q^d, q^{d-2}, q^{d-4}, \ldots, q^{-d}$.

Proof Let $E = \exp_q \left(\frac{a^{-1}}{q-q^{-1}} \psi \right)$. By (46), $\mathcal{M} = E^{-1} K E$. By construction K is diagonalizable with eigenvalues $q^d, q^{d-2}, q^{d-4}, \ldots, q^{-d}$. The result follows. □

Definition 8 For $0 \leq i \leq d$, let W_i denote the eigenspace of $\mathcal{M}$ corresponding to the eigenvalue q^{d-2i}. Note that $\{W_i\}_{i=0}^{d}$ is a decomposition of V, and that $W_i^{\Downarrow} = W_i$ for $0 \leq i \leq d$. For notational convenience, let $W_{-1} = 0$ and $W_{d+1} = 0$.

Proposition 2 *For* $0 \leq i \leq d$,

$$U_i = \exp_q \left(\frac{a^{-1}}{q-q^{-1}} \psi \right) W_i, \qquad U_i^{\Downarrow} = \exp_q \left(\frac{a}{q-q^{-1}} \psi \right) W_i, \qquad (50)$$

$$W_i = \exp_{q^{-1}} \left(-\frac{a^{-1}}{q-q^{-1}} \psi \right) U_i, \qquad W_i = \exp_{q^{-1}} \left(-\frac{a}{q-q^{-1}} \psi \right) U_i^{\Downarrow}. \quad (51)$$

Proof Define E as in the proof of Lemma 23. We show that $U_i = E W_i$. By (46), $KE = E\mathcal{M}$. Recall that U_i (resp. W_i) is the eigenspace of K (resp. $\mathcal{M}$) corresponding to the eigenvalue q^{d-2i}. By these comments $U_i = E W_i$.

Define $F = \exp_q(\frac{a}{q-q^{-1}} \psi)$. We show $U_i^{\Downarrow} = F W_i$. By (47), $BF = F\mathcal{M}$. Recall that $U_i^{\Downarrow}$ (resp. W_i) is the eigenspace of B (resp. $\mathcal{M}$) corresponding to the eigenvalue q^{d-2i}. By these comments $U_i^{\Downarrow} = F W_i$.

To obtain (51) from (50), use (42). □

Lemma 24 *For* $0 \leq i \leq d$, *the dimension of W_i is ρ_i.*

Proof This follows from Proposition 2 and the fact that U_i, $U_i^{\Downarrow}$ have dimension ρ_i. □

Recall from (6) that

$$\sum_{h=0}^{i} E_h^* V = \sum_{h=0}^{i} U_h = \sum_{h=0}^{i} U_h^{\Downarrow} \tag{52}$$

for $0 \le i \le d$.

Lemma 25 *For $0 \le i \le d$, the sum $\sum_{h=0}^{i} W_h$ is equal to the common value of (52).*

Proof Define $W = \sum_{h=0}^{i} W_h$ and let U denote the common value of (52). We show that $W = U$. By Lemma 8 and the equation on the left in (51), $W \subseteq U$. By Lemma 24, W and U have the same dimension. Thus $W = U$.   □

9   The Actions of ψ, K, B, Δ, A, A^* on $\{W_i\}_{i=0}^{d}$

We continue to discuss the situation of Assumption 1. Recall the eigenspace decomposition $\{W_i\}_{i=0}^{d}$ for M. In this section, we discuss the actions of ψ, K, B, Δ, A, A^* on $\{W_i\}_{i=0}^{d}$.

Lemma 26 *For $0 \le i \le d$,*

$$\psi W_i \subseteq W_{i-1}. \tag{53}$$

Proof Use Lemma 19.   □

Lemma 27 *For $0 \le i \le d$,*

$$(K - q^{d-2i} I) W_i \subseteq W_{i-1}, \qquad (B - q^{d-2i} I) W_i \subseteq W_{i-1}. \tag{54}$$

Proof Use Lemmas 16 and 26.   □

Lemma 28 *For $0 \le i \le d$,*

$$(\Delta - I) W_i \subseteq W_0 + W_1 + \cdots + W_{i-1}, \tag{55}$$

$$(\Delta^{-1} - I) W_i \subseteq W_0 + W_1 + \cdots + W_{i-1}. \tag{56}$$

Proof To show (55), use (25) and Lemma 25.
To show (56), use (26) and Lemma 25.   □

Lemma 29 *For $0 \le i \le d$,*

$$(A - (a + a^{-1}) q^{d-2i} I) W_i \subseteq W_{i-1} + W_{i+1}. \tag{57}$$

Proof By Lemma 22, the expression

$$(M^{-1} - q^{2i+2-d}I)(M^{-1} - q^{2i-2-d}I)(A - (a + a^{-1})q^{d-2i}I)$$

vanishes on W_i. Therefore $(M^{-1} - q^{2i+2-d}I)(M^{-1} - q^{2i-2-d}I)$ vanishes on $(A - (a + a^{-1})q^{d-2i}I)W_i$. The result follows.　　　□

Lemma 30 *For $0 \le i \le d$,*

$$(A^* - \theta_i^*I)W_i \subseteq W_0 + W_1 + \cdots + W_{i-1}. \tag{58}$$

Proof Use $(A^* - \theta_i^*I)E_i^*V = 0$ together with (25) and Lemma 25.　　　□

10　The Actions of $M^{\pm 1}$ on $\{U_i\}_{i=0}^d, \{U_i^{\Downarrow}\}_{i=0}^d, \{E_iV\}_{i=0}^d,$ $\{E_i^*V\}_{i=0}^d$

We continue to discuss the situation of Assumption 1. In Sect. 8 we saw how various operators act on the decomposition $\{W_i\}_{i=0}^d$. In this section we investigate the action of M on the first and second split decompositions of V, as well as on the eigenspace decompositions of A, A^*.

Lemma 31 *For $0 \le i \le d$,*

$$(M - q^{d-2i}I)U_i \subseteq U_0 + U_1 + \cdots + U_{i-1}, \tag{59}$$
$$(M - q^{d-2i}I)U_i^{\Downarrow} \subseteq U_0^{\Downarrow} + U_1^{\Downarrow} + \cdots + U_{i-1}^{\Downarrow}. \tag{60}$$

Proof To show (59), use Definition 5, Lemma 2, and Definition 7.
　　To show (60), use (59) applied to $\Phi^{\Downarrow}$, along with $M^{\Downarrow} = M$.　　　□

Lemma 32 *For $0 \le i \le d$,*

$$(M^{-1} - q^{2i-d}I)U_i \subseteq U_{i-1}, \qquad (M^{-1} - q^{2i-d}I)U_i^{\Downarrow} \subseteq U_{i-1}^{\Downarrow}. \tag{61}$$

Proof We first show the equation on the left in (61). By Lemma 17,

$$M^{-1} = (I - a^{-1}q^{-1}\psi)K^{-1}. \tag{62}$$

From this and Definition 5, it follows that on U_i,

$$M^{-1} - q^{2i-d}I = a^{-1}q^{2i-d-1}\psi. \tag{63}$$

The result follows from this along with Lemma 8.
　　The proof of the equation on the right in (61) follows from the equation on the left in (61) applied to $\Phi^{\Downarrow}$, along with the fact that $M^{\Downarrow} = M$.　　　□

Lemma 33 *For* $0 \le i \le d$,

$$M^{-1}E_iV \subseteq E_{i-1}V + E_iV + E_{i+1}V. \tag{64}$$

Proof We first show that $M^{-1}E_iV \subseteq \sum_{h=0}^{i+1} E_hV$. Recall from (5) that $E_iV \subseteq \sum_{h=d-i}^{d} U_h^{\Downarrow}$. By this, Lemma 32, and (5), we obtain $M^{-1}E_iV \subseteq \sum_{h=0}^{i+1} E_hV$.

We now show that $M^{-1}E_iV \subseteq \sum_{h=i-1}^{d} E_hV$. Recall from (4) that $E_iV \subseteq \sum_{h=i}^{d} U_h$. By this, Lemma 32, and (4), we obtain $M^{-1}E_iV \subseteq \sum_{h=i-1}^{d} E_hV$.

Thus $M^{-1}E_iV$ is contained in the intersection of $\sum_{h=0}^{i+1} E_hV$ and $\sum_{h=i-1}^{d} E_hV$, which is $E_{i-1}V + E_iV + E_{i+1}V$. $\qquad\square$

Lemma 34 *For* $0 \le i \le d$,

$$(M - q^{d-2i}I)E_i^*V \subseteq E_0^*V + E_1^*V + \cdots + E_{i-1}^*V,$$

$$(M^{-1} - q^{2i-d}I)E_i^*V \subseteq E_0^*V + E_1^*V + \cdots + E_{i-1}^*V.$$

Proof Note that $E_i^*V \subseteq E_0^*V + E_1^*V + \cdots + E_i^*V = W_0 + W_1 + \cdots + W_i$ by Lemma 25. The result follows from this fact along with Definition 8. $\qquad\square$

11 When Φ Is a Leonard System

We continue to discuss the situation of Assumption 1. For the rest of the paper we assume $\rho_i = 1$ for $0 \le i \le d$. In this case Φ is called a Leonard system.

We use the following notational convention. Let $\{v_i\}_{i=0}^{d}$ denote a basis for V. The sequence of subspaces $\{\mathbb{K}v_i\}_{i=0}^{d}$ is a decomposition of V said, to be *induced* by the basis $\{v_i\}_{i=0}^{d}$.

We display a basis $\{u_i\}_{i=0}^{d}$ (resp. $\{u_i^{\Downarrow}\}_{i=0}^{d}$) (resp. $\{w_i\}_{i=0}^{d}$) that induces the decomposition $\{U_i\}_{i=0}^{d}$ (resp. $\{U_i^{\Downarrow}\}_{i=0}^{d}$) (resp. $\{W_i\}_{i=0}^{d}$). We find the actions of ψ, K, B, $\Delta^{\pm 1}$, A on these bases. We also display the transition matrices between these bases.

For the rest of this section fix $0 \ne u_0 \in U_0$. Let M denote the subalgebra of End(V) generated by A. By [21, Lemma 5.1], the map $M \to V$, $X \mapsto Xu_0$ is an isomorphism of vector spaces. Consequently, the vectors $\{A^iu_0\}_{i=0}^{d}$ form a basis for V.

We now define a basis $\{u_i\}_{i=0}^{d}$ of V that induces $\{U_i\}_{i=0}^{d}$. For $0 \le i \le d$, define

$$u_i = \left(\prod_{j=0}^{i-1}(A - \theta_jI)\right)u_0. \tag{65}$$

Observe that $u_i \ne 0$. By [9, Theorem 4.6], $u_i \in U_i$. So u_i is a basis for U_i. Consequently, $\{u_i\}_{i=0}^{d}$ is a basis for V that induces $\{U_i\}_{i=0}^{d}$.

Next we define a basis $\{u_i^{\Downarrow}\}_{i=0}^d$ of V that induces $\{U_i^{\Downarrow}\}_{i=0}^d$. For $0 \le i \le d$, define

$$u_i^{\Downarrow} = \left(\prod_{j=0}^{i-1} (A - \theta_{d-j}I) \right) u_0. \tag{66}$$

Observe that $u_i^{\Downarrow} \ne 0$. By Lemma 11, $u_i^{\Downarrow} \in U_i^{\Downarrow}$. So $u_i^{\Downarrow}$ is a basis for $U_i^{\Downarrow}$. Consequently, $\{u_i^{\Downarrow}\}_{i=0}^d$ is a basis for V that induces $\{U_i^{\Downarrow}\}_{i=0}^d$.

Lemma 35 *For $0 \le i \le d$,*

$$u_i^{\Downarrow} = \Delta u_i. \tag{67}$$

Proof By Lemma 11, $\Delta U_i = U_i^{\Downarrow}$. So there exists $0 \ne \lambda \in \mathbb{K}$ such that $\Delta u_i = \lambda u_i^{\Downarrow}$. We show that $\lambda = 1$. By [4, Lemma 7.3] and (25), $\Delta u_i - A^i u$ is a linear combination of $\{A^j u\}_{j=0}^{i-1}$. Also, $u_i^{\Downarrow} - A^i u$ is a linear combination of $\{A^j u\}_{j=0}^{i-1}$. The vectors $\{A^j u\}_{j=0}^{i-1}$ are linearly independent. By these comments $\lambda = 1$. $\square$

We next define a basis $\{w_i\}_{i=0}^d$ of V that induces $\{W_i\}_{i=0}^d$. For $0 \le i \le d$, define

$$w_i = \exp_{q^{-1}} \left(-\frac{a^{-1}}{q - q^{-1}} \psi \right) u_i. \tag{68}$$

Since $\{u_i\}_{i=0}^d$ is a basis of V and $\exp_{q^{-1}}(-\frac{a^{-1}}{q-q^{-1}} \psi)$ is invertible, w_i is a basis for W_i. Consequently, $\{w_i\}_{i=0}^d$ is a basis for V that induces $\{W_i\}_{i=0}^d$.

Lemma 36 *For $0 \le i \le d$,*

$$u_i = \exp_q \left(\frac{a^{-1}}{q-q^{-1}} \psi \right) w_i, \qquad u_i^{\Downarrow} = \exp_q \left(\frac{a}{q-q^{-1}} \psi \right) w_i, \tag{69}$$

$$w_i = \exp_{q^{-1}} \left(-\frac{a^{-1}}{q-q^{-1}} \psi \right) u_i, \qquad w_i = \exp_{q^{-1}} \left(-\frac{a}{q-q^{-1}} \psi \right) u_i^{\Downarrow}. \tag{70}$$

Proof Use (68) to obtain the equations on the left in (69),(70). To obtain the equations on the right in (69),(70), use Theorem 3, Lemma 35, and (68). $\square$

We now describe the actions of ψ, K, B, M, Δ, A on the bases $\{u_i\}_{i=0}^d$, $\{u_i^{\Downarrow}\}_{i=0}^d$, $\{w_i\}_{i=0}^d$. First we recall a notion from linear algebra. Let $\mathrm{Mat}_{d+1}(\mathbb{K})$ denote the $\mathbb{K}$-algebra of $(d+1) \times (d+1)$ matrices that have all entries in $\mathbb{K}$. We index the rows and columns by $0, 1, \ldots, d$. Let $\{v_i\}_{i=0}^d$ denote a basis of V. For $T \in \mathrm{End}(V)$ and $X \in \mathrm{Mat}_{d+1}(\mathbb{K})$, we say that X *represents* T with respect to $\{v_i\}_{i=0}^d$ whenever $Tv_j = \sum_{i=0}^d X_{ij} v_i$ for $0 \le j \le d$.

By (65) and (66), the matrices that represent A with respect to $\{u_i\}_{i=0}^d$ and $\{u_i^{\Downarrow}\}_{i=0}^d$ are, respectively,

$$\begin{pmatrix} \theta_0 & & & \mathbf{0} \\ 1 & \theta_1 & & \\ & \ddots & \ddots & \\ \mathbf{0} & & 1 & \theta_d \end{pmatrix}, \qquad \begin{pmatrix} \theta_d & & & \mathbf{0} \\ 1 & \theta_{d-1} & & \\ & \ddots & \ddots & \\ \mathbf{0} & & 1 & \theta_0 \end{pmatrix}. \qquad (71)$$

By construction, the matrix $\operatorname{diag}(q^d, q^{d-2}, \ldots, q^{-d})$ represents K with respect to $\{u_i\}_{i=0}^d$, and B with respect to $\{u_i^{\Downarrow}\}_{i=0}^d$, and M with respect to $\{w_i\}_{i=0}^d$.

Definition 9 We define a matrix $\widehat{\psi} \in \operatorname{Mat}_{d+1}(\mathbb{K})$. For $1 \le i \le d$, the $(i-1, i)$-entry is $(q^i - q^{-i})(q^{d-i+1} - q^{i-d-1})$. All other entries are 0.

Proposition 3 *The matrix $\widehat{\psi}$ represents ψ with respect to each of the bases $\{u_i\}_{i=0}^d$, $\{u_i^{\Downarrow}\}_{i=0}^d$, $\{w_i\}_{i=0}^d$.*

Proof By [5, Line (23)], $\widehat{\psi}$ represents ψ with respect to $\{u_i\}_{i=0}^d$. The remaining assertions follow from Lemma 36. $\qquad\square$

Next we give the matrices that represent $M^{\pm 1}$ with respect to the bases $\{u_i\}_{i=0}^d$, $\{u_i^{\Downarrow}\}_{i=0}^d$.

Lemma 37 *We give the matrix in $\operatorname{Mat}_{d+1}(\mathbb{K})$ that represents M with respect to $\{u_i\}_{i=0}^d$. This matrix is upper triangular. For $0 \le i \le j \le d$, the (i, j)-entry is*

$$a^{i-j} q^{d-j-i} \left(q - q^{-1}\right)^{2(j-i)} \frac{[j]_q^! [d-i]_q^!}{[i]_q^! [d-j]_q^!}. \qquad (72)$$

Proof The matrix $\operatorname{diag}(q^d, q^{d-2}, \ldots, q^{-d})$ represents K with respect to $\{u_i\}_{i=0}^d$. Use this fact along with Lemma 18 and Proposition 3. $\qquad\square$

Lemma 38 *We give the matrix in $\operatorname{Mat}_{d+1}(\mathbb{K})$ that represents M^{-1} with respect to $\{u_i\}_{i=0}^d$. For $0 \le i \le d$, the (i, i)-entry is q^{2i-d}. For $1 \le i \le d$, the $(i-1, i)$-entry is*

$$-a^{-1} q^{2i-d-1} \left(q^i - q^{-i}\right) \left(q^{d-i+1} - q^{i-d-1}\right).$$

All other entries are zero.

Proof The matrix $\operatorname{diag}(q^{-d}, q^{2-d}, \ldots, q^d)$ represents K^{-1} with respect to $\{u_i\}_{i=0}^d$. Use this fact along with Lemma 17 and Proposition 3. $\qquad\square$

Lemma 39 *We give the matrix in $\operatorname{Mat}_{d+1}(\mathbb{K})$ that represents M with respect to $\{u_i^{\Downarrow}\}_{i=0}^d$. This matrix is upper triangular. For $0 \le i \le j \le d$, the (i, j)-entry is*

$$a^{j-i} q^{d-j-i} \left(q - q^{-1}\right)^{2(j-i)} \frac{[j]_q^! [d-i]_q^!}{[i]_q^! [d-j]_q^!}. \qquad (73)$$

Proof The matrix $\mathrm{diag}(q^d, q^{d-2}, \ldots, q^{-d})$ represents B with respect to $\{u_i^{\Downarrow}\}_{i=0}^d$. Use this fact along with Lemma 18 and Proposition 3. $\qquad\square$

Lemma 40 *We give the matrix in* $\mathrm{Mat}_{d+1}(\mathbb{K})$ *that represents* $\mathcal{M}^{-1}$ *with respect to* $\{u_i^{\Downarrow}\}_{i=0}^d$. *For* $0 \le i \le d$, *the* (i, i)-*entry is* q^{2i-d}. *For* $1 \le i \le d$, *the* $(i-1, i)$-*entry is*

$$-aq^{2i-d-1}\left(q^i - q^{-i}\right)\left(q^{d-i+1} - q^{i-d-1}\right).$$

All other entries are zero.

Proof The matrix $\mathrm{diag}(q^{-d}, q^{2-d}, \ldots, q^d)$ represents B^{-1} with respect to $\{u_i^{\Downarrow}\}_{i=0}^d$. Use this fact along with Lemma 17 and Proposition 3. $\qquad\square$

Next we give the matrices that represent K with respect to the bases $\{u_i^{\Downarrow}\}_{i=0}^d$, $\{w_i\}_{i=0}^d$.

Lemma 41 *We give the matrix in* $\mathrm{Mat}_{d+1}(\mathbb{K})$ *that represents* K *with respect to* $\{u_i^{\Downarrow}\}_{i=0}^d$. *For* $0 \le i \le d$, *the* (i, i)-*entry is* q^{d-2i}. *For* $0 \le i < j \le d$, *the* (i, j)-*entry is*

$$\left(1 - a^{-2}\right)a^{j-i}q^{d-j-i}\left(q - q^{-1}\right)^{2(j-i)}\frac{[j]_q^![d-i]_q^!}{[i]_q^![d-j]_q^!}. \tag{74}$$

All other entries are zero.

Proof Evaluating the equation on the right in (14) using the equation on the left in (12) we get

$$K = \left(a^{-2}I + (1 - a^{-2})\sum_{n=0}^d a^n q^n \psi^n\right)B. \tag{75}$$

The result follows from this along with Proposition 3 and the fact that the matrix $\mathrm{diag}(q^d, q^{d-2}, \ldots, q^{-d})$ represents B with respect to $\{u_i^{\Downarrow}\}_{i=0}^d$. $\qquad\square$

Lemma 42 *We give the matrix in* $\mathrm{Mat}_{d+1}(\mathbb{K})$ *that represents* K *with respect to* $\{w_i\}_{i=0}^d$. *For* $0 \le i \le d$, *the* (i, i)-*entry is* q^{d-2i}. *For* $1 \le i \le d$, *the* $(i-1, i)$-*entry is*

$$-a^{-1}q^{d-2i+1}(q^i - q^{-i})(q^{d-i+1} - q^{i-d-1}).$$

All other entries are zero.

Proof The matrix $\mathrm{diag}(q^d, q^{d-2}, \ldots, q^{-d})$ represents $\mathcal{M}$ with respect to $\{w_i\}_{i=0}^d$. Use this fact along with Proposition 3 and the equation on the left in (32). $\qquad\square$

Next we give the matrices that represent B with respect to the bases $\{u_i\}_{i=0}^d$, $\{w_i\}_{i=0}^d$.

Lemma 43 *We give the matrix in* $\mathrm{Mat}_{d+1}(\mathbb{K})$ *that represents* B *with respect to* $\{u_i\}_{i=0}^d$. *For* $0 \leq i \leq d$, *the* (i,i)-*entry is* q^{d-2i}. *For* $0 \leq i < j \leq d$, *the* (i,j)-*entry is*

$$\left(1 - a^2\right) a^{i-j} q^{d-j-i} \left(q - q^{-1}\right)^{2(j-i)} \frac{[j]_q^!\,[d-i]_q^!}{[i]_q^!\,[d-j]_q^!}. \tag{76}$$

All other entries are zero.

Proof Evaluating the equation on the left in (14) using the equation on the right in (12) we get

$$B = \left(a^2 I + (1 - a^2) \sum_{n=0}^{d} a^{-n} q^n \psi^n\right) K. \tag{77}$$

The result follows from this along with Proposition 3 and the fact that the matrix $\mathrm{diag}(q^d, q^{d-2}, \ldots, q^{-d})$ represents K with respect to $\{u_i\}_{i=0}^d$. $\square$

Lemma 44 *We give the matrix in* $\mathrm{Mat}_{d+1}(\mathbb{K})$ *that represents* B *with respect to* $\{w_i\}_{i=0}^d$. *For* $0 \leq i \leq d$, *the* (i,i)-*entry is* q^{d-2i}. *For* $1 \leq i \leq d$, *the* $(i-1,i)$-*entry is*

$$-aq^{d-2i+1}(q^i - q^{-i})(q^{d-i+1} - q^{i-d-1}).$$

All other entries are zero.

Proof The matrix $\mathrm{diag}(q^d, q^{d-2}, \ldots, q^{-d})$ represents $\mathcal{M}$ with respect to $\{w_i\}_{i=0}^d$. Use this fact along with Proposition 3 and the equation on the left in (33). $\square$

Next we consider the matrices

$$\exp_q\left(\frac{a}{q - q^{-1}} \widehat{\psi}\right), \qquad \exp_q\left(\frac{a^{-1}}{q - q^{-1}} \widehat{\psi}\right). \tag{78}$$

Their inverses are

$$\exp_{q^{-1}}\left(-\frac{a}{q - q^{-1}} \widehat{\psi}\right), \qquad \exp_{q^{-1}}\left(-\frac{a^{-1}}{q - q^{-1}} \widehat{\psi}\right) \tag{79}$$

respectively. The matrices in (78), (79) are upper triangular. We now consider the entries of (78), (79).

Lemma 45 *For* $0 \neq x \in \mathbb{K}$, *the matrix* $\exp_q(x\widehat{\psi})$ *is upper triangular. For* $0 \leq i \leq j \leq d$, *the* (i,j)-*entry is*

$$x^{j-i}q^{\binom{j-i}{2}}\left(q-q^{-1}\right)^{2(j-i)}\cdot\frac{[j]_q^![d-i]_q^!}{[i]_q^![j-i]_q^![d-j]_q^!}. \tag{80}$$

The matrix $exp_{q^{-1}}(x\widehat{\psi})$ is upper triangular. For $0 \le i \le j \le d$, the (i,j)-entry is

$$x^{j-i}q^{-\binom{j-i}{2}}\left(q-q^{-1}\right)^{2(j-i)}\cdot\frac{[j]_q^![d-i]_q^!}{[i]_q^![j-i]_q^![d-j]_q^!}. \tag{81}$$

Lemma 46 The transition matrices between the basis $\{w_i\}_{i=0}^d$ and the bases $\{u_i\}_{i=0}^d$, $\{u_i^{\Downarrow}\}_{i=0}^d$ are given in the table below.

From	To	Transition matrix
$\{u_i\}_{i=0}^d$	$\{w_i\}_{i=0}^d$	$exp_{q^{-1}}\left(-\frac{a^{-1}}{q-q^{-1}}\widehat{\psi}\right)$
$\{w_i\}_{i=0}^d$	$\{u_i\}_{i=0}^d$	$exp_q\left(\frac{a^{-1}}{q-q^{-1}}\widehat{\psi}\right)$
$\{u_i^{\Downarrow}\}_{i=0}^d$	$\{w_i\}_{i=0}^d$	$exp_{q^{-1}}\left(-\frac{a}{q-q^{-1}}\widehat{\psi}\right)$
$\{w_i\}_{i=0}^d$	$\{u_i^{\Downarrow}\}_{i=0}^d$	$exp_q\left(\frac{a}{q-q^{-1}}\widehat{\psi}\right)$

Proof Use Lemma 36 and Proposition 3.   □

We next consider the product

$$exp_q\left(\frac{a}{q-q^{-1}}\widehat{\psi}\right)exp_{q^{-1}}\left(-\frac{a^{-1}}{q-q^{-1}}\widehat{\psi}\right). \tag{82}$$

The inverse of (82) is

$$exp_q\left(\frac{a^{-1}}{q-q^{-1}}\widehat{\psi}\right)exp_{q^{-1}}\left(-\frac{a}{q-q^{-1}}\widehat{\psi}\right). \tag{83}$$

The matrices in (82), (83) are upper triangular.

Lemma 47 The transition matrices between the bases $\{u_i\}_{i=0}^d$, $\{u_i^{\Downarrow}\}_{i=0}^d$ are given in the table below.

From	To	Transition matrix
$\{u_i\}_{i=0}^d$	$\{u_i^{\Downarrow}\}_{i=0}^d$	$exp_q\left(\frac{a}{q-q^{-1}}\widehat{\psi}\right)exp_{q^{-1}}\left(-\frac{a^{-1}}{q-q^{-1}}\widehat{\psi}\right)$
$\{u_i^{\Downarrow}\}_{i=0}^d$	$\{u_i\}_{i=0}^d$	$exp_q\left(\frac{a^{-1}}{q-q^{-1}}\widehat{\psi}\right)exp_{q^{-1}}\left(-\frac{a}{q-q^{-1}}\widehat{\psi}\right)$

Proof Use Lemma 46.   □

Lemma 48 *With respect to each of the bases* $\{u_i\}_{i=0}^{d}$, $\{u_i^{\Downarrow}\}_{i=0}^{d}$, $\{w_i\}_{i=0}^{d}$, *the matrices that represent* Δ *and* Δ^{-1} *are* $\exp_q\left(\frac{a}{q-q^{-1}}\widehat{\psi}\right)\exp_{q^{-1}}\left(-\frac{a^{-1}}{q-q^{-1}}\widehat{\psi}\right)$ *and* $\exp_q\left(\frac{a^{-1}}{q-q^{-1}}\widehat{\psi}\right)\exp_{q^{-1}}\left(-\frac{a}{q-q^{-1}}\widehat{\psi}\right)$ *respectively.*

Proof Use Theorem 3 and Proposition 3.   □

We give the entries of the matrices representing Δ, Δ^{-1} in the following lemma.

Lemma 49 *The matrix in* (82) *is upper triangular. For* $0 \le i \le j \le d$, *the* (i, j)-*entry of* (82) *is*

$$\frac{(q-q^{-1})^{j-i}\,[j]_q^![d-i]_q^!}{[i]_q^![j-i]_q^![d-j]_q^!}\prod_{n=1}^{j-i}\left(aq^{n-1}-a^{-1}q^{1-n}\right). \tag{84}$$

The matrix in (83) *is upper triangular. For* $0 \le i \le j \le d$, *the* (i, j)-*entry of* (83) *is*

$$\frac{(q-q^{-1})^{j-i}\,[j]_q^![d-i]_q^!}{[i]_q^![j-i]_q^![d-j]_q^!}\prod_{n=1}^{j-i}\left(a^{-1}q^{n-1}-aq^{1-n}\right). \tag{85}$$

Proof Use Corollary 1 and Proposition 3.   □

We finish the paper by giving the matrix that represents A with respect to $\{w_i\}_{i=0}^{d}$.

Lemma 50 *We give the matrix in* $\mathrm{Mat}_{d+1}(\mathbb{K})$ *that represents* A *with respect to* $\{w_i\}_{i=0}^{d}$. *For* $1 \le i \le d$, *the* $(i, i-1)$-*entry is* 1. *For* $0 \le i \le d$, *the* (i, i)-*entry is* $(a+a^{-1})q^{d-2i}$. *For* $1 \le i \le d$, *the* $(i-1, i)$-*entry is*

$$-q^{d-2i+1}(q^i - q^{-i})(q^{d-i+1} - q^{i-d-1}).$$

All other entries are zero.

Proof Let $\mathcal{A}$ denote the matrix that represents A with respect to $\{w_i\}_{i=0}^{d}$. By Lemma 29, $\mathcal{A}$ is tridiagonal with (i, i)-entry given by $(a+a^{-1})q^{d-2i}$ for $0 \le i \le d$.

We now show that the subdiagonal entries of $\mathcal{A}$ are all 1. Let $\mathcal{A}'$ denote the matrix that represents A with respect to $\{u_i\}_{i=0}^{d}$. Recall that this matrix is displayed on the left in (71). Observe that $\mathcal{A}$ is equal to $\exp_{q^{-1}}(-\frac{a^{-1}}{q-q^{-1}}\widehat{\psi})\mathcal{A}'\exp_q(\frac{a^{-1}}{q-q^{-1}}\widehat{\psi})$. It follows from this fact that the subdiagonal entries of $\mathcal{A}$ are all 1.

We next obtain the superdiagonal entries of $\mathcal{A}$. Let $0 \le i \le d$. Apply both sides of (37) to w_i. Evaluate the result using Proposition 3 and the fact that the w_i is an eigenvector for M with eigenvalue q^{2i-d}. Analyze the result in light of the above comments concerning the entries of $\mathcal{A}$ to obtain the desired result.   □

Acknowledgement This research was partially supported by a grant from the College of Science and Health at DePaul University.

References

1. H. Alnajjar and B. Curtin. A family of tridiagonal pairs related to the quantum affine algebra $U_q(\widehat{sl}_2)$. *Electron. J. Linear Algebra* **13** (2005), 1–9.
2. G. Andrews and R. Askey and R. Roy. Special Functions. Encyclopedia of Mathematics and its Applications, Cambridge University Press, Cambridge, 1999.
3. P. Baseilhac. A family of tridiagonal pairs and related symmetric functions. *J. Phys. A* **39** (2006), 11773–11791.
4. S. Bockting-Conrad. Two commuting operators associated with a tridiagonal pair. *Linear Algebra Appl.* **437** (2012), 242–270.
5. S. Bockting-Conrad. Tridiagonal pairs of q-Racah type, the double lowering operator ψ, and the quantum algebra $U_q(sl_2)$. *Linear Algebra Appl.* **445** (2014), 256–279.
6. L. Dolan and M. Grady. Conserved charges from self-duality. *Phys. Rev. D* **25** (1982), 1587–1604.
7. D. Funk-Neubauer. Tridiagonal pairs and the q-tetrahedron algebra. *Linear Algebra Appl.* **431** (2009), 903–925.
8. G. Gasper and M. Rahman. Basic Hypergeometric Series, Second Ed. Cambridge University Press, New York, 2003.
9. T. Ito, K. Tanabe, and P. Terwilliger. Some algebra related to P- and Q-polynomial association schemes, in: *Codes and Association Schemes* (Piscataway NJ, 1999), Amer. Math. Soc., Providence RI, 2001, pp. 167–192.
10. T. Ito and P. Terwilliger. The q-tetrahedron algebra and its finite-dimensional irreducible modules. *Comm. Algebra* **35** (2007) 3415–3439.
11. T. Ito and P. Terwilliger. Tridiagonal pairs and the quantum affine algebra $U_q(\widehat{sl}_2)$. *Ramanujan J.* **13** (2007), 39–62.
12. T. Ito and P. Terwilliger. Tridiagonal pairs of q-Racah type. *J. Algebra* **322** (2009), 68–93.
13. T. Ito and P. Terwilliger. The augmented tridiagonal algebra. *Kyushu J. Math* **64** (2010), 81–144.
14. T.H. Koornwinder. The relationship between Zhedanov's algebra $AW(3)$ and the double affine Hecke algebra in the rank one case. *SIGMA* **3** (2007), 063, 15 pages.
15. T.H. Koornwinder. Zhedanov's algebra $AW(3)$ and the double affine Hecke algebra in the rank one case. II. The spherical subalgebra. *SIGMA* **4** (2008), 052, 17 pages.
16. A. Korovnichenko and A. S. Zhedanov. "Leonard pairs" in classical mechanics. *J. Phys. A* **35** (2002), 5767–5780.
17. K. Nomura and P. Terwilliger. Towards a classification of the tridiagonal pairs. *Linear Algebra Appl.* 429 (2008), 503–518.
18. S. Odake and R. Sasaki. Orthogonal polynomials from Hermitian matrices. *J. Math. Phys.* **49** (2008), 053503, 43 pages.
19. L. Onsager. Crystal statistics. I. A two-dimensional model with an order-disorder transition. *Phys. Rev.* **65** (1944), 117–149.
20. P. Terwilliger. Two linear transformations each tridiagonal with respect to an eigenbasis of the other. *Linear Algebra Appl.* 330 (2001), 149–203.
21. P. Terwilliger. Leonard pairs from 24 points of view. *Rocky Mountain J. Math.* 32 (2002), 827–888.
22. P. Terwilliger. Introduction to Leonard pairs. OPSFA Rome 2001. *J. Comput. Appl. Math.* **153**(2) (2003), 463–475.
23. P. Terwilliger. Two linear transformations each tridiagonal with respect to an eigenbasis of the other; comments on the parameter array. *Des. Codes Cryptogr.* **34** (2005), 307–332.
24. P. Terwilliger. An algebraic approach to the Askey scheme of orthogonal polynomials. Orthogonal polynomials and special functions, Lecture Notes in Math., 1883 (Berlin), Springer, 2006, 255–330.
25. P. Terwilliger and R. Vidunas. Leonard pairs and the Askey-Wilson relations. *J. Algebra Appl.* **3** (2004), 411–426.

Distance Graphs Generated by Five Primes (Research)

Daphne Der-Fen Liu, Grant Robinson, and Angel Chavez

1 Introduction

Let D be a set of positive integers, called a *distance set*. The *distance graph* generated by D, denoted $G(\mathbf{Z}, D)$, is the graph with vertex set of the integers and an edge between any pair of vertices a and b if $|a - b| \in D$. The chromatic number of distance graphs was first studied by Eggleton et al. [5] in 1985. The subject has been studied extensively since [1–4, 6, 9–15, 17–20, 22]. We denote the chromatic number of $G(\mathbf{Z}, D)$ by $\chi(D)$.

Let $\mathbf{P}$ denote the set of prime numbers. In [6] prime distance graphs were considered, that is, graphs with distance set $D \subseteq \mathbf{P}$. It was shown and easy to see that $\chi(\mathbf{P}) = 4$. Thus, given that D is a subset of $\mathbf{P}$, $\chi(D) \in \{1, 2, 3, 4\}$, since $D \subseteq D'$ implies $\chi(D) \leqslant \chi(D')$. The task considered is to classify a set of primes D according to its chromatic number. We say D is *class i* if $\chi(D) = i$. Clearly the only set that is class 1 is the empty set, and every singleton is class 2. If $|D| \geqslant 2$, then D is class 2 if and only if $2 \notin D$. Also if $2 \in D$ but $3 \notin D$, then D is class 3. A less trivial result is that $\{2, 3, p\}$ is class 4 if $p = 5$, and class 3 otherwise (see [6]). In view of these results, the remaining problem is to classify prime sets $D \supset \{2, 3\}$ with $|D| \geqslant 4$ into either class 3 or class 4.

It was shown in [6] that if $D = \{2, 3, p, p+2\}$ where p and $p+2$ are twin primes, then D is class 4. Voigt and Walther [19] classified all prime sets with cardinality 4:

D. D.-F. Liu (✉) · A. Chavez
California State University, Los Angeles, CA, USA
e-mail: dliu@calstatela.edu; achav121@calstatela.edu

G. Robinson
University of Washington, Seattle, WA, USA
e-mail: grantrob@uw.edu

© The Author(s) and the Association for Women in Mathematics 2020
B. Acu et al. (eds.), *Advances in Mathematical Sciences*, Association for
Women in Mathematics Series 21, https://doi.org/10.1007/978-3-030-42687-3_3

Theorem 1 *Let $D = \{2, 3, p, q\}$ be a set of primes with $p \geqslant 7$ and $q > p + 2$. Then D is class 4 if and only if*

$(p,q) \in \{(11, 19), (11, 23), (11, 37), (11, 41), (17, 29), (23, 31), (23, 41), (29, 37)\}.$

Since Voigt's paper in 1994, little progress has been made on the subject. It is interesting to note that, besides the potentially infinite family of distance sets containing twin primes, there are only finitely many class 4 sets of four primes. Thus it is natural to ask whether the same is true when D has five primes. A result from [6] shows that the set of potentially infinite families of class 4 distance sets will necessarily be more complicated than just those containing twin primes:

Theorem 2 *The set $\{2, 3\} \cup \{p, p + 8, 2p + 13\}$ is class 4 whenever p, $p + 8$ and $2p + 13$ are all primes.*

In this article we begin to look at prime distance sets with 5 elements that do not contain twin primes nor any of the eight minimal class 4 sets of cardinality 4 obtained in Theorem 1. We call a prime distance set D *minimal* class 4, or just minimal, if D is class 4 but every proper subset is class 3 or less. Thus we are interested in distance sets which do not contain twin primes or any of the minimal class 4 sets in Theorem 1. We present the following main result:

Theorem 3 *A prime set of the form $D = \{2, 3, 7, p, q\}$ is class 3 if none of the following is true:*

1. *D contains a proper subset that is class 4.*
2. *The pair (p, q) is one of the following 31 pairs:*

(19, 31)	(19, 37)	(19, 41)	(19, 43)	(19, 47)	(19, 53)	(19, 67)
(19, 73)	(19, 79)	(19, 83)	(19, 89)	(19, 109)	(19, 131)	(19, 151)
(19, 157)	(19, 167)	(19, 193)	(29, 41)	(29, 73)	(29, 109)	(31, 43)
(37, 59)	(41, 53)	(47, 59)	(61, 73)	(67, 79)	(71, 83)	(89, 101)
(97, 109)	(139, 151)	(181, 193).				

3. *$p \equiv 2311139, 2311163 \pmod{4622310}$ and $q = p + 8$.*

Moreover, the D sets with pairs $(19, q)$ in 2 are all class 4.

In Sect. 3, we give a proof of Theorem 3, except the moreover part, which is presented in Sect. 4. In order to show that a distance set is class 3, we will make extensive use of the number theoretic function $\kappa : \mathcal{P}(\mathbf{Z}^+) \to \mathbf{R}^+ \cup \{0\}$. For a real number x, let $||x||$ denote the minimum distance from x to an integer, that is $||x|| = \min\{\lceil x \rceil - x, x - \lfloor x \rfloor\}$. For any real t, denote by $||tD||$ the smallest value $||td||$ among all $d \in D$. The *kappa value* of D, denoted by $\kappa(D)$, is the supremum of $||tD||$ among all reals t. That is, $\kappa(D) := \sup\{||tD|| : t \in \mathbf{R}^+ \cup \{0\}\}$. The fact that the kappa value of D is always a rational number with denominator dividing a sum of two elements in D gives an effective algorithm for computing $\kappa(D)$ (see [8]).

The primary connection which we use in this paper is that (see [21])

$$\chi(D) \leqslant \left\lceil \frac{1}{\kappa(D)} \right\rceil.$$

Thus, if $\kappa(D) \geqslant 1/3$, then $\chi(D) \leqslant 3$. In particular, since we assume $\{2, 3\} \subset D$, if $\kappa(D) \geqslant 1/3$, then D is class 3.

2   Three Lemmas on $\kappa(D)$

An alternative definition of $\kappa(D)$ introduced by Gupta in [7] involves looking at the sets of "good times" for each element $d \in D$, that is, the times $t \in [0, 1)$ such that $\|td\|$ is greater than some desired value. For $\alpha \in [0, 1/2]$ and an element $d \in D$, let $I_d(\alpha) = \{t \in [0, 1) : \|td\| \geqslant \alpha\}$. Let $I_D(\alpha)$ be the intersection over D of $I_d(\alpha)$. If $I_D(\alpha)$ is not empty, then $\kappa(D) \geqslant \alpha$. Thus,

$$\kappa(D) = \sup\{\alpha \in [0, 1/2] : I_D(\alpha) \neq \varnothing\}.$$

Note that if $\kappa(D) > \alpha$, then $I_D(\alpha)$ is a union of intervals, and if $\kappa(D) = \alpha$, then $I_D(\alpha)$ is a union of singletons.

If $I_D(\alpha)$ contains a nontrivial interval or, equivalently, if $\kappa(D) > \alpha$, one might be interested in how large a number x must be to guarantee that the intersection of $I_D(\alpha)$ and $I_x(\alpha)$ is not empty, that is, $\kappa(D \cup \{x\}) \geqslant \alpha$. Note that $I_x(\alpha)$ is the union of x disjoint intervals with center $(2n + 1)/2x$ for $n \in \{0, 1, \ldots, x - 1\}$ and length $(1 - 2\alpha)/x$, that is,

$$I_x(\alpha) = \bigcup_{n=0}^{x-1} \left[\frac{n + \alpha}{x}, \frac{n + 1 - \alpha}{x} \right].$$

We call these x-*intervals*. The length of the space between any two consecutive x-intervals is $2\alpha/x$. Now let $[a, b]$ be a connected subset of $I_D(\alpha)$. If the length of the space between each pair of consecutive intervals of $I_x(\alpha)$ is less than the length of that subset, $b - a$, then it must be that one of the intervals of $I_x(\alpha)$ hits the interval $[a, b]$. This can be summarized in the following lemma:

Lemma 1 *Let* $[a, b] \subseteq I_D(\alpha)$ *with* $a < b$ *for some set* D. *If* x *is an integer,* $x \geqslant 2\alpha/(b - a)$, *then* $I_D(\alpha) \cap I_x(\alpha) \neq \varnothing$. *Consequently,* $\kappa(D \cup \{x\}) \geqslant \alpha$.

Considering two elements to be added to a set D, we describe an upper bound for the length of an interval of time in which the two sets $I_x(\alpha)$ and $I_{x+i}(\alpha)$ can be disjoint. If this bound is smaller than the length of a target interval contained in

$I_D(\alpha)$, we can similarly guarantee that the intersection of $I_D(\alpha)$, $I_x(\alpha)$ and $I_{x+i}(\alpha)$ is not empty.

Lemma 2 *Let* $1/4 \leqslant \alpha \leqslant 1/3$ *and* $[a, b] \subseteq I_D(\alpha)$ *with* $a < b$. *If* x *and* i *are integers with* $\frac{4\alpha-1}{i} + \frac{2}{x} \leqslant b - a$, *then* $I_D(\alpha) \cap I_x(\alpha) \cap I_{x+i}(\alpha) \neq \varnothing$. *Consequently,* $\kappa(D \cup \{x, x+i\}) \geqslant \alpha$.

Proof Similar to Lemma 1, it is enough to show that $I_x \cap I_{x+i} \cap I \neq \varnothing$ for any interval $I \subseteq [0, 1]$ of length $\frac{4\alpha-1}{i} + \frac{2}{x}$.

We introduce some notation to make it easier to keep track of the different intervals. As noted above,

$$I_x(\alpha) = \bigcup_{n=0}^{x-1} \left[\frac{n+\alpha}{x}, \frac{n+1-\alpha}{x} \right].$$

Fixing $1/4 \leqslant \alpha \leqslant 1/3$, let $[\frac{n+\alpha}{x}, \frac{n+1-\alpha}{x}]$ be denoted by I_x^n. Let $L(I_x^n)$ and $R(I_x^n)$ denote the left and the right endpoint of I_x^n, respectively.

Assume $i \geqslant x$. We first claim that every x-interval must intersect at least one $(x+i)$-interval. It suffices to show that the length of the gap between two consecutive $(x+i)$-intervals is less than the length of an x-interval, that is, $\frac{2\alpha}{x+i} \leqslant \frac{1-2\alpha}{x}$. This is true with the assumptions $\alpha \leqslant 1/3$ and $x \leqslant i$.

Therefore, $R(I_x^n) - L(I_x^{n-1}) = \frac{2-2\alpha}{x}$ is an upper bound on the length of an interval during which I_x and I_{x+i} are disjoint. By our assumption that $\alpha \geqslant 1/4$, the result follows, as $\frac{2-2\alpha}{x} < \frac{4\alpha-1}{i} + \frac{2}{x} \leqslant b - a$.

Now assume $i < x$. Let I_x^m be any x-interval. If $m = 0$, then with the assumptions $\alpha \leqslant 1/3$ and $i < x$, it can be shown that $L(I_x^0) \leqslant R(I_{x+i}^0)$, and therefore there is some intersection between the two intervals.

If $m \geqslant 1$, then let I_{x+i}^n be the closest $(x+i)$-interval to I_x^m such that $R(I_{x+i}^n) \leqslant L(I_x^m)$ (that is, n is the largest such integer), and set $L(I_x^m) - R(I_{x+i}^n) = \Delta$. Note that $L(I_x^m) - L(I_x^{m-1}) = 1/x$. This implies that the separation between previous pairs of x and $(x+i)$-intervals decreases until the left point of an x-interval is less than the right point of an $(x+i)$-interval. More precisely,

$$L(I_x^{m-r}) - R(I_{x+i}^{n-r}) = \left(L(I_x^m) - \frac{r}{x} \right) - \left(R(I_{x+i}^n) - \frac{r}{x+i} \right)$$

$$= L(I_x^m) - R(I_{x+i}^n) - \frac{r}{x} + \frac{r}{x+i}$$

$$= \Delta - \frac{ir}{x(x+i)}.$$

Fix $j \geqslant 0$ so that $\Delta - \frac{ij}{x(x+i)} \leqslant 0$ but $\Delta - \frac{i(j-1)}{x(x+i)} > 0$. This implies that

$$R(I_{x+i}^{n-j}) - L(I_x^{m-j}) = \frac{ij}{x(x+i)} - \Delta \leqslant \frac{i}{x(x+i)}.$$

With the assumptions that $i \leqslant x$ and $\alpha \leqslant 1/3$, it can be shown that

$$\frac{i}{x(x+i)} \leqslant \frac{1-2\alpha}{x} + \frac{1-2\alpha}{x+i}. \tag{1}$$

The right-hand side of the above inequality is the sum of the lengths of an x-interval and an $(x+i)$-interval. Therefore, since $R(I_{x+i}^{n-j}) - L(I_x^{m-j}) \leqslant \frac{1-2\alpha}{x} + \frac{1-2\alpha}{x+i}$, there must be some intersection between I_x^{m-j} and I_{x+i}^{n-j}.

Having found an intersection between an x-interval and an $(x+i)$-interval at or before I_x^m, we now move forward, looking at the right endpoint of the x-intervals. Notice,

$$\begin{aligned}
L(I_{x+i}^{n+1+r}) - R(I_x^{m+r}) &= L(I_{x+i}^{n+1}) - R(I_x^m) - \frac{ir}{x(x+i)} \\
&= R(I_{x+i}^n) + \frac{2\alpha}{x+i} - R(I_x^m) - \frac{ir}{x(x+i)} \\
&= L(I_x^m) - \Delta + \frac{2\alpha}{x+i} - R(I_x^m) - \frac{ir}{x(x+i)} \\
&= \frac{2\alpha}{x+i} - \left(\frac{1-2\alpha}{x} + \frac{ir}{x(x+i)} + \Delta\right).
\end{aligned}$$

Fix $k \geqslant 0$ so that k is the smallest such that $\frac{2\alpha}{x+i} \leqslant \frac{1-2\alpha}{x} + \frac{ik}{x(x+i)} + \Delta$, that is, the smallest with $L(I_{x+i}^{n+1+k}) \leqslant R(I_x^{m+k})$. We now show that $L_x^{m+k} \cap L_{x+i}^{n+k+1} \neq \varnothing$. Suppose $k = 0$, that is, $L(I_{x+i}^{n+1}) \leqslant R(I_x^m)$. By our choice of n as the largest such that $R(I_{x+i}^n) \leqslant L(I_x^m)$, we have $R(I_{x+i}^{n+1}) > L(I_x^m)$. This, together with the fact that $L(I_{x+i}^{n+1}) \leqslant R(I_x^m)$, implies $L_x^m \cap L_{x+i}^{n+1} \neq \varnothing$.

Assume $k \geqslant 1$. Then $R(I_x^{m+k-1}) < L(I_{x+i}^{n+k})$. The only possibility that $I_x^{m+k} \cap I_{x+i}^{n+1+k} = \varnothing$ is when the following inequality holds:

$$\begin{aligned}
\frac{1-2\alpha}{x} + \frac{1-2\alpha}{x+i} &< R(I_x^{m+k}) - L(I_{x+i}^{n+1+k}) \\
&= R(I_x^{m+k-1}) - L(I_{x+i}^{n+k}) + \frac{i}{x(x+i)} \\
&< \frac{i}{x(x+i)}.
\end{aligned}$$

This contradicts Eq. (1). Hence, $I_x^{m+k} \cap I_{x+i}^{n+1+k} \neq \varnothing$

In summary, given that $j = \lceil \frac{x(x+i)\Delta}{i} \rceil$ and $k = \lceil \frac{4\alpha x + 2\alpha i - x - i - x(x+i)\Delta}{i} \rceil$, we know that both I_x^{m-j} and I_x^{m+k} intersect an $(x+i)$-interval. Moreover, the length between these two intersections is bounded by the following:

$$R(I_x^{m+k}) - L(I_x^{m-j}) = \frac{k+j}{x} + \frac{1-2\alpha}{x}$$

$$\leqslant \frac{\frac{4\alpha x + 2\alpha i - x - i}{i} + 3 - 2\alpha}{x}$$

$$= \frac{4\alpha - 1}{i} + \frac{2}{x}.$$

Therefore, the result follows. Note that if $m + k \geqslant x$, then $R(I_x^{m+k})$ is undefined. In this case the bound $1 - L(I_x^{m-j})$ is smaller than the bound above. Similar arguments apply if $m - j < 0$. Note that $1/4 \leqslant \alpha \leqslant 1/3$ implies that $I_x \cap I_{x+i} \neq \varnothing$, since $\kappa(\{x, x+i\}) \geqslant 1/3$. $\qquad\square$

The final result of this section rationalizes the set of good times by expanding the unit circle to a circle of circumference q. This proposition will be useful because, fixing a rational point and an α, the proposition gives a finite list of residue classes of x modulo q such that the point will be in $I_x(\alpha)$.

Lemma 3 *Fix an integer x and an $\alpha \in [0, 1/2]$, and let p/q be a point in $(0, 1)$. Then $p/q \in I_x(\alpha)$ if and only if $q\alpha \leqslant xp \pmod{q} \leqslant q(1 - \alpha)$.*

Proof To say that $p/q \in I_x(\alpha)$ is equivalent to saying that there exists an $n \in \{0, 1, \ldots, x - 1\}$ such that $(n + \alpha)/x \leqslant p/q \leqslant (n + 1 - \alpha)/x$. Rearranging this inequality gives $q\alpha \leqslant px - qn \leqslant q(1 - \alpha)$. $\qquad\square$

3   Class 3 Prime Sets of the Form $\{2, 3, 7, p, q\}$

We apply the lemmas presented in the previous section to prove Theorem 3, except the moreover part, which will be shown in the next section. Recall, if $\kappa(D) \geqslant 1/3$, then $\chi(D) \leqslant 3$. Thus we fix $\alpha = 1/3$ in the following.

While the proof of Theorem 3 is conceptually simple, using nothing more sophisticated than modular arithmetic, there are many cases to check. Full verification requires a computer. For a more detailed discussion with all cases explained, see [16].

Let $D = \{2, 3, 7, x, x + i\}$ where x and $x + i$ are primes. First, Lemma 2 is applied to show that if both x and i are sufficiently large, then D must be class 3. As can be seen from Fig. 1, $[4/21, 2/9] \subseteq I_{\{2,3,7\}}(1/3)$, and the length of this interval is $2/63$. The smallest gap i such that $1/(3i) < 2/63$ is $i = 11$. Since the difference between any odd prime numbers is even, we only need to consider the cases of even integers $i \geqslant 12$. For each $i \geqslant 12$, there exists a bound M_i such that, whenever $p \geqslant M_i$ and $q \geqslant p + i$, the set $\{2, 3, 7, p, q\}$ is class 3.

For example, fixing $i = 12$, we solve the following inequality from Lemma 2 for p: $\frac{1}{3i} + \frac{2}{p} \leqslant \frac{2}{63}$. Thus if $p \geqslant 504$ and $q \geqslant p + 12$, then, by Lemma 2, $\{2, 3, 7, p, q\}$ will be class 3. Noting that as i increases the bound M_i decreases, we can repeat

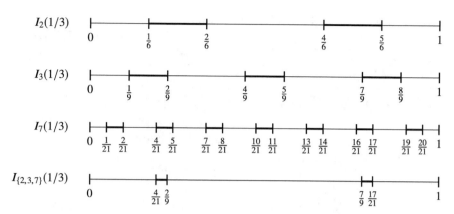

Fig. 1 The set $I_{\{2,3,7\}}(1/3)$

this process. The bound $M_{52} = 79$, and by computing the kappa value for all sets $\{2, 3, 7, p, p+i\}$ where $12 \leqslant i < 52$ and $79 \leqslant p \leqslant M_i$, we obtain the following proposition.

Proposition 1 *If $i \geqslant 12$ and $p \geqslant 79$ and $D = \{2, 3, 7, p, p+i\}$, then D is class 3 for any pair of primes $(p, p+i) \notin \{(89, 101), (97, 109), (139, 151), (181, 193)\}$.*

The next step in the process is to remove the bound that p must be greater than 79. To accomplish this, for each set $D = \{2, 3, 7, p\}$ for primes $7 < p < 79$, we apply Lemma 1 to get a bound on q such that $\{2, 3, 7, p, q\}$ is class 3 for every q exceeding the bound. Then we check whether $\kappa(D) \geqslant 1/3$ for each of the small primes q which are below the bound. This work is summarized in Table 1 and justifies the following proposition. Note that the table includes twin primes and the known results from Theorem 1.

Proposition 2 *If $i \geqslant 12$ and $7 < p < 79$ and $D = \{2, 3, 7, p, p+i\}$ does not contain a proper subset that is class 4, then D is class 3 for any pair of primes $(p, p+i)$ not listed below:*

$(19, 31)$	$(19, 37)$	$(19, 41)$	$(19, 43)$	$(19, 47)$	$(19, 53)$	$(19, 67)$
$(19, 73)$	$(19, 79)$	$(19, 83)$	$(19, 89)$	$(19, 109)$	$(19, 131)$	$(19, 151)$
$(19, 157)$	$(19, 167)$	$(19, 193)$	$(29, 41)$	$(29, 73)$	$(29, 109)$	$(31, 43)$
$(37, 59)$	$(41, 53)$	$(47, 59)$	$(61, 73)$	$(67, 79)$	$(71, 83)$.	

The fact that we switch from using Lemma 2 to Lemma 1 at $p < 79$ is arbitrary. Computationally, the hardest part of using Lemma 1 is finding the length of the longest interval in $\{2, 3, 7, p\}$, which is why Lemma 2 was used as long as it was.

Thus far we have shown that, as long as $i \geqslant 12$, there are only finitely many minimal prime sets with $\kappa(\{2, 3, 7, p, p+i\}) < 1/3$. If $i = 2$, then p and $p+2$

Table 1 Applying Lemma 1 to $\{2, 3, 7, p\}$ for primes $7 < p < 79$

p	$[a, b] \subset I_{\{2,3,7,p\}}$	Bound on q	Primes $q > p$ with $\kappa(\{2, 3, 7, p, q\}) < 1/3$
11	[7/33, 2/9]	66	13, 19, 23, 37, 41
13	[4/21, 8/39]	46	
17	[10/51, 11/51]	34	19, 29
19	[4/21, 11/57]	266	31,37,41,43,47,53, 67,73,79,83,89,109, 131,151,157,167,193
23	[4/21, 14/69]	54	31,41
29	[4/21, 17/87]	136	31,37,41,73,109
31	[19/93, 20/93]	62	43
37	[22/111, 23/111]	74	59
41	[25/123, 26/123]	82	43,53
43	[25/129, 26/129]	86	
47	[28/141, 29/141]	94	59
53	[34/159, 35/159]	106	
59	[37/177, 38/177]	118	61
61	[37/183, 38/183]	122	73
67	[43/201, 44/201]	134	79
71	[46/213, 47/213]	142	73,83
73	[46/219, 47/219]	146	

The new results from Theorem 3 are underlined; others are known results

are twin primes and the set is class 4. The last step in the process is to show that, for $i \in \{4, 6, 8, 10\}$, all prime sets of the form $\{2, 3, 7, p, p + i\}$ that do not contain one of the known class 4 sets are class 3.

Consider the case when p and $p + 4$ are both primes. Note that this implies that $p \equiv 1 \pmod 6$. We want to apply Lemma 3 to check if any rational points in the interval $[4/21, 2/9] \subset I_{\{2,3,7\}}$ are in both I_p and I_{p+4}. A natural place to start is by checking points with reduced denominator of 126, the least common multiple of 6, 21 and 9. The target interval $[4/21, 2/9] = [24/126, 28/126]$, so we will apply Lemma 3 for the points $\{n/126 : 24 \leqslant n \leqslant 28\}$. After removing the residue classes modulo 126 for which $p \not\equiv 1 \pmod 6$, we are left with Table 2.

From Table 2 we see that, for each of the rows that is not highlighted, $I_{\{2,3,7,p,p+4\}}$ will contain the point in the rightmost column, implying that $\{2, 3, 7, p, p + 4\}$ is class 3. To investigate the highlighted rows further, we increase the number of rational points to check by a factor of 5. The new denominator $q = 630$, and we must accordingly expand the undetermined list of residues to check. This gives Table 3.

From Table 3 we see that if $p \equiv 1 \pmod{630}$, then $p+4$ is not prime, if $p \equiv 625 \pmod{630}$, then p is not prime, and if $p \not\equiv 307, 319 \pmod{630}$, then $I_{\{2,3,7,p,p+4\}}$ is not empty. Iterating again, this time just increasing by a factor of 2 gives Table 4, which has no highlighted rows. This means, no matter the residue class of a prime p modulo 1260, there exists some point in $I_{\{2,3,7,p,p+4\}}$. Thus, this is the final table

Table 2 Rational points in $I_{\{2,3,7\}} \cap I_{\{p,p+4\}}$ (Round 1)

p (mod 126)	gcd$(p, 126)$	gcd$(p + 4, 126)$	Point in $I_{\{p,p+4\}}$
1			
7	7		27/126
13			25/126
19			24/126
25			28/126
31		7	27/126
37			26/126
43			28/126
49	7		27/126
55			
61			24/126
67			
73		7	27/126
79			28/126
85			26/126
91	7		27/126
97			28/126
103			24/126
109			25/126
115		7	27/126
121			

needed to finish the case when $i = 4$. Tables 2, 3, and 4 show that $\{2, 3, 7, p, p+4\}$ is class 3 for every pair of primes p and $p + 4$.

The cases when $i \in \{6, 10\}$ can be established similarly, but the case when $i = 8$ is much more difficult (see [16]). This is not surprising, as we have already have seen from Theorem 1 that $\{2, 3, 5, 13\}$, $\{2, 3, 11, 19\}$, $\{2, 3, 23, 31\}$, and $\{2, 3, 29, 37\}$ are all class 4 sets. Using similar methods to those above, we were able to show that, if $\{2, 3, 7, p, p + 8\}$ is a minimal class 4 set, it must be that $p \equiv 2311139, 2311163$ (mod 4622310). Note that $4622310 = 2 \cdot 3^2 \cdot 5 \cdot 7 \cdot 11 \cdot 23 \cdot 29$. At this point it becomes computationally intractable to inspect the millions of rational points considered by Lemma 3. From this work we obtain the following proposition.

Proposition 3 *If $i < 12$ and $D = \{2, 3, 7, p, p + i\}$ does not contain a proper subset that is class 4, then D is class 3 for any pair of primes $(p, p + i)$ where $p \not\equiv 2311139, 2311163$ (mod 4622310) when $i = 8$.*

We have shown how Lemma 2 generates Proposition 1, Lemma 1 generates Proposition 2, and Lemma 3 generates Proposition 3. Together Propositions 1 to 3 imply Theorem 3, except for the moreover part, which is the subject of the next section.

Table 3 Rational points in $I_{\{2,3,7\}} \cap I_{\{p,p+4\}}$ (Round 2)

p (mod 630)	$\gcd(p, 630)$	$\gcd(p + 4, 630)$	Point in $I_{\{p,p+4\}}$
1		5	
55	5		122/630
67			128/631
121		5	123/630
127			122/631
181		5	124/630
193			126/631
247			124/630
253			121/630
307			
319			
373			121/630
379			124/630
433			126/630
445	5		124/630
499			122/630
505	5		123/630
559			128/630
571		5	122/630
625	5		

Table 4 Rational points in $I_{\{2,3,7\}} \cap I_{\{p,p+4\}}$ (Round 3)

p (mod 1260)	$\gcd(p, 1260)$	$\gcd(p + 4, 1260)$	Point in $I_{\{p,p+4\}}$
307			253/1260
319			251/1260
937			251/1260
949			253/1260

4　Class 4 Prime Sets of the Form $\{2, 3, 7, 19, p\}$

In this section, we prove the moreover part of Theorem 3. Precisely, we show that any 3-coloring of the distance graph generated by $\{2, 3, 7, 19\}$ cannot be extended to a 3-coloring of the distance graph generated by $\{2, 3, 7, 19, p\}$ for any p in the following set:

$$\{31, 37, 41, 43, 47, 53, 67, 73, 79, 83, 89, 109, 131, 151, 157, 167, 193\}.$$

Our notation will follow that of Eggleton in [4]. Let c be a function $c \colon \mathbf{Z} \to \{0, 1, 2\}$. For a set D of positive integers, we say c is a D-consistent coloring if for every $i, j \in \mathbf{Z}$,

$$|i - j| \in D \implies c(i) \neq c(j).$$

It follows from the definition that c is a 3-coloring for a set D if and only if c is a D'-consistent coloring for any $D' \subseteq D$.

In the following we will consider a coloring c as a two-way infinite sequence, $c := \{c(i)\}_{i \in \mathbf{Z}}$. The structure of a coloring sequence c can be described by breaking it apart into the three constituent color classes. The k-color-class is defined as the set $\{i \in \mathbf{Z} : c(i) = k\}$. Let c be a $\{2, 3\}$-consistent coloring. Since each five consecutive integers in the distance graph generated by $\{2, 3\}$ contains the 5-cycle $\{i + 1, i + 3, i + 5, i + 2, i + 4\}$, the difference between any two consecutive elements in a color class is at most 5, otherwise the five cycle must be properly colored with just two colors, which is impossible. In light of this we can consider each color class as a strictly increasing sequence of integers $k := \{k_i\}_{i \in \mathbf{Z}}$ where $c(k_i) = k$ for every i and $k_i < k_{i+1}$. The structure of a color class is primarily captured by the gaps or differences between consecutive elements in the ordered color class sequence. The *gap sequence* of a k-color-class k is defined as $\Delta_k(c) = d = \{d_i\}_{i \in \mathbf{Z}}$ where $d_i = k_{i+1} - k_i$.

For any gap sequence $d = \Delta_k(c)$, let $\sigma(d)$ be the set of all partial sums of consecutive terms in d. Equivalently,

$$\sigma(d) = \{x : c(a) = c(x + a) = k \text{ for some } a \in \mathbf{Z}\}.$$

Given a coloring c, let $\sigma(c) := \bigcup_i \sigma(\Delta_i(c))$. By definition, we obtain

Proposition 4 *Let c be a function $c \colon \mathbf{Z} \to \{0, 1, 2\}$. Then c is a D-consistent coloring if and only if $\sigma(c) \cap D = \varnothing$.*

Often the colorings considered are periodic. This is denoted by enclosing the repeated block in parenthesis. As an example of these definitions, consider the periodic coloring function c defined by

$$c(i) = \begin{cases} 0 \text{ if } i \equiv 0, 1, 5, 6, 10, 11, 16 \pmod{21} \\ 1 \text{ if } i \equiv 2, 7, 8, 12, 13, 17, 18 \pmod{21} \\ 2 \text{ if } i \equiv 3, 4, 9, 14, 15, 19, 20 \pmod{21}. \end{cases}$$

The corresponding coloring sequence is $c = (001220011200112201122)$, and the three color classes are:

$$0 = \{\dots 0, 1, 5, 6, 10, 11, 16, \dots\}$$
$$1 = \{\dots 2, 7, 8, 12, 13, 17, 18, \dots\}$$
$$2 = \{\dots 3, 4, 9, 14, 15, 19, 20, \dots\}.$$

The three gap sequences are:

$$\Delta_0(c) = (1, 4, 1, 4, 1, 5, 5)$$
$$\Delta_1(c) = (5, 1, 4, 1, 4, 1, 5)$$
$$\Delta_2(c) = (1, 5, 5, 1, 4, 1, 4).$$

Since each of these gap sequences is a cyclic permutation of the others, the partial sums are the same for each:

$$\sigma(\Delta_0(c)) = \sigma(c) = \{x : x \equiv 0, \pm 1, \pm 4, \pm 5, \pm 6, \pm 9, \pm 10 \pmod{21}\}$$

Since $\{2, 3, 7, 19\} \cap \sigma(c) = \varnothing$, by Proposition 4, c is a $\{2, 3, 7, 19\}$-consistent 3-coloring.

4.1 Characterizing Gap Sequences

For either a color sequence or a gap sequence, we call any finite set of consecutive terms a *block* of the sequence. In this section we will investigate what blocks are possible for the gap sequences of a $\{2,3,7,19\}$-consistent coloring. Blocks of length l will be called *l-blocks*. In order to show that certain blocks are not possible, we will need to investigate how all three color classes interact. A gap sequence d almost completely determines a color sequence, as made precise by the following proposition from [4]:

Proposition 5 *If d is a $\{2, 3\}$-consistent gap sequence, then $d = \Delta_0(c)$ where, up to a permutation of the labels, c is given by the following rule that assigns terms of the gap sequence to blocks of a color sequence:*

$$\theta(d_i) = \begin{cases} 0 & \text{if } d_i = 1 \\ 0112 & \text{if } d_{i-1} > 1 \text{ and } d_i = 4 \\ 01z2 & \text{if } d_{i-1}d_i d_{i+1} = 141 \\ 0122 & \text{if } d_i = 4 \text{ and } d_{i+1} > 1 \\ 01122 & \text{if } d_i = 5 \end{cases}$$

where $z \in \{1, 2\}$ can be arbitrarily chosen for each 141 block in d.

The only possible gaps between consecutive elements of a color class are 1, 4 and 5. The fact that 2 or 3 cannot be gaps follows clearly from the definition, and the fact that no gap can be greater than 5 follows from existence of a 5-cycle in any block of five consecutive integers.

There are 9 possible 2-blocks of 1, 4, and 5: 11, 14, 15, 41, 44, 45, 51, 54, 55. Of these, 11 is impossible since it contains a partial sum of 2. In the following we prove that 44, 45, and 54 are also impossible.

Proposition 6 *Any $\{2, 3, 7, 19\}$-consistent gap sequence cannot contain a 2-blocks of the form* 44, 45, *or* 54.

Proof We consider each case separately.

Case 1: Let d be a $\{2, 3, 7, 19\}$-consistent gap sequence containing a 44 block. By Proposition 5, the corresponding color sequence must have the form $c = \ldots 012201120\ldots$. Without loss of generality, let $c_0 = 0$, $c_1 = 1$, $c_3 = 2$, etc. We can now make the following chain of inferences:

$$(c_0 = 0) \wedge (c_1 = 1) \wedge (c_2 = 2) \implies (c_{-2} = 2) \wedge (c_{-1} = 0)$$

$$(c_7 = 2) \wedge (c_8 = 0) \implies c_{10} = 1$$

$$(c_{-2} = 2) \wedge (c_{10} = 1) \implies c_{17} = 0$$

$$(c_6 = 1) \wedge (c_7 = 2) \wedge (c_{17} = 0) \implies (c_9 = 0) \wedge (c_{14} = 1) \implies c_{16} = 2$$

$$(c_9 = 0) \wedge (c_5 = 1) \implies c_{12} = 2$$

$$(c_{12} = 2) \wedge (c_8 = 0) \implies c_{15} = 1$$

The fact that $c_{-1} = 0$, $c_{15} = 1$ and $c_{16} = 2$ implies that c_{18} cannot be properly colored, contradicting that d is a $\{2, 3, 7, 19\}$-consistent gap sequence.

Case 2: Let d be a $\{2, 3, 7, 19\}$-consistent gap sequence containing the 2-block 45. By Proposition 5, we can assume the associated coloring sequence c contains the following block: $c_0 \ldots c_9 = 0122011220$. Then

$$(c_0 = 0) \wedge (c_1 = 1) \implies c_{-2} = 2$$

$$(c_5 = 1) \wedge (c_9 = 0) \implies c_{12} = 2$$

$$(c_0 = 0) \wedge (c_{12} = 2) \wedge (c_{11} = 1) \implies (c_{19} = 1) \wedge (c_{14} = 0) \implies c_{17} = 2.$$

This is a contradiction as $c_{-2} = c_{17}$.

Case 3: Let d be a $\{2, 3, 7, 19\}$-consistent gap sequence containing the 2-block 54. By Proposition 5, we can assume the associated coloring sequence c contains the following block: $c_0 \ldots c_9 = 0112201120$. Then

$$(c_7 = 1) \wedge (c_8 = 2) \wedge (c_9 = 0) \implies (c_{10} = 0) \wedge (c_{11} = 1) \implies c_{13} = 2$$

$$(c_9 = 0) \wedge (c_{13} = 2) \wedge (c_1 = 1) \implies (c_{16} = 1) \wedge (c_{20} = 0) \implies c_{23} = 2.$$

This is a contradiction, since $c_4 = c_{23}$.

$\square$

From the five allowable 2-blocks, nine 3-blocks can be built: 141, 151, 155, 414, 415, 514, 515, 551, 555. Of these, 151 produces a partial sum of 7, and is therefore not possible. In the following we prove 515 is also not possible.

Proposition 7 *Any* $\{2, 3, 7, 19\}$-*consistent gap sequence cannot contain the 3-block* 515.

Proof Let d be a $\{2, 3, 7, 19\}$-consistent gap sequence containing the 3-block 515. By Proposition 5, we can assume the associated coloring sequence c contains the following block: $c_0 \ldots c_{11} = 011220011220$. Then the fact that $c_1 = c_8 = 1$ contradicts the fact that c is a proper coloring.   □

Finally three larger blocks are not allowed: 5555, 14141414 and 51415. The block 14141414 contains a partial sum of 19, and therefore cannot be in a $\{2, 3, 7, 19\}$-consistent gap sequence. In the following we prove that 5555 and 51415 are also impossible.

Proposition 8 *Any* $\{2, 3, 7, 19\}$-*consistent gap sequence cannot contain the block* 5555 *nor* 51415.

Proof First, assume d is a $\{2, 3, 7, 19\}$-consistent gap sequence containing 5555. By Proposition 5, the associated color sequence contains the following block:

$$c_0 \ldots c_{19} = 01122011220112201122.$$

The fact that $c_1 = 1$ and $c_{13} = 2$ implies $c_{20} = 0$, but this together with the fact that $c_4 = 2$ and $c_{16} = 1$ means that c_{23} cannot be properly colored.

Next, assume d is a $\{2, 3, 7, 19\}$-consistent gap sequence containing the block 51415. By Proposition 5, the associated color sequence must contain the following block:

$$c_0 \ldots c_{15} = 01122001x2001122$$

where $c_8 = x \in \{1, 2\}$ is not determined by the θ-rule. But the fact that $c_1 = 1$, $c_6 = 0$ and $c_{15} = 2$ implies that c_8 cannot be properly colored.   □

With the above classification of allowable blocks, we can characterize the possible $\{2, 3, 7, 19\}$-consistent gap sequences. The fact that 151, 45, 54 and 5555 are all impossible implies that any time a 5 occurs it must be part of a 1551 or a 15551 block. The fact that 11, 44 and 14141414 are all impossible implies that a 5 must occur in all gap sequences. The fact that 515 and 51415 are impossible implies that any $\{2, 3, 7, 19\}$-consistent gap sequence can be partitioned into a sequence consisting entirely of the following four blocks:

$$C_1 = 1414155, \quad C_2 = 14141555, \quad C_3 = 141414155, \quad C_4 = 1414141555.$$
$$(2)$$

4.2 Characterizing Color Sequences

The monochromatic gap sequences are not sufficient to classify all sets $\{2, 3, 7, 19, p\}$, as $43 \notin \sigma(\boldsymbol{d})$ when $\boldsymbol{d} := (C_1 C_2)$. As we are concerned with $\{2, 3, 7, 19\}$-consistent colorings we can strengthen Proposition 5 to the following:

Lemma 4 *If $\boldsymbol{d}$ is a $\{2, 3, 7, 19\}$-consistent gap sequence, then $\boldsymbol{d} = \Delta_0(\boldsymbol{c})$ where, up to a permutation of the labels, $\boldsymbol{c}$ is given by the following rule:*

$$\eta(d_i) = \begin{cases} 0 & \text{if } d_i = 1 \\ 0112 & \text{if } d_{i-6} \cdots d_i = 5551414 \text{ or } d_i \cdots d_{i+2} = 415 \\ 01z2 & \text{if } d_{i-6} \cdots d_{i+6} = 1551414141551 \\ 0122 & \text{if } d_{i-2} \cdots d_i = 514 \text{ or } d_i \cdots d_{i+6} = 4141555 \\ 01122 & \text{if } d_i = 5 \end{cases}$$

where $z \in \{1, 2\}$ can be chosen arbitrarily for each 1551414141551 block in $\boldsymbol{d}$.

Proof By Proposition 5, we need only prove the cases where $d_i = 4$.

Case 1: Suppose $d_{i-6} \cdots d_i = 5551414$. Then by Proposition 5

$$\theta(d_{i-6} \cdots d_i) = 011220112201122001z_1 2001z_2 2.$$

The integer 19 spaces before z_2 is colored with a 2, so $z_2 = 1$ and $\eta(d_i) = 0112$.

Case 2: Suppose $d_i d_{i+1} d_{i+2} = 415$. Then

$$\theta(d_i d_{i+1} d_{i+2}) = 01z2001122.$$

The integer 7 spaces after z is colored with a 2, so $z = 1$ and $\eta(d_i) = 0112$.

Case 3: Suppose $d_{i-2} d_{i-1} d_i = 514$. Then

$$\theta(d_{i-2} d_{i-1} d_i) = 01122001z2.$$

The integer 7 spaces before z is colored with a 1, so $z = 2$ and $\eta(d_i) = 0122$.

Case 4: Suppose $d_i \cdots d_{i+6} = 4141555$. Then

$$\theta(d_i \cdots d_{i+6}) = 01z_1 2001z_2 2001122011220112201122.$$

The integer 19 spaces after z_1 is colored with a 1, so $z_1 = 2$ and $\eta(d_i) = 0122$.

Case 5: The only block that has not been covered by the previous four cases is:

$$d_{i-6} \cdots d_{i+6} = 1551414141551,$$

where the indeterminate color z in $\theta(d_i)$ can still be either 1 or 2.

$\square$

Our four gap sequence blocks can now be expanded to color sequence blocks. The strengthened η in Lemma 4 completely determines the color sequences from C_1, C_2 and C_4. The block C_3 can expand into two different color sequence blocks, depending on the choice for z.

$$A_1 := \eta(C_1) = 001220011200112201122$$

$$A_2 := \eta(C_2) = 0012200112001122011220 1122$$

$$A_3 := \eta(C_3) = 001220011200112001122 01122 \quad \text{(with } z = 1\text{)}$$

$$A'_3 := \eta(C_3) = 001220012200112001122 01122 \quad \text{(with } z = 2\text{)}$$

$$A_4 := \eta(C_4) = 00122001220011200112201122 01122.$$

It is more convenient to work with gap sequence triples rather than undifferentiated color sequences, so we unravel the above color sequences into the gap sequences for each color class.

$$\Delta_0(A_1) = 1414155 \qquad \Delta_1(A_1) = 5141415 \qquad \Delta_2(A_1) = 1551414$$

$$\Delta_0(A_2) = 14141555 \qquad \Delta_1(A_2) = 514141415 \qquad \Delta_2(A_2) = 155141414$$

$$\Delta_0(A_3) = 141414155 \qquad \Delta_1(A_3) = 514141415 \qquad \Delta_2(A_3) = 15551414$$

$$\Delta_0(A'_3) = 141414155 \qquad \Delta_1(A'_3) = 55141415 \qquad \Delta_2(A'_3) = 141551414$$

$$\Delta_0(A_4) = 1414141555 \qquad \Delta_1(A_4) = 5514141415 \qquad \Delta_2(A_4) = 14155141414.$$

Thus any color sequence c can be partitioned into a sequence of blocks $\{X_i\}$ where $X_i \in \{A_1, A_2, A_3, A'_3, A_4\}$. But we need to put some restrictions on which blocks can follow one another. Considering $\Delta_2(c)$, it is clear that A_2 cannot be followed by either A'_3 or A_4, since this would create a 14141414 block. Similarly A_4 cannot be followed by either A'_3 or A_4. Otherwise the blocks can be freely concatenated.

4.3 Guaranteed Partial Sums

Theorem 4 *If $p \in \{31, 37, 41\}$, then $\{2, 3, 7, 19, p\}$ is class 4.*

Proof Let $p \in \{31, 37, 41\}$, and assume that c is a $\{2, 3, 7, 19, p\}$-consistent 3-coloring. We know that $d := \Delta_0(c)$ must contain at least one of the blocks C_1, C_2, C_3 or C_4. Let $|C_i|$ denote the sum of all the terms in C_i. That is, $|C_1| = 21$, $|C_2| = |C_3| = 26$, and $|C_4| = 31$. By the structure of $\{2, 3, 7, 19\}$-consistent gap sequences (the blocks of (2)), we know that, regardless of what block precedes or follows C_i, the sequence must have the form

$$d = \cdots 55 C_i 14141 \cdots$$

Thus we know $\sigma(d)$ will contain the set $\{|C_i|+n : n \in \{1, 5, 6, 10, 11, 15, 16, 20, 21\}\}$. Since

$$31 = |C_1| + 10 = |C_2| + 5 = |C_3| + 5 = |C_4|,$$
$$37 = |C_1| + 16 = |C_2| + 11 = |C_3| + 11 = |C_4| + 6,$$
$$41 = |C_1| + 20 = |C_2| + 15 = |C_3| + 15 = |C_4| + 10,$$

we know that $\{21, 37, 41\} \subset \sigma(d)$, and by Proposition 4 this contradicts the claim that c is a $\{2, 3, 7, 19, p\}$-consistent 3-coloring. $\square$

Theorem 5 $\{2, 3, 7, 19, 43\}$ *is class 4.*

Proof Assume that c is a $\{2, 3, 7, 19, 43\}$-consistent 3-coloring.

Case 1: Suppose c contains the block A_1. The fact that $|\Delta_2(A_1)| = 21$ and, no matter what blocks follow, $\Delta_2(c)$ has an initial sum of 22 implies that c has a partial sum of 43, contradicting the claim that c is a consistent coloring.

Case 2: Suppose c contains A_2, A_3 or A'_3. Each of these blocks has sum 26. Thus the fact that $\Delta_2(c)$ has an initial sum of 17 no matter what block follows implies c contains a partial sum of 43, a contradiction.

Case 3: Suppose c contains A_4. Note that $|A_4| = 31$. As the block after A_4 cannot be A'_3 or A_4, $\Delta_1(c)$ must be of the form

$$\cdots 15 \Delta_1(A_4) 51 \cdots$$

This gives a partial sum of 43, a contradiction.

In all three cases c cannot be a consistent 3-coloring, and the result follows. $\square$

For the rest of the primes, the arguments only get more involved. We leave the verification that the partial sums of each color sequence of the prescribed form contain each prime listed at the beginning of this section to a computer (see [16]). To do so we construct an infinite tree colorings shown in Fig. 2. The tree is mutually recursively defined with the tree colorings' shown in Fig. 3. Any path of the tree colorings, concatenating the color sequence blocks at each vertex, will produce a color sequence of the form $\sum A_i$. Any path producing either a block A_2 or A_4 must be followed by a path producing either A_1, A_2 or A_3. This is represented by the pruned tree colorings'. Conversely, any one way infinite coloring sequence will be contained in a path of colorings. Thus it suffices to show that each path in colorings contains a partial sum of p for each prime p considered.

This is done by the pair of functions pathsToLists and check. The function pathsToLists tree n creates a list of lists of length n, representing all the paths of length n in tree. Then the function check p is a Boolean function that, when applied to a list, returns True if the list contains a pair of equal elements with

Fig. 2 The tree `colorings`

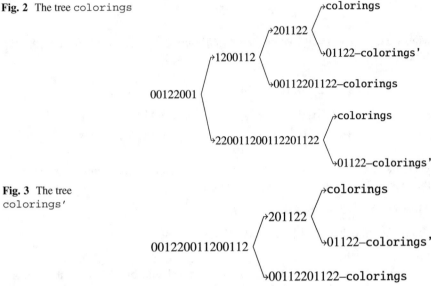

Fig. 3 The tree `colorings'`

indices differing by p. This is equivalent to checking whether the coloring block represented by the list contains a partial sum of p. In this way, running the Haskell code in [16] verifies the following proposition.

Proposition 9 *The set* $D = \{2, 3, 7, 19, p\}$ *is class 4 for any p in the following set:*

$$\{31, 37, 41, 43, 47, 53, 67, 73, 79, 83, 89, 109, 131, 151, 157, 167, 193\}.$$

5   Conclusion

By establishing Theorem 3 we completely classify the prime sets $\{2, 3, 7, 19, p\}$ and settle most of the more general family of the form $\{2, 3, 7, p, q\}$. Further we propose the following conjecture:

Conjecture 1 A prime set of the form $D = \{2, 3, 7, p, q\}$ is minimal class 4 if and only if the pair (p, q) is one of the 31 pairs listed in Theorem 3 part 2.

In order to establish this conjecture, the 14 prime pairs not covered by Proposition 9 would need to be proven class 4. While the block method developed in Sect. 4 could be extended to more general distance sets, the results of that section are very tied to the fact that both 7 and 19 are in D. This makes it seem unlikely that the method would be tractable to the other 14 sets of the form $\{2, 3, 7, p, q\}$.

To confirm the conjecture, in addition the condition that $p \not\equiv 2311139, 2311163$ (mod 4622310) when $q = p + 8$ would need to be removed. We believe that the most economical way to prove those D sets are indeed class 3 might be to find periodic 3-colorings for the associated distance graphs.

Acknowledgements The research is partially supported by the National Science Foundation under grant DMS-1600778 and the National Aeronautics and Space Administration under grant NASA MIRO NX15AQ06A.

References

1. Chang, G., Huang, L., Zhu, X.: Circular chromatic numbers and fractional chromatic numbers of distance graphs. European Journal of Combinatorics **19**(4), 423–431 (1998)
2. Chang, G.J., Liu, D.D.F., Zhu, X.: Distance graphs and t-coloring. Journal of Combinatorial Theory, Series B **75**(2), 259–269 (1999)
3. Chen, J.J., Chang, G.J., Huang, K.C.: Integral distance graphs. Journal of Graph Theory **25**(4), 287–294 (1997)
4. Eggleton, R.B.: New results on 3-chromatic prime distance graphs. Ars Combin. **26**(B), 153–180 (1988)
5. Eggleton, R.B., Erdős, P., Skilton, D.K.: Colouring the real line. J. Combin. Theory Ser. B **39**(1), 86–100 (1985)
6. Eggleton, R.B., Erdős, P., Skilton, D.K.: Colouring prime distance graphs. Graphs Combin. **6**(1), 17–32 (1990)
7. Gupta, S.: Sets of integers with missing differences. J. Combin. Theory Ser. A **89**(1), 55–69 (2000)
8. Haralambis, N.M.: Sets of integers with missing differences. J. Combinatorial Theory Ser. A **23**(1), 22–33 (1977)
9. Kemnitz, A., Kolberg, H.: Coloring of integer distance graphs. Discrete Math. **191**(1–3), 113–123 (1998). Graph theory (Elgersburg, 1996)
10. Kemnitz, A., Marangio, M.: Chromatic numbers of integer distance graphs. Discrete Math. **233**(1–3), 239–246 (2001). Graph theory (Prague, 1998)
11. Lam, P.C.B., Lin, W.: Coloring of distance graphs with intervals as distance sets. European J. Combin. **26**(8), 1216–1229 (2005)
12. Liu, D.D.F.: T-colorings and chromatic number of distance graphs. Ars Comb. **56**, 65–80 (2000)
13. Liu, D.D.F.: From rainbow to the lonely runner: a survey on coloring parameters of distance graphs. Taiwanese J. Math. **12**(4), 851–871 (2008)
14. Liu, D.D.F., Zhu, X.: Distance graphs with missing multiples in the distance sets. J. Graph Theory **30**(4), 245–259 (1999)
15. Liu, D.D.F., Zhu, X.: Fractional chromatic number and circular chromatic number for distance graphs with large clique size. J. Graph Theory **47**(2), 129–146 (2004)
16. Robinson, G.: Coloring prime distance graphs. Master's thesis, California State University, Los Angeles (2018).
17. Voigt, M.: Colouring of distance graphs. Ars Combin. **52**, 3–12 (1999)
18. Voigt, M., Walther, H.: On the chromatic number of special distance graphs. Discrete Math. **97**(1–3), 395–397 (1991)
19. Voigt, M., Walther, H.: Chromatic number of prime distance graphs. Discrete Appl. Math. **51**(1–2), 197–209 (1994). 2nd Twenty Workshop on Graphs and Combinatorial Optimization (Enschede, 1991)
20. Zhu, X.: Pattern periodic coloring of distance graphs. Journal of Combinatorial Theory, Series B **73**(2), 195–206 (1998)

21. Zhu, X.: Circular chromatic number: a survey. Discrete Math. **229**(1–3), 371–410 (2001). Combinatorics, graph theory, algorithms and applications
22. Zhu, X.: Circular chromatic number of distance graphs with distance sets of cardinality 3. J. Graph Theory **41**(3), 195–207 (2002)

Combinatorial Characterization of Queer Supercrystals (Survey)

Maria Gillespie, Graham Hawkes, Wencin Poh, and Anne Schilling

1 Introduction

The representation theory of Lie algebras is of fundamental importance, and hence combinatorial models for representations, especially those amenable to computation, are of great use. In the 1990s, Kashiwara [1] showed that integrable highest weight representations of the Drinfeld–Jimbo quantum groups $U_q(\mathfrak{g})$, where $\mathfrak{g}$ is a symmetrizable Kac–Moody Lie algebra, in the $q \to 0$ limit result in a combinatorial skeleton of the integrable representation. He coined the term crystal bases, reflecting the fact that q corresponds to the temperature of the underlying physical system. Since then, crystal bases have appeared in many areas of mathematics, including algebraic geometry, combinatorics, mathematical physics, representation theory, and number theory. One of the major advances in the theory of crystals for simply-laced Lie algebras was the discovery by Stembridge [2] of local axioms that uniquely characterize the crystal graphs corresponding to Lie algebra representations. These local axioms provide a completely combinatorial approach to the theory of crystals; this viewpoint was taken in [3].

Lie superalgebras [4] arose in physics in theories that unify bosons and fermions. They are essential in modern string theories [5] and appear in other areas of mathematics, such as the projective representations of the symmetric group. The crystal basis theory has been developed for various quantum superalgebras [6–12]. In this paper, we are in particular interested in the queer superalgebra $\mathfrak{q}(n)$ (see for example [13]). A theory of highest weight crystals for the queer superalgebra $\mathfrak{q}(n)$ was recently developed by Grantcharov et al. [7–9]. They provide an explicit combinatorial realization of the highest weight crystal bases in terms of

M. Gillespie (✉) · G. Hawkes · W. Poh · A. Schilling
Department of Mathematics, University of California at Davis, Davis, CA, USA
e-mail: anne@math.ucdavis.edu

© The Author(s) and the Association for Women in Mathematics 2020
B. Acu et al. (eds.), *Advances in Mathematical Sciences*, Association for
Women in Mathematics Series 21, https://doi.org/10.1007/978-3-030-42687-3_4

semistandard decomposition tableaux and show how these crystals can be derived
from a tensor product rule and the vector representation. They also use the tensor
product rule to derive a Littlewood–Richardson rule. Choi and Kwon [14] provide a
new characterization of Littlewood–Richardson–Stembridge tableaux for Schur P-
functions by using the theory of $q(n)$-crystals. Independently, Hiroshima [15] and
Assaf and Oguz [16, 17] defined a queer crystal structure on semistandard shifted
tableaux, extending the type A crystal structure of [18] on these tableaux.

In this paper, we provide a characterization of the queer supercrystals in analogy
to Stembridge's [2] characterization of crystals associated to classical simply-laced
root systems. Assaf and Oguz [16, 17] conjecture a local characterization of queer
crystals in the spirit of Stembridge [2], which involves local relations between the
odd crystal operator f_{-1} with the type A_{n-1} crystal operators f_i for $1 \leqslant i < n$.
However, we provide a counterexample to [17, Conjecture 4.16], which conjectures
that these local axioms uniquely characterize the queer supercrystals. Instead, we
define a new graph $G(C)$ on the relations between the type A components of the
queer supercrystal C, which together with Assaf's and Oguz' local queer axioms
and further new axioms uniquely fixes the queer crystal structure (see Theorem 3).
We provide a combinatorial description of $G(C)$ by providing the combinatorial
rules for all odd queer crystal operators f_{-i} on certain highest weight elements for
$1 \leqslant i < n$. A long version of this paper containing all proofs is available in [19].

2   Queer Supercrystals

An *(abstract) crystal* of type A_n is a nonempty set B together with the maps
$e_i, f_i : B \to B \sqcup \{0\}$ for $i \in I$ and wt: $B \to \Lambda$, where $\Lambda = \mathbb{Z}_{\geqslant 0}^{n+1}$ is the weight
lattice of the root of type A_n and $I = \{1, 2, \ldots, n\}$ is the index set, subject to several
conditions. Denote by $\alpha_i = \epsilon_i - \epsilon_{i+1}$ for $i \in I$ the simple roots of type A_n, where
ϵ_i is the i-th standard basis vector of $\mathbb{Z}^{n+1}$. Then we require:

A1. For $b, b' \in B$, we have $f_i b = b'$ if and only if $b = e_i b'$. Also wt$(b') = $
         wt$(b) - \alpha_i$.

For $b \in B$, we also define $\varphi_i(b) = \max\{k \in \mathbb{Z}_{\geqslant 0} \mid f_i^k(b) \neq 0\}$ and $\varepsilon_i(b) = $
$\max\{k \in \mathbb{Z}_{\geqslant 0} \mid e_i^k(b) \neq 0\}$. For further details, see for example [3, Definition 2.13].

There is an action of the symmetric group S_n on a type A_n crystal B given by the
operators

$$s_i(b) = \begin{cases} f_i^k(b) & \text{if } k \geqslant 0, \\ e_i^{-k}(b) & \text{if } k < 0, \end{cases} \tag{1}$$

for $b \in B$, where $k = \varphi_i(b) - \varepsilon_i(b)$. An element $b \in B$ is called *highest weight* if
$e_i(b) = 0$ for all $i \in I$. For a subset $J \subseteq I$, we say that b is J-highest weight if
$e_i(b) = 0$ for all $i \in J$. We are now ready to define an abstract queer crystal.

Fig. 1 q($n + 1$)-queer crystal
of letters $\mathcal{B}$

Definition 1 ([8, Definition 1.9]) An *abstract* q($n + 1$)-*crystal* is a type A_n crystal B together with the maps $e_{-1}, f_{-1} \colon B \to B \sqcup \{0\}$ satisfying the following conditions:

Q1. wt(B) $\subset \Lambda$;
Q2. wt($e_{-1}b$) = wt(b) + α_1 and wt($f_{-1}b$) = wt(b) − α_1;
Q3. for all $b, b' \in B$, $f_{-1}b = b'$ if and only if $b = e_{-1}b'$;
Q4. if $3 \leqslant i \leqslant n$, we have (a) the crystal operators e_{-1} and f_{-1} commute with e_i and f_i and (b) if $e_{-1}b \in B$, then $\varepsilon_i(e_{-1}b) = \varepsilon_i(b)$ and $\varphi_i(e_{-1}b) = \varphi_i(b)$.

Given two q($n + 1$)-crystals B_1 and B_2, Grantcharov et al. [8, Theorem 1.8] provide a crystal on the tensor product $B_1 \otimes B_2$, which we state here in reverse convention. It consists of the type A_n tensor product rule (see for example [3, Section 2.3]) and the *tensor product rule* for $b_1 \otimes b_2 \in B_1 \otimes B_2$

$$
e_{-1}(b_1 \otimes b_2) = \begin{cases} b_1 \otimes e_{-1}b_2 & \text{if wt}(b_1)_1 = \text{wt}(b_1)_2 = 0, \\ e_{-1}b_1 \otimes b_2 & \text{otherwise,} \end{cases} \tag{2}
$$

and similarly for f_{-1}. *Queer supercrystals* are connected components of $\mathcal{B}^{\otimes \ell}$, where $\mathcal{B}$ is the q($n + 1$)-queer crystal of letters depicted in Fig. 1.

In addition to the queer crystal operators $f_{-1}, f_1, \ldots, f_n$ and $e_{-1}, e_1, \ldots, e_n$, we define crystal operators $f_{-i} := s_{w_i}^{-1} f_{-1} s_{w_i}$ and $e_{-i} := s_{w_i}^{-1} e_{-1} s_{w_i}$ for $1 < i \leqslant n$, where $s_{w_i} = s_2 \cdots s_i s_1 \cdots s_{i-1}$ with s_i as in (1). By [8, Theorem 1.14], with all operators e_i, f_i for $i \in \{\pm 1, \pm 2, \ldots, \pm n\}$ each connected component of $\mathcal{B}^{\otimes \ell}$ has a unique highest weight vector.

The operators f_i for $i \in I_0$ have an easy combinatorial description on $b \in \mathcal{B}^{\otimes \ell}$ given by the *signature rule*, which can be directly derived from the tensor product rule (see for example [3, Section 2.4]). One can consider b as a word in the alphabet $\{1, 2, \ldots, n+1\}$. Consider the subword of b consisting only of the letters i and $i+1$. Pair any consecutive letters $i + 1, i$ in this order, remove this pair, and repeat. Then f_i changes the rightmost unpaired i to $i + 1$; if there is no such letter $f_i(b) = 0$. Similarly, e_i changes the leftmost unpaired $i + 1$ to i; if there is no such letter $e_i(b) = 0$.

Remark 1 From (2), one may also derive a simple combinatorial rule for f_{-1} and e_{-1}. Consider the subword v of $b \in \mathcal{B}^{\otimes \ell}$ consisting of the letters 1 and 2. The crystal operator f_{-1} on b is defined if the leftmost letter of v is a 1, in which case it turns it into a 2. Otherwise $f_{-1}(b) = 0$. Similarly, e_{-1} on b is defined if the leftmost letter of v is a 2, in which case it turns it into a 1. Otherwise $e_{-1}(b) = 0$.

We now give explicit descriptions of $\varphi_{-i}(b)$ and $f_{-i}b$ for J-highest-weight elements $b \in \mathcal{B}^{\otimes \ell}$ for certain $J \subseteq I_0 := \{1, 2, \ldots, n\}$ (see Proposition 1 and

Theorem 1). We will need these results in Sect. 4 when we characterize certain graphs on the type A components of the queer crystal.

Definition 2 The *initial k-sequence* of a word $b = b_1 \ldots b_\ell \in \mathcal{B}^{\otimes \ell}$, if it exists, is the sequence of letters $b_{p_k}, b_{p_{k-1}}, \ldots, b_{p_1}$, where b_{p_k} is the leftmost k and b_{p_j} is the leftmost j to the right of $b_{p_{j+1}}$ for all $1 \leqslant j < k$.

Let $i \in I_0$ and $b \in \mathcal{B}^{\otimes \ell}$ be $\{1, 2, \ldots, i\}$-highest weight with $\mathrm{wt}(b)_{i+1} > 0$, where $\mathrm{wt}(b)_{i+1}$ is the $(i + 1)$-st entry in $\mathrm{wt}(b) \in \mathbb{Z}^{n+1}$. Then note that b has an initial $(i + 1)$-sequence, say $b_{p_{i+1}}, b_{p_i}, \ldots, b_{p_1}$. Also let $b_{q_i}, b_{q_{i-1}}, \ldots, b_{q_1}$ be the initial i-sequence of b. Note that $p_{i+1} < p_i < \cdots < p_1$ and $q_i < q_{i-1} < \cdots < q_1$ by the definition of initial sequence. Furthermore either $q_j = p_j$ or $q_j < p_{j+1}$ for all $1 \leqslant j \leqslant i$.

Proposition 1 *Let* $b \in \mathcal{B}^{\otimes \ell}$ *be* $\{1, 2, \ldots, i\}$*-highest weight for* $i \in I_0$. *Then* $\varphi_{-i}(b) = 1$ *if and only if* $\mathrm{wt}(b)_i > 0$ *and either* $\mathrm{wt}(b)_{i+1} = 0$ *or* $p_j \neq q_j$ *for all* $j \in \{1, 2, \ldots, i\}$.

Example 1 Take $b = 1331242312111$ and $i = 3$. Then $p_4 = 6, p_3 = 8, p_2 = 10, p_1 = 11$ and $q_3 = 2, q_2 = 5, q_1 = 9$. We indicate the chosen letters p_j by underlines and q_j by overlines: $b = 13\overline{3}1\underline{2}\overline{4}2\underline{3}\overline{1}\underline{2}111$. Since no letter has a both an overline and underline (meaning $p_j \neq q_j$ for all j), we have $\varphi_{-3}(b) = 1$.

Recall that in a queer crystal B an element $b \in B$ is *highest-weight* if $e_i(b) = 0$ for all $i \in I_0 \cup I_-$, where $I_0 = \{1, 2, \ldots, n\}$ and $I_- = \{-1, -2, \ldots, -n\}$.

Proposition 2 ([8, Prop.1.13]) *Let* $b \in \mathcal{B}^{\otimes \ell}$ *be highest weight. Then* $\mathrm{wt}(b)$ *is a strict partition.*

Next, we provide an explicit description of $f_{-i}(b)$ for $i \in I_0$, when b is $\{1, 2, \ldots, i\}$-highest weight. Recall that the sequence $b_{q_i}, b_{q_{i-1}}, \ldots, b_{q_1}$ is the leftmost sequence of letters $i, i - 1, \ldots, 1$ from left to right. Set $r_1 = q_1$ and recursively define $r_j < r_{j-1}$ for $1 < j \leqslant i$ to be maximal such that $b_{r_j} = j$. Note that by definition $q_j \leqslant r_j$. Let $1 \leqslant k \leqslant i$ be maximal such that $q_k = r_k$.

Theorem 1 *Let* $b \in \mathcal{B}^{\otimes \ell}$ *be* $\{1, 2, \ldots, i\}$*-highest weight for* $i \in I_0$ *and* $\varphi_{-i}(b) = 1$ *(see Proposition 1). Then* $f_{-i}(b)$ *is obtained from* b *by changing* $b_{q_j} = j$ *to* $j - 1$ *for* $j = i, i - 1, \ldots, k + 1$ *and* $b_{r_j} = j$ *to* $j + 1$ *for* $j = i, i - 1, \ldots, k$.

Example 2 Let us continue Example 1 with $b = 1331242312111$ and $i = 3$. We overline b_{q_j} and underline b_{r_j}, so that $b = 1\overline{3}\underline{3}1\overline{2}4\underline{2}3\overline{1}\underline{2}111$. From this we read off $q_3 = 2, q_2 = 5, q_1 = 9, r_3 = 3, r_2 = 7, r_1 = 9, k = 1$ and $f_{-3}(b) = 1241143322111$.

As another example, take $b = 545423321211$ in the q(6)-crystal $\mathcal{B}^{\otimes 12}$ and $i = 5$. Again, we overline b_{q_j} and underline b_{r_j}, so that $b = \overline{5}4\underline{5}4\overline{2}3\overline{3}\underline{2}\overline{1}211$. This means that $q_5 = 1, q_4 = 2, q_3 = 6, q_2 = 8, q_1 = 9, r_5 = 3, r_4 = 4, r_3 = 7, r_2 = 8, r_1 = 9, k = 2$, and $f_{-5}(b) = 436522431211$.

Corollary 1 *Let $b \in \mathcal{B}^{\otimes \ell}$ be J-highest weight for $\{1, 2, \ldots, i\} \subseteq J \subseteq I_0$ and $\varphi_{-i}(b) = 1$ for some $i \in I_0$. Then:*

1. *Either $f_{-i}(b) = f_i(b)$ or $f_{-i}(b)$ is J-highest weight.*
2. *$f_{-i}(b)$ is I_0-highest weight only if $b = f_{i+1} f_{i+2} \cdots f_{h-1} u$ for some $n + 1 \geqslant h > i$ and u a I_0-highest weight element.*

3 Local Axioms

In [17, Definition 4.11], Assaf and Oguz give a definition of regular queer crystals. In essence, their axioms are rephrased in the following definition, where $\tilde{I} := I_0 \cup \{-1\}$.

Definition 3 (Local Queer Axioms) Let C be a graph with labeled directed edges given by f_i for $i \in I_0$ and f_{-1}. If $b' = f_j b$ for $j \in \tilde{I}$ define e_j by $b = e_j b'$.

LQ1. The subgraph with all vertices but only edges labeled by $i \in I_0$ is a type A_n Stembridge crystal.

LQ2. $\varphi_{-1}(b), \varepsilon_{-1}(b) \in \{0, 1\}$ for all $b \in C$.

LQ3. $\varphi_{-1}(b) + \varepsilon_{-1}(b) > 0$ if $\mathrm{wt}(b)_1 + \mathrm{wt}(b)_2 > 0$.

LQ4. Assume $\varphi_{-1}(b) = 1$ for $b \in C$.

    (a) If $\varphi_1(b) > 2$, we have $f_1 f_{-1}(b) = f_{-1} f_1(b)$, $\varphi_1(b) = \varphi_1(f_{-1}(b)) + 2$, and $\varepsilon_1(b) = \varepsilon_1(f_{-1}(b))$.

    (b) If $\varphi_1(b) = 1$, we have $f_1(b) = f_{-1}(b)$.

LQ5. Assume $\varphi_{-1}(b) = 1$ for $b \in C$.

    (a) If $\varphi_2(b) > 0$, we have $f_2 f_{-1}(b) = f_{-1} f_2(b)$, $\varphi_2(b) = \varphi_2(f_{-1}(b)) - 1$, and $\varepsilon_2(b) = \varepsilon_2(f_{-1}(b))$.

    (b) If $\varphi_2(b) = 0$, we have

$$\varphi_2(b) = \varphi_2(f_{-1}(b)) - 1 = 0, \quad \text{or} \quad \varphi_2(b) = \varphi_2(f_{-1}(b)) = 0,$$
$$\varepsilon_2(b) = \varepsilon_2(f_{-1}(b)), \qquad\qquad \varepsilon_2(b) = \varepsilon_2(f_{-1}(b)) + 1.$$

LQ6. Assume that $\varphi_{-1}(b) = 1$ and $\varphi_i(b) > 0$ with $i \geqslant 3$ for $b \in C$. Then $f_i f_{-1}(b) = f_{-1} f_i(b)$, $\varphi_i(b) = \varphi_i(f_{-1}(b))$, and $\varepsilon_i(b) = \varepsilon_i(f_{-1}(b))$.

Axioms **LQ4** and **LQ5** are illustrated in Fig. 2.

Proposition 3 ([17]) *The queer crystal of words $\mathcal{B}^{\otimes \ell}$ satisfies the axioms in Definition 3.*

In [17, Conjecture 4.16], Assaf and Oguz conjecture that every regular queer crystal is a normal queer crystal. In other words, every connected graph satisfying the local queer axioms of Definition 3 is isomorphic to a connected component in some $\mathcal{B}^{\otimes \ell}$. We provide a counterexample to this claim in [19, Figure 3]. This

Fig. 2 Illustration of axioms
LQ4 (left) and **LQ5** (right).
The (-1)-arrow at the bottom
of the right figure might or
might not be there

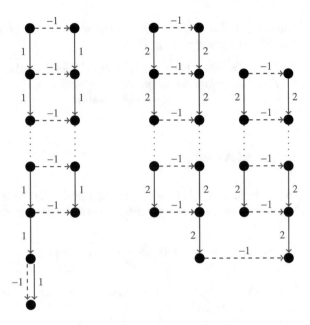

counterexample is based on the I_0-components of the q(3)-crystal of highest weight
$(4, 2, 0)$. In addition to the usual queer crystal, there is another choice of arrows that
does not violate the conditions of Definition 3.

The problem with Axiom **LQ5** illustrated in Fig. 2 is that the (-1)-arrow at
the bottom of the 2-strings is not closed at the top. Hence, as demonstrated by
the counterexample in switching components with the same I_0-highest weights can
cause non-uniqueness.

4 Graph on Type A Components

Definition 4 Let C be a crystal with index set $I_0 \cup \{-1\}$ that is a Stembridge crystal
of type A_n when restricted to the arrows labeled I_0. We define the *component graph*
of C, denoted by $G(C)$, as the following simple directed graph. The vertices of
$G(C)$ are the type A_n components of C (typically labeled by their highest weight
elements). There is a directed edge from vertex C_1 to vertex C_2, if there is an element
b_1 in component C_1 and an element b_2 in component C_2 such that $f_{-1}b_1 = b_2$.

Example 3 Let C be the connected component in the q(3)-crystal $\mathcal{B}^{\otimes 6}$ with highest
weight element $1 \otimes 2 \otimes 1 \otimes 1 \otimes 2 \otimes 1$ of highest weight $(4, 2, 0)$. The graph $G(C)$
is given in Fig. 3 on the left. The graph $G(C')$ for the counterexample C' in [19,
Figure 3] is given in Fig. 3 on the right. Since the two graphs are not isomorphic
as unlabeled graphs, this confirms that the purple dashed arrows in [19, Figure 3]

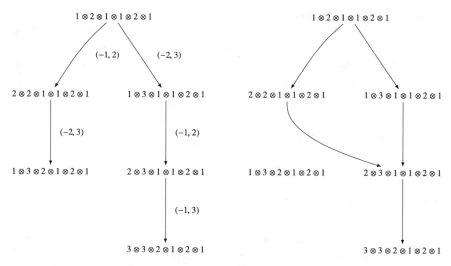

Fig. 3 **Left:** $G(C)$ for the crystals of Example 3. **Right:** $G(C')$ for the crystals of Example 3

do not give the queer crystal even though the induced crystal satisfies the axioms in Definition 3.

Next we show that the arrows in $G(C)$, where C is a connected component in $\mathcal{B}^{\otimes \ell}$, can be modeled by e_{-i} on type A highest weight elements.

Proposition 4 *Let C be a connected component in the $\mathfrak{q}(n+1)$-crystal $\mathcal{B}^{\otimes \ell}$. Let C_1 and C_2 be two distinct type A_n components in C and let u_2 be the I_0-highest weight element in C_2. Then there is an edge from C_1 to C_2 in $G(C)$ if and only if $e_{-i}u_2 \in C_1$ for some $i \in I_0$.*

By Proposition 4, there is an edge from component C_1 to component C_2 in $G(C)$ if and only if $e_{-i}u_2 \in C_1$ for some $i \in I_0$, where u_2 is the I_0-highest weight element of C_2.

We call the arrow *combinatorial* if $e_{-i}u_2$ is $\{1, 2, \ldots, i\}$-highest weight. Define $f_{(-i,h)} := f_{-i}f_{i+1}f_{i+2}\cdots f_{h-1}$.

Theorem 2 *Let C be a connected component in $\mathcal{B}^{\otimes \ell}$. Then each combinatorial edge in $G(C)$ can be obtained by $f_{(-i,h)}$ for some $i \in I_0$ and $h > i$ minimal such that $f_{(-i,h)}$ applies.*

In [19], we showed that it suffices to know the combinatorial edges to construct all vertices in $G(C)$. By Theorem 2, every combinatorial edge in the graph is labeled by the operator $f_{(-i,h)}$, where f_{-i} is given by the combinatorial rules stated in Theorem 1 and connects an I_0-highest weight element to another I_0-highest weight element. Hence, all vertices of $G(C)$ can be constructed from the $\mathfrak{q}(n+1)$-highest weight element u by the application of these combinatorial arrows.

Remark 2 The construction of the component graph of C with highest weight λ produces a Schur expansion of the Schur-P polynomial $P_\lambda(x_1, \ldots, x_{n+1})$. This expansion is obtained by counting the multiplicities of highest weights for all type A_n components that are present in $G(C)$. For example, the component graph in Example 3 yields the expansion $P_{42} = s_{42} + s_{33} + s_{411} + 2s_{321} + s_{222}$.

5 Characterization of Queer Crystals

Our main theorem gives a characterization of the queer supercrystals.

Theorem 3 *Let C be a connected component of a generic abstract queer crystal (see Definition 1). Suppose that C satisfies the following conditions:*

1. *C satisfies the local queer axioms of Definition 3.*
2. *C satisfies the connectivity axioms of [19, Definition 4.4].*
3. *$G(C)$ is isomorphic to $G(\mathcal{D})$, where $\mathcal{D}$ is some connected component of $\mathcal{B}^{\otimes \ell}$.*

Then the queer supercrystals C and $\mathcal{D}$ are isomorphic.

Acknowledgements We are grateful to Sami Assaf, Dan Bump, Zach Hamaker, Ezgi Oguz, and Travis Scrimshaw for helpful discussions. We would also like to thank Dimitar Grantcharov, Ji-Hye Jung, and Masaki Kashiwara for answering our questions about their work. This work benefited from experimentations in SageMath. This work was partially supported by NSF grants DMS–1500050, DMS–1760329, DMS–1764153 and NSF MSPRF grant PDRF 1604262.

References

1. M. Kashiwara. On crystal bases of the Q-analogue of universal enveloping algebras. *Duke Math. J.*, 63(2):465–516, 1991.
2. J. R. Stembridge. A local characterization of simply-laced crystals. *Trans. Amer. Math. Soc.*, 355(12):4807–4823, 2003.
3. D. Bump and A. Schilling. *Crystal bases*. World Scientific Publishing Co. Pte. Ltd., Hackensack, NJ, 2017. Representations and combinatorics.
4. V. G. Kac. Lie superalgebras. *Advances in Math.*, 26(1):8–96, 1977.
5. S. J. Gates, Jr., M. T. Grisaru, M. Roček, and W. Siegel. *Superspace*, volume 58 of *Frontiers in Physics*. Benjamin/Cummings Publishing Co., Inc., Advanced Book Program, Reading, MA, 1983. One thousand and one lessons in supersymmetry, With a foreword by David Pines.
6. G. Benkart, S.-J. Kang, and M. Kashiwara. Crystal bases for the quantum superalgebra $U_q(\mathfrak{gl}(m, n))$. *J. Amer. Math. Soc.*, 13(2):295–331, 2000.
7. D. Grantcharov, J. H. Jung, S.-J. Kang, M. Kashiwara, and M. Kim. Quantum queer superalgebra and crystal bases. *Proc. Japan Acad. Ser. A Math. Sci.*, 86(10):177–182, 2010.

8. D. Grantcharov, J. H. Jung, S.-J. Kang, M. Kashiwara, and M. Kim. Crystal bases for the quantum queer superalgebra and semistandard decomposition tableaux. *Trans. Amer. Math. Soc.*, 366(1):457–489, 2014.

9. D. Grantcharov, J. H. Jung, S.-J. Kang, M. Kashiwara, and M. Kim. Crystal bases for the quantum queer superalgebra. *J. Eur. Math. Soc. (JEMS)*, 17(7):1593–1627, 2015.

10. D. Grantcharov, J. H. Jung, S.-J. Kang, and M. Kim. A categorification of q(2)-crystals. *Algebr. Represent. Theory*, 20(2):469–486, 2017.

11. J.-H. Kwon. Super duality and crystal bases for quantum ortho-symplectic superalgebras. *Int. Math. Res. Not. IMRN*, (23):12620–12677, 2015.

12. J.-H. Kwon. Super duality and crystal bases for quantum ortho-symplectic superalgebras II. *J. Algebraic Combin.*, 43(3):553–588, 2016.

13. S.-J. Cheng and W. Wang. *Dualities and representations of Lie superalgebras*, volume 144 of *Graduate Studies in Mathematics*. American Mathematical Society, Providence, RI, 2012.

14. S.-I. Choi and J.-H. Kwon. Crystals and Schur *P*-positive expansions. *Electron. J. Combin.*, 25(3):Paper 3.7, 27, 2018.

15. T. Hiroshima. q-crystal structure on primed tableaux and on signed unimodal factorizations of reduced words of type *B*. *Publ. Res. Inst. Math. Sci.*, 55(2):369–399, (2019).

16. S. Assaf and E. Kantarci Oguz. Crystal graphs for shifted tableaux. *Sém. Lothar. Combin.*, 80B:Art. 26, 12, 2018.

17. S. Assaf and E. Kantarci Oguz. Toward a local characterization of crystals for the quantum queer superalgebra. *Ann. Combin.*, 24(1):3–36, 2020. preprint arXiv:1803.06317v1.

18. G. Hawkes, K. Paramonov, and A. Schilling. Crystal analysis of type *C* Stanley symmetric functions. *Electron. J. Combin.*, 24(3):Paper 3.51, 32, 2017.

19. M. Gillespie, G. Hawkes, W. Poh, and A. Schilling. Characterization of queer supercrystals. *J. Combin. Theo.*, Series A, 173:105235, 2020. https://doi.org/10.1016/j.jcta.2020.105235.

Enumerating in Coxeter Groups (Survey)

Bridget Eileen Tenner

1 Introduction

A *Coxeter group* is defined by a set of generators S and relations of the form

$$(ss')^{m(s,s')} = e$$

for $s, s' \in S$, with $m(s, s) = 1$. There are many so-called "types" of Coxeter groups, with, perhaps, the most well-studied being the *finite Coxeter group of type A*, also known as the *symmetric group*. Due to the length of this article, we focus our discussion on the symmetric group and, as appropriate, cite analogous results for Coxeter groups of other types. The other types referenced will most often be B and D, which have interpretations as signed permutations and signed permutations with restrictions, respectively. The reader will notice that several of the enumerative problems discussed here do not have such analogues, and we close this paper by highlighting a selection of these for future research.

The symmetric group S_n consists of all permutations of $\{1, \ldots, n\}$, and it is generated by the *adjacent transpositions* $\{s_i \ : \ 1 \leq i < n\}$, where s_i is the permutation that swaps i and $i + 1$, and fixes all other elements. (Note that there are other generating sets for S_n, as well, but the one of interest to us here is the set of adjacent transpositions.) In addition to being involutions, these generators satisfy the *commutation relation*

$$s_i s_j = s_j s_i \, when \, |i - j| > 1$$

B. E. Tenner (✉)
Department of Mathematical Sciences, DePaul University, Chicago, IL, USA
e-mail: bridget@math.depaul.edu

© The Author(s) and the Association for Women in Mathematics 2020
B. Acu et al. (eds.), *Advances in Mathematical Sciences*, Association for
Women in Mathematics Series 21, https://doi.org/10.1007/978-3-030-42687-3_5

and the *braid relation*

$$s_i s_{i+1} s_i = s_{i+1} s_i s_{i+1}.$$

Further information about general Coxeter groups and their combinatorial properties can be found in the aptly titled [2].

Despite the great interest in Coxeter groups from a variety of mathematical perspectives, many questions about them remain unanswered. Combinatorial aspects of these objects are no outlier in this sense, and open combinatorial questions range from an understanding of intricate structural features to fundamental enumerative issues.

Counting questions about Coxeter groups can take a range of forms, including the enumeration of Coxeter group elements that possess certain properties, and the quantification of particular features of the group elements themselves. In this article, we present problems in both of these categories. We also hint at large classes of open questions. In this way, we hope to attract and inspire new work in this area, where this is much yet to be done and much potential interest in the results.

2 Main Tools

The main tools for our work are two theorems from the literature. Before we can state these, we make a few important definitions. We phrase these in terms of the symmetric group because that is the focus of this work, but analogous objects exist for Coxeter groups of other types, too.

Definition 1 A *reduced decomposition* of a permutation $w \in S_n$ is a decomposition of w into minimally many generators: $w = s_{i_1} \cdots s_{i_{\ell(w)}}$. This minimal value $\ell(w)$ is the *length* of w. The set of all reduced decompositions of w is denoted $R(w)$.

A permutation can have many reduced decompositions, as demonstrated below. The number of reduced decompositions of a permutation was calculated in [10] in terms of standard Young tableaux.

Example 1 Let $w \in S_4$ be such that $w(1) = 3$, $w(2) = 2$, $w(3) = 4$, and $w(4) = 1$. This permutation has three reduced decompositions:

$$w = s_2 s_1 s_2 s_3 = s_1 s_2 s_1 s_3 = s_1 s_2 s_3 s_1,$$

where we think of the adjacent transpositions as maps, and thus compose them from right to left.

The Coxeter relations described above can act on reduced decompositions. They do this by replacing $s_i s_j$ by $s_j s_i$ when $|i - j| > 1$, and $s_i s_{i+1} s_i$ by $s_{i+1} s_i s_{i+1}$. Each of these actions suggests an equivalence relation on the elements of $R(w)$.

Definition 2 Let w be a permutation and $R(w)$ its set of reduced decompositions. This set has two natural partitions, arising from the Coxeter relations:

- the *commutation classes* of w are $C(w) := R(w)/(s_i s_j \sim s_j s_i)$ when $|i - j| > 1$, and
- the *braid classes* of w are $B(w) := R(w)/(s_i s_{i+1} s_i \sim s_{i+1} s_i s_{i+1})$.

In [1, §3], the authors consider such relation-based partitions more generally, for arbitrary Coxeter groups.

That is, any two reduced decompositions that are in the same commutation class $C \in C(w)$ (respectively, braid class $B \in B(w)$) can be obtained from each other by a sequence of commutation (respectively, braid) moves. We demonstrate these partitions by continuing the previous example.

Example 2 Let w be as in Example 1. Then

$$C(w) = \left\{ \{s_2 s_1 s_2 s_3\}, \ \{s_1 s_2 s_1 s_3, \ s_1 s_2 s_3 s_1\} \right\} \text{ and}$$

$$B(w) = \left\{ \{s_2 s_1 s_2 s_3, \ s_1 s_2 s_1 s_3\}, \ \{s_1 s_2 s_3 s_1\} \right\}.$$

Up to now, we have written permutations as products of adjacent transpositions. In fact, there are many ways to represent permutations, including as products of different generating sets, as products of cycles, as graphs, as arrow diagrams, and in one-line notation. The final definition that we need at this point concerns a seemingly (but not for long!) unrelated feature of permutations, related to the one-line notation for a permutation.

Definition 3 Let $w \in S_n$ be a permutation and write w, in *one-line notation*, as the word $w(1) \cdots w(n)$. Let $p \in S_k$ be a permutation written similarly, with $k \leq n$. The permutation w contains a *p-pattern* if there exist $j_1 < j_2 < \cdots < j_k$ such that the subword $w(j_1) \cdots w(j_k)$ is in the same relative order as the word for p. If this is the case, then we write $p \prec w$. If not, then w *avoids* p, written $p \not\prec w$.

Example 3 The permutation w from Example 1 is written in one-line notation as 3241. This permutation has a 231-pattern (in fact, it has two: the subwords $w(1)w(3)w(4) = 341$ and $w(2)w(3)w(4) = 241$ are both *order isomorphic* to 231). On the other hand, $123 \not\prec w$.

Although Definition 3 is specific to the symmetric group S_n, there is a notion of *signed* pattern for the finite Coxeter groups of types B and D. Despite obvious parallels between patterns and signed patterns, however, the (unsigned) pattern literature is notably richer than the literature for signed patterns. The reader is referred to [6, 7], among many other works.

The two theorems that have been most useful for tackling the problems discussed here each have a number of technical details. While important, pausing to define and characterize those details could be distracting in an article of this length. As a compromise, we give "big picture" statements of the theorems here, with citations to their full statements in other works.

Dictionary.   There is a way to translate between statements about permutation patterns and statements about reduced decompositions. (See [11, Theorem 3.8] and its generalization [13, Theorem 3.9].)

Rhombic Tilings.   There is a bijection between rhombic tilings of certain polygons and commutation classes of reduced decompositions. (See [3, Theorem 2.2].)

Elnitsky developed analogous tiling-to-commutation class bijections for types B and D, as well [3, §§6–7]. In those settings, the tilings have reflective requirements to account for sign and, in the case of type D, so-called "megatiles" are permitted.

3   Counting Special Elements: An Example

One type of enumerative problem about Coxeter groups would be to count the elements with a particular property or feature. We given an example of this type of work here.

There is a natural partial ordering on Coxeter group elements defined in terms of reduced decompositions.

Definition 4 Let G be a Coxeter group with elements v and w. Then $v \leq w$ in the *(strong) Bruhat order* if a reduced decomposition of v is a subword of a reduced decomposition of w.

Despite the fact that both v and w in Definition 4 can have multiple reduced decompositions, this ordering is well-defined. (The *weak Bruhat order*, which we do not study here, requires that the reduced decomposition of v appear as a prefix (or suffix) of a reduced decomposition of w, whereas the Bruhat order does not even require the reduced decomposition of v to appear as a consecutive subword in the reduced decomposition of w.)

Viewed as posets under the Bruhat order, Coxeter groups can have quite snarly structure. The *principal order ideal* of an element w is the set of all elements that are less than or equal to w in the poset, and even the principal order ideals in the Bruhat order need not be well-behaved. To get a sense of which elements might have "nice" principal order ideals, we consider the following classification.

Definition 5 In a Coxeter group, an element w is a *boolean* element if its principal order ideal $\{v : v \leq w$ in the Bruhat order$\}$ is a boolean poset.

We demonstrate Definition 5 by looking at boolean elements in the Coxeter group S_4. This group has 13 boolean elements. The poset structure of S_4, with those 13 boolean elements highlighted, is shown in Fig. 1.

To support interest in these boolean elements, we note that they describe a structure with beautiful topology. Indeed, the collection of boolean elements in any

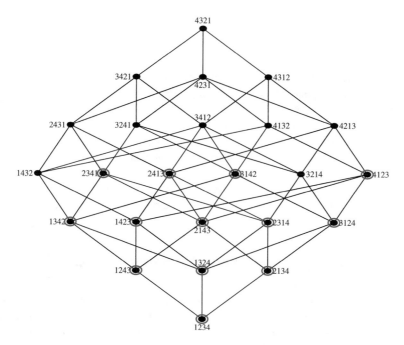

Fig. 1 The Coxeter group S_4, drawn as a poset under the Bruhat order. Group elements are written as permutations in one-line notation, and the 13 boolean elements of the group are circled

Coxeter group forms a simplicial poset. This poset, then, is the face poset of a regular cell complex called the *boolean complex*. If the Coxeter group has rank n, then that boolean complex is homotopy equivalent to a wedge of $(n-1)$-dimensional spheres. This topology is discussed, and in more depth, in [8, 9].

It turns out (see [12]) that, in any Coxeter group, boolean elements can be identified by whether or not their reduced decompositions have repeated letters. In the case of the symmetric group, among others, this can also be characterized by pattern avoidance.

Theorem 1 ([12]) *A permutation is boolean iff it avoids* 321 *and* 3412.

Boolean elements in the finite Coxeter groups of types B and D can also be characterized by (signed, in these cases) pattern avoidance [12, §7].

Having identified these elements, and with such attractive characterizations, it is enticing to try to enumerate them. In fact, they can be enumerated, both overall and by length.

Theorem 2 ([4, 14]) *The number of boolean permutations in* S_n *is* F_{2n-1}, *the odd-indexed Fibonacci number.*

Theorem 3 ([12]) *The number of boolean permutations in* S_n *of length* k *is*

$$\sum_{i=1}^{k} \binom{n-i}{k+1-i}\binom{k-1}{i-1}.$$

Boolean elements in the finite Coxeter groups of types B and D can be enumerated [4], and enumerated by length [12, §7]. For type B, these enumerations closely resemble the type A results cited above, while the results for type D are notably different and more complicated to state.

Boolean elements are just one example of a noteworthy class of elements in a Coxeter group whose enumeration might be of interest. Moreover, in enumerating such a class of objects, one might develop a new characterization for them that could shed light on other topics or unanswered questions. Depending on the enumerative technique used, this might even hint at a deeper structure in the group.

4   Counting an Element's Special Features: An Example

The second category of enumerative problems that we present for Coxeter groups is to calculate the size of a particular feature of a group element. To demonstrate this, we look at the reduced decompositions $R(w)$, the commutation classes $C(w)$, and the braid classes $B(w)$ of a permutation w.

The natural first approach to evaluating the sizes of these sets—to evaluating the size of anything, really—is a straightforward calculation. Indeed, as discussed above, Stanley computes $|R(w)|$ by counting Young tableaux of particular shape(s), with the special case that just one shape is needed iff w is 2143-avoiding [10]. On the other hand, outside of special cases like classifying the permutations for which $|C(w)| = 1$ or those for which $|B(w)| = 1$, there are no known similar results for $|C(w)|$ or $|B(w)|$.

This leads us to a second approach, which is not to evaluate the absolute sizes of these sets, but to determine *relative* sizes. "Relative to what?" one might ask. Recall the Dictionary mentioned above, and the link it provides between reduced decompositions and permutation patterns. Inspired by that result, we consider pattern containment as a possible yardstick against which to measure these set sizes. (Note that the idea of ordering the set of all permutations—of any size—by pattern containment is not new. Indeed, this leads to a poset whose Möbius function has been the subject of great interest since at least [15].)

Not only does pattern containment seem to be an appropriate yardstick, but it yields more information about the sizes of these sets than was previously known.

Theorem 4 ([13])

(a) If $p \prec w$ then $|R(p)| \leq |R(w)|$. Moreover, if $p \prec w$ and $|R(w)| > 1$, then $|R(p)| = |R(w)|$ iff $\ell(p) = \ell(w)$; equivalently, iff p and w have equally many 21-patterns.

(b) If $p \prec w$ then $|C(p)| \leq |C(w)|$. Moreover, if $p \prec w$, then $|C(p)| = |C(w)|$ iff p and w have equally many 321-patterns.

Any analysis of commutation classes is greatly helped by the Rhombic Tilings result mentioned above. Unfortunately, similar machinery has not (yet) been developed for studying braid classes. This seems to be more an issue of oversight than the result of any great complexity to braid classes that might prevent such machinery's existence. Braid classes have received some attention (see [1, 16]), but not nearly as much as commutation classes. Recently, in an attempt to begin to remedy this, important strides were made in understanding the set $B(w)$ by considering it *simultaneously* with $C(w)$ [5].

The strength of that technique is that it recognizes that $B(w)$ and $C(w)$ are both partitions of the same set, $R(w)$, and so knowledge about one of the partitions might imply knowledge about the other. In fact, this turns out to be the case, and the sets can be leveraged against each other to good effect.

In an important sense, these two sets are orthogonal to each other: as shown in [5], for any permutation w and any $B \in B(w)$ and $C \in C(w)$,

$$|B \cap C| \leq 1.$$

Thus, one can index the reduced decompositions of $R(w)$ by ordered pairs (B, C) representing their braid and commutation classes, and each possible pair appears at most once in this list. This gives an upper bound to $|R(w)|$ in terms of $|B(w)|$ and $|C(w)|$. A lower bound follows from the fact (see, for example, [2]) that any reduced decomposition for w can be obtained from any other by a sequence of braid and commutation moves. These bounds are combined in the following theorem.

Theorem 5 ([5]) *For any permutation w,*

$$|B(w)| + |C(w)| - 1 \leq |R(w)| \leq |B(w)| \cdot |C(w)|.$$

With so little known about the structure and behavior of braid classes, it is instructive to try to understand the bounds of Theorem 5. As shown in [5], those bounds are sharp, and the permutations achieving them can be characterized and enumerated.

We have used this section to give a sense of this category of enumerative questions about Coxeter group elements, and the results and implications that they might have. Certainly there is a substantial range of topics still to be studied, both related to the discussion above and independent of it.

5 Directions for Future Research

The goal of this article is to demonstrate the different ways that enumerative combinatorialists might approach the study of Coxeter groups. Although we have listed many results and cited many sources, the reader should not assume that this is a "closed" field of study. There is much still to be uncovered about the combinatorics of these objects, including questions that have been studied for many years and others that, themselves, have not yet been identified.

In particular, while some of the results described for the symmetric group have analogues in Coxeter groups of other types, much remains to be uncovered. There is every reason to expect, for example, a type B analogue of Theorem 5 relating braid classes, commutation classes, and the classes $R(w)/(s_0s_1s_0s_1 \sim s_1s_0s_1s_0)$. Similarly, just as Enlitsky's tiling bijections can be constructed for types B and D, there may well be a Dictionary for groups of other types. Finally, we reiterate that while much has studied about commutation classes of reduced decompositions, much less attention has been given to partitions based on other Coxeter relations.

Acknowledgement Research partially supported by Simons Foundation Collaboration Grant for Mathematicians 277603.

References

1. N. Bergeron, C. Ceballos, and J.-P. Labbé, Fan realizations of type A subword complexes and multi-associahedra of rank 3, *Discrete Computational Geom.* **54** (2015), 195–231.
2. A. Björner and F. Brenti, *Combinatorics of Coxeter Groups*, Graduate Texts in Mathematics 231, Springer, New York, 2005.
3. S. Elnitsky, Rhombic tilings of polygons and classes of reduced words in Coxeter groups, *J. Comb. Theory, Ser. A* **77** (1997), 193–221.
4. C. K. Fan, Schubert varieties and short braidedness, *Transform. Groups* **3** (1998), 51–56.
5. S. Fishel, E. Milićević, R. Patrias, and B. E. Tenner, Enumerations relating braid and commutation classes, *European J. Comb.* **74** (2018), 11–26.
6. S. Kitaev, *Patterns in Permutations and Words*, Monographs in Theoretical Computer Science, Springer-Verlag, Berlin, 2011.
7. S. Linton, N. Ruškuc, V. Vatter, *Permutation Patterns*, London Mathematical Society Lecture Note Series 376, Cambridge University Press, Cambridge, 2010.
8. K. Ragnarsson and B. E. Tenner, Homotopy type of the boolean complex of a Coxeter system, *Adv. Math.* **222** (2009), 409–430.
9. K. Ragnarsson and B. E. Tenner, Homology of the boolean complex, *J. Algebraic Comb.* **34** (2011), 617–639.
10. R. P. Stanley, On the number of reduced decompositions of elements of Coxeter groups, *European J. Comb.* **5** (1984), 359–372.
11. B. E. Tenner, Reduced decompositions and permutation patterns, *J. Algebraic Comb.* **24** (2006), 263–284.
12. B. E. Tenner, Pattern avoidance and the Bruhat order, *J. Comb. Theory, Ser. A* **114** (2007), 888–905.
13. B. E. Tenner, Reduced word manipulation: patterns and enumeration, *J. Algebraic Comb.* **46** (2017), 189–217.
14. J. West, Generating trees and forbidden subsequences, *Discrete Math.* **157** (1996), 363–374.
15. H. Wilf, The patterns of permutations, *Discrete Math.* **257** (2002), 575–583.
16. D. M. Zollinger, Equivalence classes of reduced words, Master's thesis, University of Minnesota, 1994.

Part III
Algebraic Biology

From Chaos to Permanence Using Control Theory (Research)

Sherli Koshy-Chenthittayil and Elena Dimitrova

1 Introduction

Different aspects of stable coexistence and survival of species in an ecological system have been studied by biologists and mathematicians. Ecologists have also considered whether chaotic behavior occurs in biological systems. Schaffer and Kot [28] give examples of real world chaos in systems such as the Canadian lynx cycle, outbreaks of *Thrips imaginis*, and others. Costantino et al. [8] famously observed chaotic behavior in cultures of flour beetles (*Tribolium castaneum*) in the 1990s. Around the same time, Sugihara and May [33] investigated the chaotic trajectories of real world time series data on measles, chickenpox and marine phytoplankton. Becks et al. [4] also observed chaotic behavior for certain values of dilution rates of a chemostat experiment. They studied a predator-prey system consisting of a bacterivorous ciliate and two bacterial prey species.

Furthermore, models of two prey and one predator [12], a three species food chain [13], microbial systems [19], multi-trophic ecological systems [32], and plankton models [10, 16, 23] have exhibited chaos for biologically relevant parameters. In [39], instances in the biomedical field are presented where chaos has been observed. Some examples are the response of cardiac and neural tissue to pacing stimuli, fluctuations in leukocyte counts in patients with chronic myelogenous leukemia, and ventricular tachycardia. It is also observed that there is no real insight on how to prevent or eliminate chaotic arrhythmia and the concept of making small perturbations to the chaotic system might be more useful. In [9], it is suggested that

S. Koshy-Chenthittayil (✉)
Center for Quantitative Medicine, UConn Health, Farmington, CT, USA
e-mail: koshychenthittayil@uchc.edu

E. Dimitrova
Department of Mathematics, California State Polytechnic University, San Luis Obispo, CA, USA
e-mail: edimitro@calpoly.edu

© The Author(s) and the Association for Women in Mathematics 2020
B. Acu et al. (eds.), *Advances in Mathematical Sciences*, Association for
Women in Mathematics Series 21, https://doi.org/10.1007/978-3-030-42687-3_6

85

chaotic behavior of populations carries evolutionary advantages. They argue that since very small changes in the initial conditions of a chaotic system can greatly alter the system's trajectory, one doesn't need to change the system completely to obtain a desired outcome.

However, survival and chaos have mostly been studied as separate topics. The goal of this work is to demonstrate how chaotic behavior can indeed be used to ensure the survival and thriving of the species involved in a system by employing a practical control such as harvesting of the predator. We provide such a control algorithm that takes advantage of chaotic trajectories to lead a system from a situation where one or more of the species may not survive to a state where in the long term all the species are sufficiently far from extinction.

The outline of the remainder of the paper is as follows: Sect. 2 presents a mathematical definition of survival of the species in a system and discusses methods for control through chaos. Section 3 presents the control algorithm we propose. Section 4 contains applications of the algorithm to predator-prey models which include harvesting of species.

2 Survival and Chaos

During the 1960s, there were doubts among biologists that either local or global asymptotic stability was not enough to describe population behavior. R.C. Lewontin [21] and J. Maynard Smith [22] give an intuitive approach to *dynamic boundedness* and *permanence*, where permanence means that the species remain at a safe threshold from extinction. There were, however, no mathematical ideas that seemed helpful in treating these concepts. To remedy this, the idea of *persistence*, that is

$$\limsup_{t \to \infty} x_i(t) > 0 \text{ for all species } x_i$$

was introduced in [11] for the autonomous model $\dot{x}_i = x_i f_i(x)$ for $i = 1, \ldots, n$.

A disadvantage of this concept is that orbits of a persistent system may approach the boundary $\partial \mathbb{R}^n_+$, thus bringing species dangerously close to extinction. In his paper, Schreiber [29] discusses multiple definitions of persistence based on [6, 7] and their dependence on perturbations.

The stronger condition of *permanence* that avoids this difficulty was introduced in [30] and is based on the boundary $\partial \mathbb{R}^n_+$ being repelling. Proving that a system is permanent is far from trivial. Below we present several approaches for showing that a system has the property of permanence.

In [2], a system is said to be permanent if the boundary (including infinity) is an unreachable repeller or, equivalently, if there exists a compact subset in the interior of the state space where all orbits starting from the interior eventually end up. The focus in [2] is primarily on ecological systems of the type

$$\dot{x}_i = x_i f_i(x) \text{ on } \mathbb{R}^n_+$$

and replicator equations

$$\dot{x}_i = x_i \left(f_i(x) - \sum x_i f_i \right)$$

which have been widely investigated in population genetics, population ecology, the theory of prebiotic evolution of self-replicating polymers and socio-biological studies of evolution. The concept of permanence, however, can be applied to a wide range of systems of differential and difference equations.

A sufficient condition for permanence is also presented in [2] with the help of an average Lyapunov function P. The function P is defined on the state space, vanishing on the boundary and strictly positive in the interior, such that $\dot{P} = P\psi$ where ψ is a continuous function with the property that for some $T > 0$,

$$\frac{1}{T} \int_0^T \psi(x(t))dt > 0 \text{ for all } x \text{ on the boundary.}$$

Another method for proving that a system is permanent is given in [15] and will be restated here as Theorem 1. We first need the following definitions.

Definition 1 ([15]) Consider a system of the type

$$\dot{x}_i = x_i f_i(x) \text{ for } i = 1, \ldots, n. \tag{1}$$

- A rest point (steady state) $\bar{x}$ is *saturated* if $f_i(\bar{x}) \leq 0$ for all i (the equality sign must hold whenever $\bar{x}_i > 0$). A rest point in the interior is trivially saturated. The quantities $f_i(\bar{x})$ are the eigenvalues of the Jacobian at $\bar{x}$ whose eigenvectors are transversal to the boundary face of $\bar{x}$. Thus they are called *transversal eigenvalues*.
- A *degenerate saturated rest point* is one which has a zero transversal eigenvalue.
- A *regular rest point* is one which has non-zero eigenvalues.
- If $\bar{x}$ is a regular rest point, then the *index* $\mathbf{i}(\bar{x})$ is the sign of the Jacobian $D_{\bar{x}} f$. Hence

$$i(\bar{x}) = (-1)^k$$

where k is the number of real negative eigenvalues of the Jacobian. For $n = 2$, for example, the index of a center, a sink, or a source is $+1$, while that of a saddle is -1.
- A *boundary rest point* is a rest point with at least one zero coordinate.

Theorem 1 ([15]) *If the system (1), does not have regular saturated boundary rest points, then it is permanent.*

Our goal is to take advantage of the chaotic behavior of a non-permanent system in order to make it permanent. To this end, we employ control theory on the chaotic orbits to push the system into permanence.

We first notice the following connection between survival and chaos.

Theorem 2 *Three-dimensional continuous chaotic systems are persistent.*

Proof If a system is not persistent and one of the species dies out, then the dimension is reduced to two and thus the system cannot be chaotic by the Poincaré–Bendixson Theorem.   □

Among the first proponents of using a chaotic attractor to control a system were E. Ott et al. [25]. They observed that a chaotic attractor has typically embedded within it an infinite number of unstable periodic orbits. In their method (OGY), the approach was to first determine some of the unstable low-period periodic orbits that are embedded in the chaotic attractor and then choose one which yields improved system performance. They then tailored their small time-dependent parameter perturbations so as to stabilize the existing orbit.

T. L. Vincent [37] first provides a motivation of why chaotic behavior in nonlinear systems is useful to set up a control design. He mentions that chaotic behavior is useful in moving a system to various points in the state space without changing the system drastically. The reasoning is very similar to the one used for the OGY method in [25]. The chaotic control algorithm in [37] requires two ingredients: a chaotic attractor and a controllable target. If chaos does not exist, it can be created using *open loop control* (where the control function is a function of time). A controllable target is any subset of the domain of attraction to an equilibrium point, under a corresponding feedback control law, that has a non-empty intersection with the chaotic attractor. The controllable target should be large enough so that one does not have to wait too long for the system to reach it. The algorithm is applied on three different systems: the Hénon map, bouncing ball system, two-link pendulum system. The first two are discrete and the third is a four-dimensional continuous system. Further detailed examples on how to control an inverted pendulum and a bouncing ball were provided in [38].

Several other control methods were described in [24, 26, 27, 31], to name a few. Here, we adopt the algorithm in [37] as a foundation of our control method due to ease of computation and adaptability to different models.

The next two sections provide details on the control algorithm we propose and its applications to predator-prey models.

3   The Chaotic Control Algorithm (CCA)

We are introducing a control algorithm, CCA (*chaotic control algorithm*), which is based on the algorithm presented in [37, 38]. CCA is designed to be applied on systems which already have chaotic orbits but are not permanent. The output of the algorithm is the required closed loop control (i.e. a control function of the state of the system) which pushes the system into permanence.

Consider the system of non-linear differential equations given by

$$\dot{X} = F[X, U] \tag{2}$$

where $F = [F_1, \ldots F_{N_X}]$ is an N_X-dimensional vector function of the state vector $X = [X_1, \ldots X_{N_X}]$, and control vector $U = [U_1, \ldots U_{N_U}]$. The control will, in general, be bounded. Assume that for all t there exists a control $\hat{U}(t)$ such that (2) has a chaotic attractor and is non-permanent. Also assume that for a specified constant control, $\bar{U}$, there is a corresponding rest point of interest which is near the chaotic attractor where the system is permanent. The rest point $\bar{X}$ is such that

$$F(\bar{X}, \bar{U}) = 0 \tag{3}$$

Notice that the rest point need not be stable after optimal control is applied since we are only interested in making the system permanent.

The steps of the CCA are outlined next.

Input: Rest point $\bar{X}$ and control $\bar{U}$ such that (3) is satisfied. In particular, $\bar{U}$ is chosen such that (2) is permanent.

Output: Optimal closed loop control which pushes the chaotic, non-permanent system into permanence.

1. Linearizing $\dot{X} = F[X, U]$ about the rest point $\bar{X}$ we have

$$\dot{x} = Ax + Bu \tag{4}$$

where

$$x = X - \bar{X}, \quad u = U - \bar{U}, \quad A = \frac{\partial F}{\partial X}\Big|_{\bar{X}, \bar{U}}, \text{ and } B = \frac{\partial F}{\partial U}\Big|_{\bar{X}, \bar{U}}.$$

Now the origin is the rest point for (4) with control $u(t) \equiv 0$. We are going to obtain a control such that the origin becomes stable.

2. The *linear quadratic regulator* (LQR) method determines gains K such that under full state feedback of the form

$$u(x) = -Kx \tag{5}$$

a quadratic performance index is minimized. The performance index is the infinite integral of the quadratic form $x^t Q x + u^t R u$ where Q and R are symmetric positive definite matrices to be chosen as part of the control design process.

The gain matrix K is given by

$$K = R^{-1} B^T S$$

and the matrix S is determined by solving the Riccati equation given by

$$SA + A^T S - SBR^{-1}B^T S = -Q.$$

3. Under full state feedback control given by (5), the linearized system is given by

$$\dot{x} = \hat{A}x \qquad (6)$$

where

$$\hat{A} = A - BK.$$

A Lyapunov function of the form

$$V(x) = x^t P x \qquad (7)$$

may now be determined for the linear stable controlled system (6) using the continuous Lyapunov equation

$$P\hat{A} + \hat{A}^T P = -\hat{Q}$$

where $\hat{Q}$ is a positive definite matrix.

For the stable linear system, starting from any point in state space, the solution obtained for P will result in the property that $\dot{V} < 0$ for every point of the linear system (6) except at the origin where $V = 0$. This will prove that the origin is asymptotically stable for (4). This, in turn, implies that for the nonlinear system (2), the rest point will be asymptotically stable in some neighborhood containing the rest point.

Figure 1 is a flowchart depicting CCA.

We next apply the CCA and demonstrate how using the chaotic nature of a system, one can transform it from non-permanent to permanent.

4   Applications of the CCA to Predator-Prey Models

We now investigate two different types of predator-prey models. The models are of Lotka-Volterra type and Leslie-Gower type. In each case harvesting of one species is introduced and chaotic behavior is observed for certain parameter values. For the same parameter values, non-permanence of the species is also observed. Using the CCA from Sect. 3, we construct a closed loop control (i.e. a control which is a function of the state of the system) which steers the system towards permanence. The systems were chosen for their different types of functional responses and since harvesting of one species was relatively easy to introduce.

Fig. 1 Flowchart for the
chaotic control algorithm
(CCA) described in Sect. 3

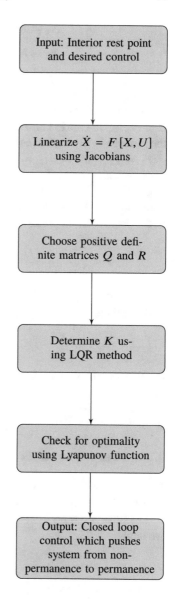

Input: Interior rest point
and desired control

Linearize $\dot{X} = F[X, U]$
using Jacobians

Choose positive defi-
nite matrices Q and R

Determine K us-
ing LQR method

Check for optimality
using Lyapunov function

Output: Closed loop
control which pushes
system from non-
permanence to permanence

4.1   Control Through Harvesting in a Predator-Prey Model

First we apply the control algorithm to a Lotka-Volterra type two-prey, one-predator
model from [3] where the predator is harvested at a constant rate. Here the
harvesting of the predator will act as a control.

The population dynamics model involves three interacting species, namely the
prey N_1 and N_2 and the predator P. The harvesting is given by a harvesting function
$H(P)$. The dynamics is described by Lotka-Volterra type equations given by

$$\dot{N}_1 = N_1(r_1 - a_{11}N_1 - a_{12}N_2 - a_{13}P)$$
$$\dot{N}_2 = N_2(r_2 - a_{21}N_1 - a_{22}N_2 - a_{23}P) \tag{8}$$
$$\dot{P} = P(-r_3 + a_{31}N_1 + a_{32}N_2) - H(P).$$

We consider the harvesting function $H(P) = H_p$ where H_p is a constant and $0 < H_p < 1$. This is known as constant harvest quota. The parameters chosen are the same as in [3] and are as follows:

$r_1 = r_2 = r_3 = a_{11} = a_{12} = a_{22} = a_{23} = 1, a_{21} = 1.5, a_{32} = 0.5, a_{13} = 5, a_{31} = 2.5$.

Since $a_{12} < a_{21}$, the first prey has a competitive advantage, i.e. N_1 is the dominant and N_2 the sub-dominant prey. Consider the two-dimensional subsystem of the preys without predation:

$$\dot{N}_1 = N_1(r_1 - a_{11}N_1 - a_{12}N_2)$$
$$\dot{N}_2 = N_2(r_2 - a_{21}N_1 - a_{22}N_2).$$

The relation

$$a_{11}a_{22} - a_{12}a_{21} < 0$$

implies that the system does not have an interior rest point, that is, the species N_1 and N_2 cannot coexist. This shows that without predation, either one of the prey species N_1 or N_2 will die out [40].

4.1.1 Boundedness of the Solutions

Following methods similar to those in [1, 5, 34], we now prove that all solutions of (8) which initiate in $\mathbb{R}_+^3$ are uniformly bounded.

Let $W = N_1 + N_2 + 2P$. Then $\dot{W} = \dot{N}_1 + \dot{N}_2 + 2\dot{P}$. Along the solutions of (8), we have

$$\dot{W} = N_1(1 - N_1 - N_2 - 5P)$$
$$+ N_2(1 - 1.5N_1 - N_2 - P)$$
$$+ 2P(-1 + 2.5N_1 + 0.5N_2) - 2H_p$$
$$= N_1(1 - N_1) + N_2(1 - N_2) - 2.5N_1N_2 - 2P - 2H_p$$
$$\leq N_1(1 - N_1) + N_2(1 - N_2) - 2P.$$

For each constant $D > 0$, the following inequality holds:

$$\dot{W} + DW \leq N_1(1 - N_1 + D) + N_2(1 - N_2 + D) + 2P(D - 1).$$

Now if we take D such that $0 < D < 1$ and the maximum value $\dfrac{1+D}{2}$ of both the expressions $N_1(1 - N_1 + D)$ and $N_2(1 - N_2 + D)$ with respect to N_1 and N_2 respectively, we see that $\dot{W} + DW \leq 1 + D = K$ which implies that $0 \leq W(N_1, N_2, P) \leq \dfrac{K}{D} + W(N_1(0), N_2(0), P(0))e^{-Dt}$ and so $0 < W \leq \dfrac{K}{D}$ as $t \to \infty$. Therefore, all solutions of (8) that initiate in $\mathbb{R}_+^3$ are confined in the region.

4.1.2 The Harvesting Model

We now consider the harvesting model

$$\dot{N}_1 = N_1(1 - N_1 - N_2 - 5P)$$
$$\dot{N}_2 = N_2(1 - 1.5N_1 - N_2 - P) \qquad (9)$$
$$\dot{P} = P(-1 + 2.5N_1 + 0.5N_2) - H_p$$

with control $U = H_p$. The control $\hat{U}$ for which chaos is observed is $\hat{U} = H_p = 0.02$ [3]. The chaos is indicated by a positive Lyapunov exponent 0.0474 at the initial conditions $(0.4899, 0.2040, 0.0612)$. This is the interior rest point.

Since we want to demonstrate how the system can be steered towards permanence, we first check that the original system is non-permanent for these parameter values. The system is of the form $\dot{x}_i = x_i f(x_i)$ where $x_1 = N_1, x_2 = N_2, x_3 = P$ and

$$f_1(N_1, N_2, P) = 1 - N_1 - N_2 - 5P$$
$$f_2(N_1, N_2, P) = 1 - 1.5N_1 - N_2 - P$$
$$f_3(N_1, N_2, P) = -1 + 2.5N_1 + 0.5N_2 - \frac{0.02}{P}.$$

According to Theorem 1, the system is permanent if it does not have regular, saturated boundary rest points, i.e for the rest point $\bar{x}$, $f_i(\bar{x}) > 0$ for some i when $\bar{x}_i = 0$.

For this particular system, consider the biologically valid rest points $\bar{x} = (\bar{N}_1, \bar{N}_2, \bar{P})$ and the values of $f_i(\bar{x}) > 0$ for some i when $\bar{x}_i = 0$.

We see in Table 1 that the system does indeed have a saturated rest point, namely $B = (0.9236067978, 0, 0.1527864045e - 1)$, so it is not permanent.

Figure 2 is a representation of the chaotic manifold with the chosen set of parameters and initial condition set to the interior rest point.

Table 1 Check for permanence of (9) using boundary rest points

Rest point	Value of f_i
$A = (0.4763932022, 0, 0.1047213596)$	$f_2(A) = 0.18 > 0$
$B = (0.9236067978, 0, 0.1527864045e - 1)$	$f_2(B) = -0.4013 < 0$

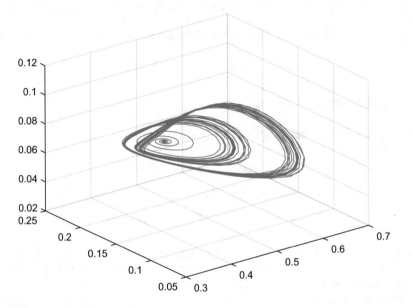

Fig. 2 Chaotic manifold for $h = 0.02$ [3] and the initial conditions ($N_1 = 0.4899, N_2 = 0.2040, P = 0.0612$)

4.1.3 Application of CCA [18]

The specific control that will lead the harvesting model towards permanence is $\bar{U} = H_p = 0.035$ and the interior rest point of the system

$$\dot{N}_1 = N_1(1 - N_1 - N_2 - 5P)$$

$$\dot{N}_2 = N_2(1 - 1.5N_1 - N_2 - P)$$

$$\dot{P} = P(-1 + 2.5N_1 + 0.5N_2) - 0.035$$

is $\bar{X} = (N_1, N_2, P) = (0.5816, 0.0549, 0.0727)$. The system is found to be permanent by checking the boundary rest points using the MATLAB code [18] given in the Appendix. In the code, the boundary rest points are calculated and checked to see if they are regular saturated rest points. If the rest points are not saturated, it is concluded that the system is permanent by Theorem 1.

So $\bar{X} = (0.5816, 0.0549, 0.0727)$.

For the linearization step we calculate the matrices A and B as

$$A = \frac{\partial F}{\partial X}|_{\bar{X},\bar{U}} = \begin{bmatrix} -0.5816 & -0.5816 & -2.9080 \\ -0.0824 & -0.0549 & -0.0549 \\ 0.1817 & 0.0363 & 0.4814 \end{bmatrix},$$

$$B = \frac{\partial F}{\partial U}|_{\bar{X},\bar{U}} = \begin{bmatrix} 0 \\ 0 \\ -1 \end{bmatrix}.$$

We choose the matrices $Q = I_3$ and R=[1] which are positive definite.

Applying the *lqr* routine of MATLAB, the gains matrix K is obtained as

$$K = \begin{bmatrix} 0.2761 & -0.2112 & -2.1591 \end{bmatrix}.$$

Thus our feedback control given by (5) is

$$u(x) = -Kx$$

$$= -0.2761N_1 - 0.2112N_2 - 2.1591P.$$

To confirm that the origin is asymptotically stable, the Lyapunov function has also been calculated. From Step 3 of the algorithm, we have

$$\hat{A} = A - BK$$

$$= \begin{bmatrix} -0.5816 & -0.5816 & -2.9080 \\ -0.0824 & -0.0549 & -0.0549 \\ 0.4579 & -0.1749 & -1.6777 \end{bmatrix}.$$

We choose $\hat{Q} = Q = I_3$ and we obtain P using the *lyap* function in MATLAB:

$$P = \begin{bmatrix} 1.4209 & -4.0725 & 0.7023 \\ -4.0725 & 17.5405 & -2.3263 \\ 0.7023 & -2.3263 & 0.7322 \end{bmatrix}.$$

For the above matrix, $V(x) = x^t P x$ will satisfy $\dot{V} < 0$ by construction of P. The chaotic nature disappears when $h = 0.035$ as seen in Fig. 3.

The system is now permanent, with harvesting still possible thanks to the chaotic orbit which we took advantage of as we applied the CCA.

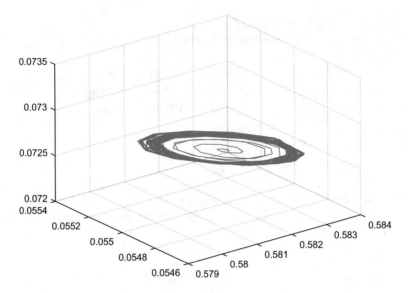

Fig. 3 Permanence and absence of a chaotic manifold when $h = 0.035$, initial conditions ($N_1 = 0.5816$, $N_2 = 0.0549$, $P = 0.0727$)

4.2 Control Through Harvesting in a Food-Chain System

The next system consider is of a simple prey-specialist predator-generalist predator (for example, plant-insect pest-spider) interaction based on the model found in [17, 20, 35]. In the system below, harvesting of the prey is considered as a control.

$$\dot{x} = a_1 x - b_1 x^2 - \frac{wxy}{x + D} - hx$$
$$\dot{y} = -a_2 y + \frac{w_1 xy}{x + D_1} - \frac{w_2 yz}{y + D_2} \tag{10}$$
$$\dot{z} = cz^2 - \frac{w_3 z^2}{y + D_3}.$$

In this model, a prey population of size x serves as the only food for the specialist predator population of size y. This population, in turn, serves as favorite food for the generalist predator population of size z. The equations for rate of change of population size for prey and specialist predator are according to the Volterra scheme (predator population dies out exponentially in absence of its prey). The interaction between this predator y and the generalist predator z is modeled by the Leslie-Gower scheme where the loss in a predator population is proportional to the reciprocal of per capita availability of its most favorite food. The basic characteristic of the Leslie-Gower model is that it leads to a solution which is asymptotically independent of the initial conditions and depends only on the intrinsic attributes of the interacting system, that is, the parameters w, w_1, and so on [36].

After introducing harvesting to the existing model, we observed chaos and non-permanence [18]. The hx term models the harvesting function being proportional to the population of the prey (constant harvest effort).

The constants are all positive and are described as follows.

a_1:	intrinsic growth rate of the prey population x;
b_1:	strength of intra-specific competition among the prey species;
w, w_1, w_2, w_3:	the maximum values which per capita growth can attain;
D, D_1:	the extent to which the environment provides protection to the prey x;
a_2:	intrinsic death rate of the predator y in the absence of the only food x;
D_2:	the value of y at which the per capita removal rate of y becomes $w_2/2$;
D_3:	the residual loss in z population due to severe scarcity of its favorite food y;
c:	the rate of self-reproduction of the generalist predator z. The square term signifies the mating frequency is directly proportional to the number of males and females;
h:	harvesting rate of the prey x.

The parameter values (except for h) are taken as in [35] and are given below.

$$a_1 = 1.93, b_1 = 0.06, w = 1, D = 10, a_2 = 1, w_1 = 2$$
$$D_1 = 10, w_2 = 0.405, D_2 = 10, c = 0.03, w_3 = 1, D_3 = 20.$$

The above parameter choices are so that the system is bounded and there is possibility of chaotic behavior for different values of h.

4.2.1 Equilibrium Analysis [18]

The possible biologically viable equilibria are $E_0 = (0, 0, 0)$, $E_1 = (\dfrac{a_1 - h}{b_1}, 0, 0)$, $E_2 = (\bar{x}, \bar{y}, 0)$, and the interior rest point $E_3 = (x^*, y^*, z^*)$.

For E_1 to be biologically relevant, we need

$$a_1 - h > 0 \implies h < a_1 = 1.93. \tag{11}$$

E_2 is obtained by solving the subsystem

$$a_1 - b_1 x - \frac{wy}{x + D} - h = 0$$

$$-a_2 + \frac{w_1 x}{x + D_1} = 0.$$

Thus $E_2 = (\bar{x}, \bar{y}, 0) = (10, 20(1.33 - h), 0)$. Again for E_2 to be biologically viable we need

$$1.33 - h > 0 \implies h < 1.33. \tag{12}$$

$E_3 = (x^*, y^*, z^*)$ is the solution of the following system.

$$a_1 - b_1 x - \frac{wy}{x + D} - h = 0 \tag{13a}$$

$$-a_2 + \frac{w_1 x}{x + D_1} - \frac{w_2 z}{y + D_2} = 0 \tag{13b}$$

$$cz - \frac{w_3 z}{y + D_3} = 0. \tag{13c}$$

From (13c), we have

$$cz - \frac{w_3 z}{y + D_3} = 0$$

$$c - \frac{w_3}{y + D_3} = 0$$

$$\implies y* = \frac{w_3}{c} - D_3 = 13.33 > 0.$$

From (13a),

$$(a_1 - h - b_1 x)(x + D) = wy^*$$

$$0.06x^2 - (1.33 - h)x + (-5.97 + 10h) = 0.$$

For real roots, we need

$$(1.33 - h)^2 - 4(0.06)(-5.97 + 10h) \geq 0.$$

Solving for h using Maple's *solve* command (up to four significant figures), we get

$$h \leq 0.7411 \text{ or } h \geq 4.3189.$$

For the rest point to be biologically valid, we need at least one positive root to the equation $0.06x^2 - (1.33 - h)x + (-5.97 + 10h)$.

According to Descartes' Rule of Signs, if $h \leq 0.7411$, then we have one sign change of the coefficients. So there is at least one positive root.

Therefore, x^* exists if

$$h \leq 0.7411. \tag{14}$$

From (13b),

$$\frac{w_2 z^*}{y^* + D_2} = -a_2 + \frac{w_1 x^*}{x^* + D_1}$$

$$z^* = \frac{y^* + D_2}{w_2}(-a_2 + \frac{w_1 x^*}{x^* + D_1})$$

$$z^* = 57.605(\frac{x^* - 10}{x^* + 10})$$

z^* exists if $x^* > 10$.

4.2.2 Conditions for Permanence

We use Lyapunov functions to derive conditions for permanence [14, 18]. Assume
the boundary rest points $E_0 = (0, 0, 0)$, $E_1 = (\frac{a_1 - h}{b_1}, 0, 0)$, $E_2 = (10, 20(1.33 - h), 0)$ exist and there are no periodic orbits on the boundary. We need $h < 1.33$ for
the system (10) to be permanent. To see this, let the Lyapunov function be $\sigma(X) = x^{p_1} y^{p_2} z^{p_3}$, where p_1, p_2, p_3 are positive constants. Clearly $\sigma(X)$ is a non-negative
C^1 function defined in $\mathbb{R}_+^3$. Consider

$$\psi(X) = \frac{\dot{\sigma}(X)}{\sigma(X)}$$

$$= p_1 \frac{\dot{x}}{x} + p_2 \frac{\dot{y}}{y} + p_3 \frac{\dot{z}}{z}$$

$$= p_1 \left(a_1 - b_1 x - \frac{wy}{x + D} - h \right)$$

$$+ p_2 \left(-a_2 + \frac{w_1 x}{x + D_1} - \frac{w_2 z}{y + D_2} \right)$$

$$+ p_3 \left(cz - \frac{w_3 z}{y + D_3} \right).$$

To show permanence, we need $\psi(X) > 0$ for all equilibria $X \in \mathrm{bd}\mathbb{R}_+^3$, i.e. the
following conditions have to be satisfied.

$$\psi(E_0) = p_1(a_1 - h) - p_2 a_2 > 0 \tag{15a}$$

$$\psi(E_1) = p_2\left(-a_2 + \frac{w_1(a_1 - h)/b_1}{(a_1 - h)/b_1 + D_1}\right) > 0 \tag{15b}$$

$$\psi(E_2) = 0. \tag{15c}$$

We note that by (11) and by increasing p to a sufficiently large value, $\psi(E_0)$ can be made positive.

From (15b) we have the following requirement.

$$-a_2 + \frac{w_1(a_1 - h)/b_1}{(a_1 - h)/b_1 + D_1} > 0$$

$$-a_2 + \frac{w_1(a_1 - h)}{(a_1 - h) + b_1 D_1} > 0$$

$$\frac{2(1.93 - h)}{(1.93 - h) + 0.06 * 10} > 1.$$

Solving we get

$$h < 1.33. \tag{16}$$

Therefore, from inequalities (11), (12), (14), and (16) we see that we need $h \leq 0.7411$ for the existence of an interior rest point and permanence.

4.2.3 Control Algorithm Using Harvesting [18]

Now suppose the harvesting coefficient $h = 0.93$. This violates the condition for permanence and we also notice that the system is chaotic by the presence of a positive Lyapunov exponent 1.4427. We can use the chaos to bring the system back to permanence with final control $U(t) = h = 0.1$. The system

$$\dot{x} = 1.93x - 0.06x^2 - \frac{xy}{x + 10} - 0.1x$$

$$\dot{y} = -y + \frac{2xy}{x + 10} - \frac{0.405yz}{y + 10} \tag{17}$$

$$\dot{z} = 0.03z^2 - \frac{z^2}{y + 20}$$

has interior rest point $\bar{X} = (x = 23.955, y = 13.333, z = 23.679)$. The system is in fact permanent using the boundary rest points and the analysis in Sect. 4.2.2.

Applying CCA, we find the matrices A and B:

$$A = \frac{\partial F}{\partial X}\Big|_{\bar{X},\bar{U}} = \begin{bmatrix} -1.1603 & -0.7055 & 0 \\ 0.2313 & 0.2349 & -0.2314 \\ 0 & 0.5046 & 0 \end{bmatrix}$$

$$B = \frac{\partial F}{\partial U}\Big|_{\bar{X},\bar{U}} = \begin{bmatrix} -23.9555 \\ 0 \\ 0 \end{bmatrix}.$$

We again choose the matrices $Q = I_3$ and $R = [1]$.
Applying the *lqr* routine of MATLAB, the gains matrix K is obtained as

$$K = \begin{bmatrix} -0.9630 & -1.0716 & -0.9891 \end{bmatrix}.$$

Thus our feedback control given by (5) is

$$u(x) = -Kx = -0.9630x - 1.0716y - 0.9891z.$$

To confirm that the origin as asymptotically stable, the Lyapunov function is also been calculated. From Step 3 of the CCA, we have

$$\hat{A} = A - BK$$

$$= \begin{bmatrix} -24.2299 & -26.3759 & -23.6950 \\ 0.2313 & 0.2349 & -0.2314 \\ 0 & 0.5046 & 0 \end{bmatrix}.$$

We choose $\hat{Q} = Q = I_3$ and we obtain P using the *lyap* function in MATLAB.

$$P = \begin{bmatrix} 179.7359 & -85.1218 & -89.0194 \\ -85.1218 & 80.7233 & -0.9909 \\ -89.0194 & -0.9909 & 90.3612 \end{bmatrix}.$$

For the above matrix, $V(x) = x^t Px$ will satisfy $\dot{V} < 0$ by construction of P.

In [35], harvesting was introduced and chaos was also observed. Non-permanence was also observed but with the CCA, we calculated an optimal harvesting level that led to permanence.

5 Conclusions

By demonstrating that the chaotic behavior of a system can be used as a control to
obtain permanence for the system, we shed more light over the significance of chaos
in ecological systems.

We investigated two predator-prey models in which instances of chaos and non-
permanence were observed for different values of a harvesting parameter. To take
advantage of the chaos present in the system, we applied a control algorithm (CCA)
which used the chaotic orbits in the system to obtain a closed loop control which
pushed the system into a permanent state. Thus chaos enabled the species to remain
at a safe threshold value from extinction.

6 MATLAB Code to Determine Permanence Using Boundary Rest Points [18]

This code determines the permanence of the systems considered in Sect. 4 using
boundary rest points. The system is said to be permanent if it does not have regular,
saturated rest points. The code first finds the boundary rest points and then checks
to see if they are saturated.

```
%Check for permanence
function [r,check] = Perm_Check2(system, p)
sys_harvest=1;%From the harvesting paper by Azar et.al
sys_Upad=3;%Multiple attractors and crisis route
 -Upadhyay

if system == sys_harvest
 syms x1 x2 x3;
%parameters
r1 = p(1);
r2 = p(2);
r3 = p(3);

a_11=p(4);
a_12=p(5);
a_13=p(6);
a_21 = p(7);
a_22=p(8);
a_23=p(9);
a_31=p(10);
a_32=p(11);
H = p(12); %Harvesting function
```

```
f1=(r1-a_11*x1-a_12*x2-a_13*x3);
f2=(r2-a_21*x1-a_22*x2-a_23*x3);
f3=(-r3+a_31*x1+a_32*x2 - H/x3);
xp1 = x1*f1== 0;
xp2 = x2*f2== 0;
xp3 = x3*f3== 0;
 S = solve([xp1,xp2,xp3]);
V=double([S.x1 S.x2 S.x3]);%gives us the rest points
F= double([subs(f1,S) subs(f2,S) subs(f3,S) ]);
%F gives f_i values at the rest points
end


%Multiple attractors and crisis route -Upadhyay
if system ==sys_Upad
syms x1 x2 x3 real;
%parameters
a1=p(1);
b1=p(2);
w=p(3);
D=p(4);
a2=p(5);
w1=p(6);
D1=p(7);
w2=p(8);
D2=p(9);
c=p(10);
w3=p(11);
D3=p(12);
h=p(13); %harvesting coefficient
f1= a1-b1*x1-(w*x2/(x1+D))-h;
f2= -a2+w1*x1/(x1+D1)-w2*x3/(x2+D2);
f3= c*x3-w3*x3/(x2+D3);
xp1 = x1*f1== 0;
xp2 = x2*f2== 0;
xp3 = x3*f3== 0;
 S = solve([xp1,xp2,xp3]);
V=double([S.x1 S.x2 S.x3]);%gives us the rest points
F= double([subs(f1,S) subs(f2,S) subs(f3,S) ]);
%F gives f_i values at the rest points
end

[m,n]=size(V);
for i = 1:m  %To get the interior rest point
    if all(V(i,:)>0)
```

```
            r=V(i,:);
            break;
        else
            r=[0  0  0];
        end
        if any(V(i,:)<0)
            V1=V([1:i-1,i+1:end],:); %To make sure rest
                                 pts are valid biologically
            F1=F([1:i-1,i+1:end],:);
        else
            V1=V;
        end
    end
    flag=0;
    [m1,n1]=size(V1);
    if all(r>0) %doing the check for permanence if there
                             is an interior rest point
        for i=1:m1
            for j=1:n1
                if V1(i,j)==0
                    if F1(i,j) >=0
                        check = 1;
                        flag=1;
                        break;
                    else
                        check =0;
                    end
                end
                if (flag==1)
                    break;
                end
            end
            if (flag==1)
                    break;
            end
        end
    else
        check=0;
    end
```

Acknowledgements We thank Oleg Yordanov for sharing ideas and productive discussions and the reviewer for the thoughtful comments which improved the manuscript.

References

1. O. Arino, A. El Abdllaoui, J. Mikram, and J. Chattopadhyay. Infection in prey population may act as a biological control in ratio-dependent predator-prey models. *Nonlinearity*, 17(3):1101, 2004.
2. Jean-Pierre Aubin and Karl Sigmund. Permanence and viability. *Journal of Computational and Applied Mathematics*, 22:203–209, 1988.
3. Christian Azar, John Holmberg, and Kristian Lindgreen. Stability analysis of harvesting in a predator-prey model. *Journal of Theoretical Biology*, 174(1):13–19, 1995.
4. Lutz Becks, Frank M. Hilker, Horst Malchow, Kalus Jürgens, and Hartmut Arndt. Experimental demonstration of chaos in a microbial food web. *Nature03627*, 435, 2005.
5. J. Chattopadhyay, R.R. Sarkar, and G. Ghosal. Removal of infected prey prevent limit cycle oscillations in an infected prey-predator system: a mathematical study. *Ecological Modelling*, 156(2):113–121, 2002.
6. Peter L Chesson. The stabilizing effect of a random environment. *Journal of Mathematical Biology*, 15(1):1–36, 1982.
7. PL Chesson and S Ellner. Invasibility and stochastic boundedness in monotonic competition models. *Journal of Mathematical Biology*, 27(2):117–138, 1989.
8. R.F. Costantino, R.A. Desharnais, J.M. Cushing, and B. Dennis. Chaotic dynamics in an insect population. *American Association for the Advancement of Science*, 275(5298):389–391, 1997.
9. Michael Doebeli. The evolutionary advantage of controlled chaos. 254(1341):281–285, 1993.
10. Francesco Doveri, M Scheffer, S Rinaldi, S Muratori, and Yu Kuznetsov. Seasonality and chaos in a plankton fish model. *Theoretical Population Biology*, 43(2):159–183, 1993.
11. H.I. Freedman and Paul Waltman. Mathematical analysis of some three-species food-chain models. *Mathematical Biosciences*, 33(3):257–276, 1977.
12. Michael E. Gilpin. Spiral chaos in a predator-prey model. *The American Naturalist*, 1979.
13. Alan Hastings and Thomas Powell. Chaos in a three-species food chain. *Ecology*, 72(3):896–903, 1991.
14. J. Hofbauer and K. Sigmund. *Evolutionary Games and Population Dynamics*. Cambridge University Press, 1998.
15. Josef Hofbauer. Saturated equilibria, permanence and stability for ecological systems. *Mathematical Ecology*, pages 625–642, 1988.
16. Jef Huisman, Nga N Pham Thi, David M Karl, and Ben Sommeijer. Reduced mixing generates oscillations and chaos in the oceanic deep chlorophyll maximum. *Nature*, 439(7074):322, 2006.
17. S.R.K. Iyengar, R.K. Upadhyay, and Vikas Rai. Chaos: An ecological reality. *International Journal of Bifurcation and Chaos*, 08(06):1325–1333, 1998.
18. Sherli Koshy-Chenthittayil. *Chaos to permanence- through control theory*. PhD thesis, Clemson University, 2017.
19. Mark Kot, Gary S. Sayler, and Terry W. Schultz. Complex dynamics in a model microbial system. *Bulletin of Mathematical Biology*, Vol No. 54(No. 4):Pg. 619–648, 1992.
20. Christophe Letellier and M.A. Aziz-Alaoui. Analysis of the dynamics of a realistic ecological model. *Chaos, Solitons and Fractals*, 13(1):95–107, 2002.
21. R. Lewontin. The meaning of stability. In *Brookhaven Symposium Biology*, volume 22, pages 13–24, 1969.
22. J. Maynard-Smith. *The status of Neo-Darwinism, Towards a Theoretical Ecology*. Edinburgh University Press, 1969.
23. AB Medvinsky, SV Petrovsk, IA Tikhonova, E Venturino, and H Malchow. Chaos and regular dynamics in model multi-habitat plankton-fish communities. *Journal of biosciences*, 26(1):109–120, 2001.
24. K.A. Mirus and J.C. Sprott. Controlling chaos in low and high dimensional systems with periodic parametric perturbations. *Physical Review E*, 59(5):5313–5324, 1999.

25. Edward Ott, Celso Grebogi, and James A. Yorke. Controlling chaos. *Phys. Rev. Lett.*, 64:1196–1199, 1990.
26. K. Pyragas. Continuous control of chaos by self-controlling feedback. *Physics Letters A*, 170(6):421–428, 1992.
27. Filipe J. Romeiras, Celso Grebogi, Edward Ott, and W.P. Dayawansa. Controlling chaotic dynamical systems. *Physica D: Nonlinear Phenomena*, 58(1):165–192, 1992.
28. William M Schaffer and M Kot. Chaos in ecological systems: the coals that newcastle forgot. *Trends in Ecology & Evolution*, 1(3):58–63, 1986.
29. Sebastian J. Schreiber. Persistence despite perturbations for interacting populations. *Journal of Theoretical Biology*, 242(4):844–852, 2006.
30. P Schuster, K Sigmund, and R Wolff. Dynamical systems under constant organization. iii. cooperative and competitive behavior of hypercycles. *Journal of Differential Equations*, 32(3):357–368, 1979.
31. Luke Shulenburger, Ying-Cheng Lai, Tolga Yalcinkaya, and Robert D Holt. Controlling transient chaos to prevent species extinction. *Physics Letters A*, 260(1):156–161, 1999.
32. Lewi Stone and Daihai He. Chaotic oscillations and cycles in multi-trophic ecological systems. *Journal of Theoretical Biology*, 248(2):382–390, 2007.
33. George Sugihara and Robert M May. Nonlinear forecasting as a way of distinguishing chaos from measurement error in time series. *Nature*, 344(6268):734, 1990.
34. Yasuhiro Takeuchi and Norihiko Adachi. Existence and bifurcation of stable equilibrium in two-prey,one-predator communities. *Bulletin of Mathematical Biology*, 45(6):877–900, 1983.
35. R.K. Upadhyay. Multiple attractors and crisis route to chaos in a model food-chain. *Chaos, Solitons and Fractals*, 16(5):737–747, 2003.
36. R.K. Upadhyay and S.R.K. Iyengar. *Introduction to Mathematical Modeling and Chaotic Dynamics*. CRC Press, 2014.
37. T.L. Vincent. Control using chaos. *IEEE Control Systems Magazine*, 17(6):65–76, 1997.
38. T.L. Vincent. Chaotic control systems. *Nonlinear Dynamics and Systems Theory*, 1(2):205–218, 2001.
39. James N Weiss, Alan Garfinkel, Mark L Spano, and William L Ditto. Chaos and chaos control in biology. *The Journal of clinical investigation*, 93(4):1355–1360, 1994.
40. Peter Yodzis. *Introduction to Theoretical Ecology*, chapter 6, pages 166–167. Harper and Row Publishers, Inc, 1989.

Classification on Large Networks: A Quantitative Bound via Motifs and Graphons (Research)

Andreas Haupt, Thomas Schultz, Mohammed Khatami, and Ngoc Tran

1 Introduction

This paper concerns classification problems when each data point is a large network. In neuroscience, for instance, the brain can be represented by a structural connectome or a functional connectome, both are large graphs that model connections between brain regions. In ecology, an ecosystem is represented as a species interaction network. On these data, one may want to classify diseased vs healthy brains, or a species network before and after an environmental shock. Existing approaches for graph classification can be divided broadly into three groups: (1) use of graph parameters such as edge density, degree distribution, or densities of motifs as features, (2) parametric models such as the stochastic k-block model [1], and (3) graph kernels [18], and graph embeddings [29]. Amongst these methods, motif counting is perhaps the least rigorously studied. Though intuitive, only small motifs are feasible to compute, and thus motif counting is often seen as an ad-hoc method with no quantitative performance guarantee.

A. Haupt (✉)
Hausdorff Center for Mathematics, Bonn, Germany

Institute for Data Systems and Society, Massachusetts Institute of Technology,
Cambridge, MA, USA
e-mail: a.haupt@me.com

T. Schultz · M. Khatami
Department of Computer Science, University of Bonn, Bonn, Germany
e-mail: schultz@cs.uni-bonn.de; khatami@informatik.uni-bonn.de

N. Tran
The University of Texas at Austin, Austin, TX, USA
e-mail: ntran@math.utexas.edu

© The Author(s) and the Association for Women in Mathematics 2020   107
B. Acu et al. (eds.), *Advances in Mathematical Sciences*, Association for
Women in Mathematics Series 21, https://doi.org/10.1007/978-3-030-42687-3_7

1.1 Contributions

In this paper, we formalize the use of motifs to distinguish graphs using graphon theory, and give a tight, explicit quantitative bound for its performance in classification (cf. Theorem 1). Furthermore, we use well-known results from graph theory to relate the spectrum (eigenvalues) of the adjacency matrix one-to-one to cycle homomorphism densities, and give an analogous quantitative bound in terms of the spectrum (cf. Theorem 2). These results put motif counting on a firm theory, and justify the use of spectral graph kernels for counting a family of motifs. We apply our method to detect the autoimmune disease *Lupus Erythematosus* from diffusion tensor imaging (DTI) data, and obtain competitive results to previous approaches (cf. Sect. 4).

Another contribution of our paper is the first study of a general model for random weighted graphs, *decorated graphons*, in a machine learning context. The proof technique can be seen as a broad tool for tackling questions on generalisations of graphons. There are three key ingredients. The first is a generalization of the Counting Lemma [see 22, Theorem 10.24], on graphons to decorated graphons. It allows one to lower bound the cut metric by homomorphism densities of motifs, a key connection between motifs and graph limits. The second is Kantorovich duality [see 37, Theorem 5.10], which relates optimal coupling between measures and optimal transport over a class of functions and which is used in relating spectra to homomorphism densities. In this, Duality translates our problem to questions on function approximation, to which we use tools from approximation theory to obtain tight bounds. Finally, we use tools from concentration of measure to deal with sampling error an generalise known sample concentration bounds for graphons [see 6, Lemma 4.4].

Our method extends results for discrete edge weights to the continuous edge weight case. Graphs with continuous edge weights naturally arise in applications such as neuroscience, as demonstrated in our dataset. The current literature for methods on such graphs is limited [16, 26], as many graph algorithms rely on discrete labels [10, 34].

1.2 Related Literature

Graphons, an abbreviation of the words "graph" and "function", are limits of large vertex exchangeable graphs under the cut metric. For this reason, graphons and their generalizations are often used to model real-world networks [8, 12, 36]. Originally appeared in the literature on exchangeable random arrays [4], it was later rediscovered in graph limit theory and statistical physics [14, 22].

There is an extensive literature on the inference of graphons from one observation, i.e. *one* large but finite graph [3, 9, 21, 40]. This is distinct from our classification setup, where one observes multiple graphs drawn from several graphons. In our setting, the graphs might be of different sizes, and crucially, they are unlabelled: There is no *a priori* matching of the graph nodes. That is, if we think of the underlying graphon as an infinitely large random graph, then the graphs in our i.i.d sample could be glimpses into entirely different neighborhoods of this graphon, and they are further corrupted by noise. A naïve approach would be to estimate one graphon for each graph, and either average over the graphs or over the graphons obtained. Unfortunately, our graphs and graphons are only defined up to relabelings of the nodes, and producing the optimal labels between a pair of graphs is NP-complete (via subgraph isomorphism). Thus, inference in our setting is not a mere "large sample" version of the graphon estimation problem, but an entirely different challenge.

A method closer to our setup is graph kernels for support-vector machines [18, 38]. The idea is to embed graphs in a high-dimensional Hilbert space, and compute their inner products via a kernel function. This approach has successfully been used for graph classification [39]. Most kernels used are transformations of homomorphism densities/motifs as feature vectors for a class of graphs [cf 41, subsection 2.5]: [33] propose so-called *graphlet counts* as features. These can be interpreted as using *induced homomorphism densities* [cf 22, (5.19)] as features which can be linearly related to homomorphism densities as is shown in [22, (5.19)]. The random walk kernel from [18, p. 135 center] uses the homomorphism densities of all paths as features. Finally [28, Prop. 5 and discussion thereafter] uses homomorphism densities of trees of height $\leq k$ as features.

However, as there are many motifs, this approach has the same problem as plain motif counting: In theory, performance bounds are difficult, in practice, one may need to make *ad hoc* choices. Due to the computational cost [18], in practice, only small motifs of size up to 5 have been used for classification [33]. Other approaches chose a specific class of subgraphs such as paths [18] or trees [34], for which homomorphism densities or linear combinations of them can be computed efficiently. In this light, our Theorem 2 is a theoretical advocation for cycles, which can be computed efficiently via the graph spectrum.

1.3 Organization

We recall the essentials of graphon theory in Sect. 2. For an extensive reference, see [22]. Main results are in Sect. 3, followed by applications in Sect. 4. Our proofs can be found in the appendix.

2 Background

A graph $G = (V, E)$ is a set of vertices V and set of pairs of vertices, called edges E. A label on a graph is a one-to-one embedding of its vertices onto $\mathbb{N}$. Say that a random labelled graph is vertex exchangeable if its distribution is invariant under relabelings.

A labelled graphon W is a symmetric function from $[0, 1]^2$ to $[0, 1]$. A relabelling ϕ is an invertible, measure-preserving transformation on $[0, 1]$. An unlabelled graphon is a graphon up to relabeling. For simplicity, we write "a graphon W" to mean *an unlabelled graphon equivalent to the labelled graphon W*. Similarly, by *a graph G* we mean *an unlabelled graph which, up to vertex permutation, equals to the labelled graph G*.

The cut metric between two graphons W, W' is

$$\delta_\square(W, W') = \inf_{\phi, \varphi} \sup_{S, T} \left| \int_{S \times T} W(\varphi(x), \varphi(y)) - W'(\phi(x), \phi(y)) \, dx \, dy \right|,$$

where the infimum is taken over all relabelings φ of W and ϕ of W', and the supremum is taken over all measurable subsets S and T of $[0, 1]$. That is, $\delta_\square(W, W')$ is the largest discrepancy between the two graphons, taken over the best relabeling possible. A major result of graphon theory is that the space of unlabelled graphons is compact and complete w.r.t. $\delta_\square$. Furthermore, the limit of any convergent sequence of finite graphs in $\delta_\square$ is a graphon [see 22, Theorem 11.21]. In this way, graphons are truly *limits of large graphs*.

A motif is an unlabelled graph. A graph homomorphism $\phi \colon F \to G$ is a map from $V(F)$ to $V(G)$ that preserves edge adjacency, that is, if $\{u, v\} \in E(F)$, then $\{\phi(u), \phi(v)\} \in E(G)$. Often in applications, the count of a motif F in G is the number of different embeddings (subgraph isomorphisms) from F to G. However, homomorphisms have much nicer theoretical and computational properties [22, par. 2.1.2]. Thus, in our paper, "motif counting" means "computation of homomorphism densities". The homomorphism density $t(F, G)$ is the number of homomorphisms from F to G, divided by $|V(G)|^{|V(F)|}$, the number of mappings $V(F) \to V(G)$. Homomorphisms extend naturally to graphons through integration with respect to the kernel W [22, subsec. 7.2.]. That is, for a graph F with $e(F)$ many edges,

$$t(F, W) = \int_{[0, 1]^{e(F)}} \prod_{\{x, y\} \in E(F)} W(x, y) \, dx dy.$$

The homomorphism density for a weighted graph G on k nodes is defined by viewing G as a step-function graphon, with each vertex of G identified with a set on the interval of Lebesgue measure $1/k$. For a graph G and a graphon W, write $t(\bullet, G)$ and $t(\bullet, W)$ for the sequence of homomophism densities, defined over all possible finite graphs F.

A finite graph G is uniquely defined by $t(\bullet, G)$. For graphons, homomorphism densities distinguish them as well as the cut metric, that is, $\delta_\square(W, W') = 0$ iff $t(\bullet, W) = t(\bullet, W')$ [22, Theorem 11.3]. In other words, if one could compute the homomorphism densities of all motifs, then one can distinguish two convergent sequences of large graphs. Computationally this is not feasible, as $(t(\bullet, W))_{F \text{ finite graph}}$ is an infinite sequence. However, this gives a sufficient condition test for graphon inequality: If $t(F, W) \neq t(F, W')$ for some motif F, then one can conclude that $\delta_\square(W, W') > 0$. We give a quantitative version of this statement in the appendix, which plays an important part in our proof. Theorem 1 is an extension of this result that accounts for sampling error from estimating $t(F, W)$ through the empirical distribution of graphs sampled from W.

2.1 Decorated graphons

Classically, a graphon generates a random unweighted graph $\mathbb{G}(k, W)$ via uniform sampling of the nodes,

$$U_1, \ldots, U_k \overset{\text{iid}}{\sim} \text{Unif}_{[0,1]}$$

$$(\mathbb{G}(k, W)_{ij} | U_1, \ldots, U_k) \overset{\text{iid}}{\sim} \text{Bern}(W(U_i, U_j)), \forall i, j \in [k].$$

Here, we extend this framework to decorated graphons, whose samples are random *weighted* graphs.

Definition 1 Let $\Pi([0, 1])$ be the set of probability measures on $[0, 1]$. A decorated graphon is a function $\mathcal{W} : [0, 1]^2 \to \Pi([0, 1])$.

For $k \in \mathbb{N}$, the k-sample of a measure-decorated graphon $\mathbb{G}(k, \mathcal{W})$ is a distribution on unweighted graphs on k nodes, generated by

$$U_1, \ldots, U_k \overset{\text{iid}}{\sim} \text{Unif}_{[0,1]}$$

$$(\mathbb{G}(k, \mathcal{W})_{ij} | U_1, \ldots, U_k) \overset{\text{iid}}{\sim} \mathcal{W}(U_i, U_j), \forall i, j \in [k].$$

We can write every decorated graphon $\mathcal{W}$ as $\mathcal{W}_{W,\mu}$ with $W(x, y)$ being the expectation of $\mathcal{W}(x, y)$, and $\mu(x, y)$ being the centered measure corresponding to $\mathcal{W}(x, y)$. This decomposition will be useful in formulating our main results, Theorems 1 and 2.

One important example of decorated graphons are *noisy* graphons, that is, graphons perturbed by an error term whose distribution does not vary with the latent parameter: Given a graphon $W : [0, 1]^2 \to [0, 1]$ and a centered noise measure $\nu \in \Pi([0, 1])$, the ν-noisy graphon is the decorated graphon $\mathcal{W}_{W,\mu}$, where $\mu(x, y) = \nu$ is constant, i.e. the same measure for all latent parameters. Hence, in the noisy graphon, there is no dependence of the noise term on the latent parameters.

As weighted graphs can be regarded as graphons, one can use the definition of homomorphisms for graphons to define homomorphism numbers of samples from a decorated graphon (which are then random variables). The k-sample from a decorated graphon is a distribution on weighted graphs, unlike that from a graphon, which is a distribution on unweighted (binary) graphs. The latter case is a special case of a decorated graphon, where the measure at (x, y) is a centered variable taking values $W(x, y)$ and $1 - W(x, y)$. Hence, our theorems generalise results for graphons.

2.2 Spectra and Wasserstein Distances

The spectrum $\lambda(G)$ of a weighted graph G is the set of eigenvalues of its adjacency matrix, counting multiplicities. Similarly, the spectrum $\lambda(W)$ of a graphon W is its set of eigenvalues when viewed as a symmetric operator [22, (7.18)]. It is convenient to view the spectrum $\lambda(G)$ as a counting measure, that is, $\lambda(G) = \sum_\lambda \delta_\lambda$, where the sum runs over all λ's in the spectrum. All graphs considered in this paper have edge weights in $[0, 1]$. Therefore, the support of its spectrum lies in $[-1, 1]$. This space is equipped with the Wasserstein distance (a variant of the earth-movers distance)

$$\mathcal{W}^1(\mu, \nu) = \inf_{\gamma \in \Pi([-1,1]^2)} \int_{(x,y) \in [-1,1]^2} |x - y| \mathrm{d}\gamma(x, y) \tag{1}$$

for $\mu, \nu \in \Pi([-1, 1])$, where the first (second) marginal of γ should equal μ (ν). Analogously, equip the space of random measures $\Pi(\Pi([-1, 1]))$ with the Wasserstein distance

$$\mathcal{W}^1(\bar{\mu}, \bar{\nu}) = \inf_{\gamma \in \Pi(\Pi([-1,1])^2)} \int_{(\mu,\nu) \in \Pi([-1,1])^2} \mathcal{W}^1(\mu, \nu) \mathrm{d}\gamma(\mu, \nu). \tag{2}$$

where again the first (second) marginal of γ should equal $\bar{\mu}$ ($\bar{\nu}$).

Equation (2) says that one must first find an optimal coupling of the eigenvalues for different realisations of the empirical spectrum and then an optimal coupling of the random measures. Equation (1) is a commonly used method for comparing point clouds, which is robust against outliers [25]. Equation (2) is a natural choice of comparison of measures on a continuous space. Similar definitions have appeared in stability analysis of features for topological data analysis [11].

3 Graphons for Classification: Main Results

Consider a binary classification problem where in each class, each data point is a finite, weighted, unlabelled graph. We assume that in each class, the graphs are i.i.d realizations of some underlying decorated graphon $\mathcal{W} = \mathcal{W}_{W,\mu}$ resp. $\mathcal{W}' = \mathcal{W}_{W',\mu'}$. Theorem 1 says that if the empirical homomorphism densities are sufficiently different in the two groups, then the underlying graphons W and W' are different in the cut metric. Theorem 2 gives a similar bound, but replaces the empirical homomorphism densities with the empirical spectra. Note that we allow for the decorated graphons to have different noise distributions and that noise may depend on the latent parameters.

Here is the model in detail. Fix constants $k, n \in \mathbb{N}$. Let $\mathcal{W}_{W,\mu}$ and $\mathcal{W}_{W',\mu'}$ be two decorated graphons. Let

$$G_1, \ldots, G_n \overset{\text{iid}}{\sim} \mathbb{G}(k, \mathcal{W}_{W,\mu})$$

$$G'_1, \ldots, G'_n \overset{\text{iid}}{\sim} \mathbb{G}(k, \mathcal{W}_{W',\mu'})$$

be weighted graphs on k nodes sampled from these graphons. Denote by $\delta_\bullet$ the Dirac measure on the space of finite graphs. For a motif graph F with $e(F)$ edges, let

$$\bar{t}(F) := \frac{1}{n} \sum_{i=1}^{n} \delta_{t(F, G_i)}$$

be the empirical measure of the homomorphism densities of F with respect to the data $(G_1, \ldots, G_n)$ and analogously $\bar{t}'(F)$ the empirical measure of the homomorphism densities of $(G'_1, \ldots, G'_n)$.

Theorem 1 *There is an absolute constant c such that with probability*

$$1 - 2\exp\left(\frac{kn^{-\frac{2}{3}}}{2e(F)^2}\right) - 2e^{-.09cn^{\frac{2}{3}}}$$

and weighted graphs $G_i, G'_i, i = 1, \ldots, n$ generated by decorated graphons $\mathcal{W}_{W,\mu}$ and $\mathcal{W}_{W',\mu'}$,

$$\delta_\square(W, W') \geq e(F)^{-1}(\mathcal{W}^1(\bar{t}, \bar{t}') - 9n^{-\frac{1}{3}}). \tag{3}$$

Note that the number of edges affect both the distance of the homomorphism densities $\mathcal{W}^1(\bar{t}, \bar{t}')$ and the constant $e(F)^{-1}$ in front, making the effect of $e(F)$ on the right-hand-side of the bound difficult to analyze. Indeed, for any fixed $v \in \mathbb{N}$, one can easily construct graphons where the lower bound in Theorem 1 is attained for $k, n \to \infty$ by a graph with $v = e(F)$ edges. Note furthermore, that the bound is given in terms of the expectation of the decorated graphon, W, unperturbed by variations due to μ resp. μ'. Therefore, in the large-sample limit, motifs as features characterise exactly the expectation of decorated graphons.

Our next result utilizes , Theorem 1 and Kantorovich duality to give a bound on $\delta_\square$ with explicit dependence on v. Let $\bar{\lambda}, \bar{\lambda}'$ be the empirical random spectra in the decorated graphon model, that is, $\bar{\lambda} = \frac{1}{n} \sum_{i=1}^{n} \lambda(G_i), \bar{\lambda}' = \frac{1}{n} \sum_{i=1}^{n} \lambda(G'_i)$.

Theorem 2 *There is an absolute constant c such that the following holds: Let $v \in$ $\mathbb{N}$. With probability $1 - 2v \exp\left(\frac{kn^{-\frac{2}{3}}}{2v^2}\right) - 2ve^{-.09cn^{\frac{2}{3}}}$, for weighted graphs generated by decorated graphons $\mathcal{W}_{W,\mu}$ and $\mathcal{W}_{W',\mu'}$,*

$$\delta_\square(W, W') \geq v^{-2}2^{-1}(4e)^{-v}\left(\mathcal{W}^1_{\mathcal{W}^1}(\bar{\lambda}', \bar{\lambda}) - \frac{3}{\pi v} - 18v(4e)^v n^{-\frac{1}{3}}\right)$$

Through the parameter v, Theorem 2 defines a family of lower bounds for the cut distance between the underlying graphons. The choice of v depends on the values of n and the Wasserstein distance of the empirical spectra. The parameter v can be thought of as a complexity control of transformations of eigenvalues that are used in a lower bound: If one restricts to differences in distribution of low-degree polynomials, the approximation with respect to the measure μ, the sampling of edge weights, is good, implying a small additive error. In this case, however, the sampling of the nodes from a graphon, i.e. of the latent node features U_i, $i \in [n]$ has a large error, which is multiplicative. We refer to the appendix for further details.

Theorems 1 and 2 give a test for graphon equality. Namely, if $\mathcal{W}^1(\bar{\lambda}', \bar{\lambda})$ is large, then the underlying graphons W and W' of the two groups are far apart. This type of sufficient condition is analogous to the result of [11, Theorem 5.5] from topological data analysis. It should be stressed that this bound is purely nonparametric. In addition, we do not make any regularity assumption on either the graphon or the error distribution μ. The theorem is stable with respect to transformations of the graph: A bound analogous to Theorem 2 holds for the spectrum of the graph Laplacian and the degree sequence, as we show in the appendix in Sect. 8. In addition, having either k or n fixed is merely for ease of exposition. We give a statement with heterogenous k and n in the appendix in Sect. 9.

We conclude with a remark on computational complexity. The exact computation of eigenvalues of a real symmetric matrix through state-of-the-art numerical linear algebra takes $O(n^3)$ time, where n is the number of the rows of the matrix [13]. In these algorithms, a matrix is transformed in $O(n^3)$ in tridiagonal form. Eigenvalues are computed in $O(kn^2)$ time, where k is the number of largest eigenvalues that are sought.

This is competitive with other graph kernels in the literature. The random walk kernel [18] has a runtime of $O(n^3)$, the subtree kernel from [28] enumerates all possible subtrees in both graphs and hence has, depending on the depth of the trees, doubly exponential runtime. The same holds for the graphlet kernels [32], which are computable in quadratic time for a constant bound on the size of the subgraph homomorphism taken into consideration, but have doubly exponential dependency on this parameter as well. Finally, the paper [16] has, in the sparse, but not ultra-sparse regime ($|E(G)| \in \Theta(|V(G)|)$ and without node labels (corresponding to the Kronecker node kernel) a runtime of $\Omega(n^3)$ for graphs with small diameter $(O(\sqrt{|V(G)|}))$ and for a graph with high diameter a runtime of $\Omega(n^4)$.

4 An Application: Classification of Lupus Erythematosus

Systemic Lupus Erythematosus (SLE) is an autoimmune disease of connective tissue. Between 25–70% of patients with SLE have neuropsychiatric symptoms (NPSLE) [15]. The relation of neuropsychiatric symptoms to other features of the disease is not completely understood. Machine learning techniques in combination with expert knowledge have successfully been applied in this field [20].

We analyse a data set consisting of weighted graphs. The data is extracted from diffusion tensor images of 56 individuals, 19 NPSLE, 19 SLE without neuropsychiatric symptoms and 18 human controls (HC) from the study [30]. The data was preprocessed to yield 6 weighted graphs on 1106 nodes for each individual. Each node in the graphs is a brain region of the hierarchical Talairach brain atlas by [35].

The edge weights are various scalar measures commonly used in DTI, averaged or integrated along all fibres from one brain region to another as in the pipeline depicted in Fig. 1. These scalar measures are the *total number* (of fibers between two regions), the *total length* (of all fibers between two regions), *fractional anisotropy* (FA), *mean diffusivity* (MD), *axial diffusivity* (AD) and *radial diffusivity* (RD) [cf 5].

The paper [20] used the same dataset [30], and considered two classification problems: HC vs NPSLE, and HC vs SLE. Using 20 brain fibers selected from all over the brain (such as the fornix and the anterior thalamic radiation) they used manifold learning to track the values AD, MD, RD and FA along fibers in the brain. Using nested cross-validation, they obtain an optimal disretisation of the bundles, and use average values on parts of the fibers as features for support-vector classification. They obtained an accuracy of 73% for the HC vs. NPSLE and 76% for HC vs. SLE, cf Table 1.

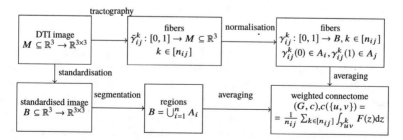

Fig. 1 Preprocessing pipeline for weighted structural connectomes. A brain can be seen as a tensor field $B : \mathbb{R}^3 \to \mathbb{R}^{3\times3}$ of flows. The support of this vector field is partitioned into regions $A_1, \ldots, A_n$, called brain regions. Fibers are parametrized curves from one region to another. Each scalar function $F : \mathbb{R}^3 \to \mathbb{R}$ (such as average diffusivity (AD) and fractional anisotropy (FA)) converts a brain into a weighted graph on n nodes, where the weight between regions i and j is F averaged or integrated over all fibers between these regions

Table 1 Result comparison. Our spectral method performs comparable to [20], who used manifold learning and expert knowledge to obtain the feature vectors. Our method is significantly simpler computationally and promises to be a versatile tool for graph classification problems

	HC vs. NPSLE	HC vs. SLE
[20]	76%	73%
Eigenvalues	78.3%	67.5%

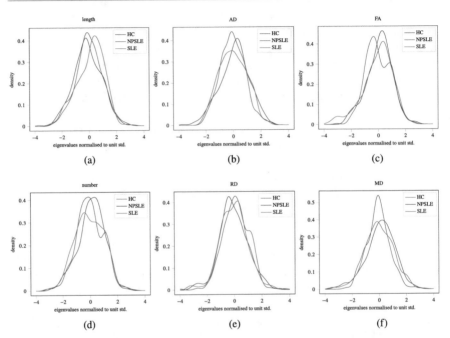

Fig. 2 Density of first and last ten eigenvalues (normalised to zero mean unit standard deviation) of the graph Laplacian for all six values. (**a**) Length. (**b**) Average diffusivity. (**c**) Fractional anisotropy. (**d**) Number. (**e**) Radial diffusivity. (**f**) Mean diffusivity

To directly compare ourselves to [20], we consider the same classification problems. For each weighted graph we reduce the dimension of graphs by averaging edge weights of edges connecting nodes in the same region on a coarser level of the Talairach brain atlas [35]. Inspired by Theorem 2, we compute the spectrum of the adjacency matrix, the graph Laplacian and the degree sequence of the dimension-reduced graphs. We truncate to keep the eigenvalues smallest and largest in absolute value, and plotted the eigenvalue distributions for the six graphs, normalized for comparisons between the groups and graphs (see Fig. 2). We noted that the eigenvalues for graphs corresponding to length and number of fibers show significant differences between HC and NPSLE. Thus, for the task HC vs NPSLE, we used the eigenvalues from these two graphs as features (this gives a total of 40 features), while in the HC vs SLE task, we use all 120 eigenvalues from the six graphs. Using a leave-one-out cross validation with ℓ^1-penalty and a linear support-vector kernel, we arrive at classification rates of 78% for HC vs. NPSLE and 67.5% for HC vs. SLE both for the graph Laplacian. In a permutation test as proposed in [27], we can reject the hypothesis that the results were obtained by pure chance at 10% accuracy. Table 1 summarises our results.

5 Conclusion

In this paper, we provide estimates relating homomorphism densities and distribution of spectra to the cut metric without any assumptions on the graphon's structure. This allows for a non-conclusive test of graphon equality: If homomorphism densities or spectra are sufficiently different, then also the underlying graphons are different. We study the *decorated graphon* model as a general model for random weighted graphs. We show that our graphon estimates also hold in this generalised setting and that known lemmas from graphon theory can be generalised. In a neuroscience application, we show that despite its simplicity, our spectral classifier can yield competitive results. Our work opens up a number of interesting theoretical questions, such as restrictions to the stochastic k-block model.

6 Proof of Theorem 1

Theorem 1 *There is an absolute constant c such that with probability*

$$1 - 2\exp\left(\frac{kn^{-\frac{2}{3}}}{2e(F)^2}\right) - 2e^{-.09cn^{\frac{2}{3}}}$$

and weighted graphs G_i, G'_i, $i = 1, \ldots, n$ generated by decorated graphons $\mathcal{W}_{W,\mu}$ and $\mathcal{W}_{W',\mu'}$,

$$\delta_\square(W, W') \geq e(F)^{-1}(\mathcal{W}^1(\bar{t}, \bar{t}') - 9n^{-\frac{1}{3}}). \tag{3}$$

6.1 Auxiliary Results

The following result is a generalisation of [6, Lemma 4.4] to weighted graph limits.

Lemma 1 *Let $\mathcal{W} = \mathcal{W}_{W,\mu}$ be a decorated graphon, $G \sim \mathbb{G}(k, \mathcal{W})$. Let F be an unweighted graph with v nodes. Then with probability at least $1 - 2\exp\left(\frac{k\varepsilon^2}{2v^2}\right)$,*

$$|t(F, G) - t(F, W)| < \varepsilon. \tag{4}$$

Proof We proceed in three steps. First, give a different formulation of $t(F, W)$ in terms of an expectation. Secondly, we show that this expectation is not too far from the expectation of $t(F, G)$. Finally, we conclude by the method of bounded differences that concentration holds.

1. Let $t_{\mathrm{inj}}(F, G)$ be the injective homomorphism density, which restricts the homomorphisms from F to G to all those ones that map distinct vertices of F to distinct vertices in G [cf 22, (5.12)]. Let $G \sim \mathbb{G}(k, \mathcal{W})$ and X be G's adjacency matrix. As a consequence of exchangeability of X, it is sufficient in the computation of t_{inj} to consider one injection from $V(F)$ to $V(G)$ instead of the average of all such. Without loss, we may assume that $V(F) = [v]$ and $V(G) = [k]$. Hence, for the identity injection $[k] \hookrightarrow [n]$,

$$\mathbb{E}[t_{\mathrm{inj}}(F, X_n)] = \mathbb{E}\left[\prod_{\{i,j\} \in E(G)} X_{ij} \right].$$

Let $U_1, \ldots, U_n$ be the rows and columns in sampling X from G. Then

$$\mathbb{E}\left[\prod_{\{i,j\} \in E(G)} X_{ij} \right] = \mathbb{E}\left[\mathbb{E}\left[\prod_{\{i,j\} \in E(G)} X_{ij} \,\middle|\, U_1, \ldots, U_n \right] \right]$$
$$= \mathbb{E}\left[\prod_{\{i,j\} \in E(G)} (W(U_i, U_j) + \mu(U_i, U_j)) \right]$$

We multiply out the last product, and use that $\mu(U_i, U_j)$ are independent and centered to see that all summands but the one involving only terms from the expectation graphon vanish, i.e.

$$\mathbb{E}\left[\prod_{\{i,j\} \in E(G)} X_{ij} \right] = \mathbb{E}\left[\prod_{\{i,j\} \in E(G)} W(U_i, U_j) \right] = t(F, W)$$

2. Note that the bound in the theorem is trivial for $\varepsilon^2 \le \ln 2 \frac{2k^2}{n} = 4 \ln 2 \frac{k^2}{2n}$. Hence, in particular, $\varepsilon \le 4 \ln 2 \frac{k^2}{2n}$.

   Furthermore, $|t(F, X) - t(F, W)| \le \frac{1}{k}\binom{v}{2} + |t(F, X) - \mathbb{E}[t(F, X)]| \le \frac{v^2}{2k} + |t(F, X) - \mathbb{E}[t(F, X)]|$ by the first part and the bound on the difference of injective homomorphism density and homomorphism density [23, Lemma 2.1]. Hence

$$\mathbb{P}[|t(F, X_n) - t(F, \mathbb{E}\mathcal{W})| \geq \varepsilon] \leq \mathbb{P}\left[|t(F, X_n) - \mathbb{E}[t(F, X_n)]| \geq \varepsilon + \frac{1}{n}\binom{k}{2}\right]$$

$$\leq \mathbb{P}\left[|t(F, X_n) - \mathbb{E}[t(F, X_n)]| \geq \varepsilon\left(1 - \frac{1}{4\ln 2}\right)\right].$$

Set $\varepsilon' = \varepsilon\left(1 - \frac{1}{4\ln 2}\right)$. Let X be the adjacency matrix of $G \sim \mathbb{G}(n, \mathcal{W})$ sampled with latent parameters $U_1, \ldots, U_n$. Define a function depending on n vectors where the i-th vector consists of all values relevant to the i-th column of the array X_n, that is $U_i, X_1, \ldots, X_n$. In formulas,

$$f: \underset{i=1}{\overset{n}{\times}}[0, 1]^{i+1} \to [0, 1],$$

$$(a_1, \ldots, a_n) = ((u_1, x_{11}), (u_2, x_{12}, x_{22}), \ldots, (u_n, x_{1n}, \ldots, x_{nn}))$$

$$\mapsto \mathbb{E}[t(F, (X_{ij})_{1 \leq i, j \leq n}) | U_1 = u_1, \ldots, U_n = u_n, X_{11} = x_{11}, \ldots, X_{nn} = x_n n].$$

We note that the random vectors $(U_i, X_{1i}, X_{2i}, \ldots, X_{ni})$ are mutually independent for varying i. Claim:

$$|f((a_1, \ldots, a_n)) - f((b_1, \ldots, b_n))| \leq \sum_{i=1}^{n} \frac{k}{n} 1_{a_i \neq b_i}$$

If this claim is proved, then we have by McDiarmid's inequality [24, (1.2) Lemma],

$$\mathbb{P}[|t(F, X_n) - t(F, \mathbb{E}\mathcal{W})| \geq \varepsilon']$$

$$\leq 2\exp\left(-\frac{2\varepsilon'^2}{n\left(\frac{k}{n}\right)^2}\right) \leq 2\exp\left(-\frac{2\varepsilon'^2 n}{k^2}\right) = 2\exp\left(-\frac{2n\varepsilon'^2}{k^2}\right),$$

Which implies the theorem by basic algebra.

Let us now prove the claim: It suffices to consider a, b differing in one coordinate, say n. By the definition of the homomorphism density of a weighted graph, $t(F, X)$ can be written as

$$\int g(x_1, \ldots, x_k) d\mathrm{Unif}_{[n]}^k((x_i)_{i \in [k]})$$

for $g(x_1, \ldots, x_k) = \prod_{\{i,k\} \in E(G)} X_{x_i x_k}$. We observe $0 \leq g \leq 1$ (in the case of graphons, one has $g \in \{0, 1\}$). It hence suffices to bound the measure where the integrand g depends on a_i by $\frac{k}{n}$. This is the case only if if $x_\ell = i$ at least for one $\ell \in [k]$. But the probability that this happens is upper bounded by,

$$1 - \left(1 - \frac{1}{n}\right)^k \le \frac{k}{n},$$

by the Bernoulli inequality. This proves the claim and hence the theorem.

□

Lemma 2 ([22, Lemma 10.23]) *Let W, W' be graphons and F be a motif. Then*

$$|t(F, W) - t(F, W')| \le e(F)\delta_\square(W, W')$$

Lemma 3 *Let $\mu \in \Pi([0, 1])$ and let μ_n be the empirical measure of n iid samples of μ. Then*

$$\mathbb{E}[\mathcal{W}^1(\mu, \mu_n)] \le 3.6462n^{-\frac{1}{3}}$$

The strategy of prove will be to adapt a proof in [19, Theorem 1.1] to the 1-Wasserstein distance.

Proof Let $X \sim \mu$, $Y \sim N(0, 1)$ and $\mu^\sigma = \text{Law}(X + Y)$. Then for any $v \in \Pi([0, 1])$, by results about the standard normal distribution, $W(v, v^\sigma) \le \mathbb{E}[|Y|] = \sigma\sqrt{\frac{2}{\pi}}$. Hence, by the triangle inequality

$$\mathcal{W}^1(\mu, \mu_n) \le 2\sqrt{\frac{2}{\pi}}\sigma + \mathcal{W}^1(\mu^\sigma, \mu_n^\sigma).$$

As the discrete norm dominates the absolute value metric on $[0, 1]$, $\mathcal{W}^1(\mu^\sigma, \mu_n^\sigma) \le \|\mu^\sigma - \mu_n^\sigma\|_{\text{TV}}$. Note that μ_n^σ and μ^σ have densities f^σ, f_n^σ. This means, as $\|\mu^\sigma - \mu_n^\sigma\|_{\text{TV}} = \int |f_n^\sigma(x) - f^\sigma(x)|\mathrm{d}x$,

$$\mathcal{W}^1(\mu^\sigma, \mu_n^\sigma) \le \int |f_n^\sigma(x) - f^\sigma(x)|\mathrm{d}x \le \sqrt{2\pi}\sqrt{\int (|x|^2 + 1)|f_n^\sigma(x) - f^\sigma(x)|^2\mathrm{d}x},$$

where the last inequality is an application of [19, (2.2)]. Now observe that by the definitions of f^σ and f_n^σ, $\mathbb{E}[|f_n^\sigma(x) - f^\sigma(x)|^2 \le n^{-1} \int \phi_\sigma^2(x - y)\mathrm{d}\mu(y)$, where ϕ_σ is the standard normal density. Hence

$$\mathbb{E}[\mathcal{W}^1(\mu_n^\sigma, \mu^\sigma)] \le \sqrt{2\pi}n^{-\frac{1}{2}}\sqrt{\int (|x|^2 + 1) \int \phi_\sigma^2(x - y)\mathrm{d}\mu(y)\mathrm{d}x}$$

By basic algebra, $\phi_\sigma^2(x) = \frac{1}{2\sigma}\pi^{-\frac{1}{2}}\phi_{\frac{\sigma}{\sqrt{2}}}(x)$. This implies for $Z \sim N(0, 1)$ by a change of variables

$$\int (|x|^2 + 1) \int \phi_\sigma^2(x - y) d\mu(y) dx$$

$$\leq \frac{1}{2\sigma\sqrt{\pi}}(1 + 2(\sigma^2\mathbb{E}[Z^2] + \int |y|^2 d\mu(y))) \leq \sigma^{-1}2^{-1}\pi^{-\frac{1}{2}}(1 + 2(\sigma^2 x + 1))$$

$$\leq \sigma^{-1}\pi^{-\frac{1}{2}}\frac{3}{2}$$

Hence $\mathbb{E}[\mathcal{W}^1(\mu_n^\sigma, \mu^\sigma)] \leq \frac{3}{2}\sqrt{2}n^{-\frac{1}{2}}\sigma^{-\frac{1}{2}} = \frac{3}{\sqrt{2}}n^{-\frac{1}{2}}\sigma^{-\frac{1}{2}}$ and

$$\mathbb{E}[\mathcal{W}^1(\mu_n, \mu)] \leq 2\sqrt{\frac{2}{\pi}}\sigma + \frac{3}{\sqrt{2}}n^{-\frac{1}{2}}\sigma^{-\frac{1}{2}}.$$

Choosing σ optimally by a first-order condition, one arrives at the lemma.   □

Lemma 4 ([17, Theorem 2]) *Let* $\mu \in \mathcal{P}(\mathbb{R})$ *such that for* $X \sim \mu$, $\ell = \mathbb{E}[e^{\gamma X^\alpha}] < \infty$ *for some choice of* γ *and* α. *Then one has with probability at least* $1 - e^{-cn\varepsilon^2}$

$$\mathcal{W}^1(\mu_n, \mu) \leq \varepsilon$$

for any $\varepsilon \in [0, 1]$ *and* c *only depending on* ℓ, γ *and* α.

6.2 Proof of Theorem 1

Proof (Proof of Theorem 1) Let $G \sim \mathbb{G}(k, \mathcal{W})$ and $G' \sim \mathbb{G}(k, \mathcal{W}')$. By combining Lemmas 3 and 4, we get that with probability at least $1 - 2e^{-.09cn^{\frac{2}{3}}}$,

$$\mathcal{W}^1(\bar{t}, \bar{t}') \leq \mathcal{W}^1(t(F, G), t(F, G')) + 8n^{-\frac{1}{3}}$$

In addition, by Lemma 1, with probability at least $1 - 2\exp\left(\frac{kn^{-\frac{2}{3}}}{2v^2}\right) - 2e^{-.09cn^{\frac{2}{3}}}$ one also has

$$\mathcal{W}^1(t(F, G), t(F, G')) \leq |t(F, W) - t(F, W')| + n^{-\frac{1}{3}}$$

Upon application of Lemma 2 and rearranging, one arrives at the theorem.   □

7 Proof of Theorem 2

Theorem 2 *There is an absolute constant c such that the following holds: Let $v \in$ $\mathbb{N}$. With probability $1 - 2v \exp\left(\frac{kn^{-\frac{2}{3}}}{2v^2}\right) - 2ve^{-.09cn^{\frac{2}{3}}}$, for weighted graphs generated by decorated graphons $\mathcal{W}_{W,\mu}$ and $\mathcal{W}_{W',\mu'}$,*

$$\delta_{\square}(W, W') \geq v^{-2}2^{-1}(4e)^{-v}\left(\mathcal{W}^1_{\mathcal{W}^1}(\bar{\lambda}, \bar{\lambda}') - \frac{3}{\pi v} - 18v(4e)^v n^{-\frac{1}{3}}\right)$$

7.1 Auxiliary Results

Lemma 5 ([7, (6.6)]) *Let G be a weighted graph and λ the spectrum interpreted as a point measure. Let C_k be the cycle of length kem. Then*

$$t(C_k, G) = \sum_{w \in \lambda} w^k.$$

Lemma 6 (Corollary of [2, p. 200]) *Let f be a 1-Lipschitz function on $[-1, 1]$. Then there is a polynomial p of degree v such that $\|f - p\|_\infty \leq \frac{3}{\pi v}$.*

Lemma 7 ([31, Lemma 4.1]) *Let $\sum_{i=0}^{v} a_i x^i$ be a polynomial on $[-1, 1]$ bounded by M. Then*

$$|a_i| \leq (4e)^v M.$$

7.2 Proof of Theorem 2

Proof (Proof of Theorem 2) Consider any coupling (λ, λ') of $\bar{\lambda}$ and $\bar{\lambda}'$. One has by the definition of the Wasserstein distance $\mathcal{W}^1_{\mathcal{W}^1}$ and Kantorovich duality

$$\mathcal{W}^1_{\mathcal{W}^1}(\bar{\lambda}', \bar{\lambda}) \leq \mathbb{E}\left[\mathcal{W}^1(\lambda, \lambda')\right] = \mathbb{E}\left[\sup_{\text{Lip}(f) \leq 1} \int f(x)\mathrm{d}(\lambda - \lambda')\right] \tag{5}$$

Fix any $\omega \in \Omega$. By Lemma 6 one can approximate Lipschitz functions by polynomials of bounded degree,

$$\sup_{\substack{f:\,[-1,1] \to \mathbb{R} \\ \text{Lip}(f) \leq 1}} \int f(x)\mathrm{d}(\lambda - \lambda')(\omega) \leq \sup_{\substack{\deg(f) \leq v \\ |f| \leq 2}} \int f(x)\mathrm{d}(\lambda - \lambda')(\omega) + \frac{3}{\pi v}.$$

Here, $|f| \leq 2$ can be assumed as f is defined on $[-1, 1]$ and because of its 1-Lipschitz continuity.

Hence, by Lemma 7 and the triangle inequality

$$\sup_{\substack{\deg(f) \leq v \\ |f| \leq 2}} \int f(x)\mathrm{d}(\lambda - \lambda')(\omega) \leq \sum_{i=1}^{v} 2(4e)^v \left| \left| \int x^k \mathrm{d}(\lambda - \lambda') \right| (\omega) \right.$$

$$= \sum_{i=1}^{v} 2(4e)^v \left| \sum_{w \in \lambda} w^i - \sum_{w' \in \lambda'} w^i \right| (\omega)$$

Tanking expectations, one gets

$$\mathcal{W}^1_{\mathcal{W}^1}(\bar{\lambda}, \bar{\lambda}') \leq \frac{3}{\pi v} + \sum_{i=1}^{v} 2(4e)^v \mathbb{E}\left[\left| \sum_{w \in \lambda} w^i - \sum_{w' \in \lambda'} w^i \right| \right]$$

for any coupling (λ, λ') of $\bar{\lambda}$ and $\bar{\lambda}'$. Now consider a coupling (λ, λ') of $\bar{\lambda}$ and $\bar{\lambda}'$ such that $\bar{t}, \bar{t}'$ (which are functions of λ, λ' by Lemma 5) are optimally coupled. Then by the definition of $\bar{\lambda}, \bar{\lambda}', \bar{t}$ and $\bar{t}'$,

$$\mathcal{W}^1(\bar{t}_k, \bar{t}'_k) = \mathbb{E}\left[\left| \sum_{w \in \lambda} w^k - \sum_{w \in \bar{\lambda}'} w'^k \right| \right]$$

where $\bar{t}_i = \frac{1}{n} \sum_{j=1}^{n} \delta_{t(C_i, G_j)}$ and $\bar{t}'_i = \frac{1}{n} \sum_{j=1}^{n} \delta_{t(C_i, G_j)}$. Hence,

$$\mathcal{W}^1_{\mathcal{W}^1}(\bar{\lambda}', \bar{\lambda}) \leq \sum_{i=1}^{v} 2(4e)^v \mathcal{W}^1(\bar{t}_i, \bar{t}'_i) + \frac{3}{\pi v}. \tag{6}$$

$$\leq \frac{3}{\pi v} + v^2 2(4e)^v \delta_{\square}(W, W') + 18v(4e)^v n^{-\frac{1}{3}}.$$

The first equality follows by (5) and the second with probability at least $1 - 2v \exp\left(\frac{kn^{-\frac{2}{3}}}{2v^2}\right) - 2ve^{-.09cn^{\frac{2}{3}}}$ from Theorem 1.   □

8   A Similar Bound for Degree Features

Let G be a graph and (d_i) be its degree sequence. Consider the point measure $d = \sum_i \delta_{d_i}$ of degrees. Denote by $\bar{d}$ resp. $\bar{d}'$ the empirical measure of degree point measures of $G_1, \ldots, G_n$ resp. $G'_1, \ldots, G'_n$.

Proposition 1 *Theorem 2 holds with the same guarantee with $\bar{\lambda}$, $\bar{\lambda}'$ replaced by $\bar{d}$, $\bar{d}'$.*

Lemma 8 *Let S_v be the star graph on v nodes and G be a weighted graph. Then*

$$t(S_v, G) = \sum_{w \in d} w^v$$

The proof of Proposition 1 is along the same lines as the one of Theorem 2, but using Lemma 8 instead of 5.

9   Heterogenous Sample Sizes

Our bounds from Theorems 1 and 2 can also be formulated in a more general setting of heterogenous sizes of graphs. In the following, we give an extension in two dimensions. First, we allow for heterogenous numbers of observations n. Secondly, we allow for random sizes of graphs k. Here is the more general model in details: There is a measure $\nu \in \Pi(\mathbb{N})$ such that $G_1, \ldots, G_{n_1}$ are sampled iid as

$$k \sim \nu \qquad\qquad\qquad G_i \sim \mathbb{G}(k, \mathcal{W}_{W,\mu}); \qquad\qquad (7)$$

sampling of $G'_1, \ldots, G'_{n_2}$ is analogously. Hence the samples G_i are sampled from a mixture over the measures $\mathbb{G}(k, \mathcal{W}_{W',\mu'})$. We can define $\bar{t}$, $\bar{t}'$, $\bar{\lambda}$ and $\bar{\lambda}'$ using the same formulas as we did in the main text. Then the following result holds.

Corollary 1 *There is an absolute constant c such that the following holds: Let $n_1, n_2 \in \mathbb{N}$ and $G_i, i = 1, \ldots, n_1$, $G'_i, i = 1, \ldots, n_2$ sampled as in (7). Then with probability at least $1 - \exp\left(\dfrac{kn_1^{-\frac{2}{3}}}{2e(F)^2}\right) - e^{-.09cn_1^{\frac{2}{3}}} - \exp\left(\dfrac{kn_2^{-\frac{2}{3}}}{2e(F)^2}\right) - e^{-.09cn_2^{\frac{2}{3}}}$,*

$$\delta_{\square}(W, W') \geq e(F)^{-1}(\mathcal{W}^1(t, \bar{t}) - 5n_1^{-\frac{1}{3}} + 5n_2^{-\frac{1}{3}}).$$

Corollary 2 *In the setting of Corollary 1 and with the same absolute constant, the following holds: Let $v \in \mathbb{N}$. With probability $1 - v \exp\left(\dfrac{kn_1^{-\frac{2}{3}}}{2v^2}\right) - ve^{-.09cn_1^{\frac{2}{3}}} - v \exp\left(\dfrac{kn_2^{-\frac{2}{3}}}{2v^2}\right) - ve^{-.09cn_2^{\frac{2}{3}}}$,*

$$\delta_{\square}(W, W') \geq v^{-2}2^{-1}(4e)^{-v}\left(\mathcal{W}^1_{\mathcal{W}^1}(\bar{\lambda}, \bar{\lambda}') - \frac{3}{\pi v} - 18v(4e)^v(n_1^{-\frac{1}{3}} + n_2^{-\frac{1}{3}})\right)$$

The proofs are very similar to the ones in the main text. For the differences in n_1 and n_2, the concentration results Lemmas 3 and 4 will have to be applied separately with different values of n. For the random values k, we can choose a coupling that couples random graphs of similar sizes, leading to the expressions in the Corollaries.

References

1. Emmanuel Abbe. "Community detection and stochastic block models: recent developments". In: *arXiv preprint arXiv:1703.10146 (2017)*.
2. Naum I Achieser. *Theory of approximation. Courier Corporation, 2013*.
3. Edo M Airoldi, Thiago B Costa, and Stanley H Chan. "Stochastic blockmodel approximation of a graphon: Theory and consistent estimation". In: *Advances in Neural Information Processing Systems*. 2013, pp. 692–700.
4. David J Aldous. "Representations for partially exchangeable arrays of random variables". In: *Journal of Multivariate Analysis* 11.4 (1981), pp. 581–598.
5. Peter J Basser et al. "In vivo fiber tractography using DT-MRI data". In: *Magnetic resonance in medicine* 44.4 (2000), pp. 625–632.
6. Christian Borgs, Jennifer Chayes, and Adam Smith. "Private graphon estimation for sparse graphs". In: *Advances in Neural Information Processing Systems*. 2015, pp. 1369–1377.
7. Christian Borgs et al. "An L^p theory of sparse graph convergence I: limits, sparse random graph models, and power law distributions". In: *arXiv preprint arXiv:1401.2906 (2014)*.
8. Christian Borgs et al. "Convergent sequences of dense graphs I: Subgraph frequencies, metric properties and testing". In: *Advances in Mathematics* 219.6 (2008), pp. 1801-1851.
9. Christian Borgs et al. "Convergent sequences of dense graphs II. Multiway cuts and statistical physics". In: *Annals of Mathematics* 176.1 (2012), pp. 151–219.
10. Karsten M Borgwardt and Hans-Peter Kriegel. "Shortest-path kernels on graphs". In: *Data Mining, Fifth IEEE International Conference on*. IEEE. 2005, 8-pp.
11. Peter Bubenik. "Statistical topological data analysis using persistence landscapes". In: *Journal of Machine Learning Research* 16.1 (2015), pp. 77–102.
12. Diana Cai, Trevor Campbell, and Tamara Broderick. "Edge-exchangeable graphs and sparsity". In: *Advances in Neural Information Processing Systems*. 2016, pp. 4242–4250.
13. Inderjit S Dhillon and Beresford N Parlett. "Multiple representations to compute orthogonal eigenvectors of symmetric tridiagonal matrices". In: *Linear Algebra and its Applications* 387 (2004), pp. 1–28.
14. Persi Diaconis and Svante Janson. "Graph limits and exchangeable random graphs". In: *arXiv preprint arXiv:0712.2749 (2007)*.
15. Edward J Feinglass et al. "Neuropsychiatric manifestations of systemic lupus erythematosus: diagnosis, clinical spectrum, and relationship to other features of the disease". In: *Medicine* 55.4 (1976), pp. 323–339.
16. Aasa Feragen et al. "Scalable kernels for graphs with continuous attributes". In: *Advances in Neural Information Processing Systems*. 2013, pp. 216–224.
17. N. Fournier and A. Guillin. "On the rate of convergence in Wasserstein distance of the empirical measure". In: *ArXiv e-prints* (Dec. 2013). arXiv: 1312.2128 [math.PR].
18. Thomas Gärtner, Peter Flach, and Stefan Wrobel. "On graph kernels: Hardness results and efficient alternatives". In: *Learning Theory and Kernel Machines*. Springer, 2003, pp. 129–143.
19. Joseph Horowitz and Rajeeva L Karandikar. "Mean rates of convergence of empirical measures in the Wasserstein metric". In: *Journal of Computational and Applied Mathematics* 55.3 (1994), pp. 261–273.

20. Mohammad Khatami et al. "BundleMAP: anatomically localized features from dMRI for detection of disease". In: *International Workshop on Machine Learning in Medical Imaging*. Springer. 2015, pp. 52–60.

21. Olga Klopp, Alexandre B Tsybakov, Nicolas Verzelen, et al. "Oracle inequalities for network models and sparse graphon estimation". In: *The Annals of Statistics* 45.1 (2017), pp. 316–354.

22. László Lovász. *Large networks and graph limits*. Vol. 60. American Mathe- matial Soc., 2012.

23. László Lovász and Balázs Szegedy. "Limits of dense graph sequences". In: *Journal of Combinatorial Theory, Series B* 96.6 (2006), pp. 933–957.

24. Colin McDiarmid. "On the method of bounded differences". In: *Surveys in combinatorics* 141.1 (1989), pp. 148–188.

25. Patrick Mullen et al. "Signing the unsigned: Robust surface reconstruction from raw pointsets". In: *Computer Graphics Forum*. Vol. 29. Wiley Online Library. 2010, pp. 1733–1741.

26. Marion Neumann et al. "Efficient graph kernels by randomization". In: *Machine Learning and Knowledge Discovery in Databases* (2012), pp. 378–393.

27. Markus Ojala and Gemma C Garriga. "Permutation tests for studying classifier performance". In: *Journal of Machine Learning Research* 11.Jun (2010), pp. 1833–1863.

28. Jan Ramon and Thomas Gärtner. "Expressivity versus efficiency of graph kernels". In: *Proceedings of the first international workshop on mining graphs, trees and sequences*. 2003, pp. 65–74.

29. Kaspar Riesen and Horst Bunke. *Graph classification and clustering based on vector space embedding*. Vol. 77. World Scientific, 2010.

30. Tobias Schmidt-Wilcke et al. "Diminished white matter integrity in patients with systemic lupus erythematosus". In: *NeuroImage: Clinical* 5 (2014), pp. 291–297.

31. Alexander A Sherstov. "Making polynomials robust to noise". In: *Proceedings of the forty-fourth annual ACM symposium on Theory of computing*. ACM. 2012, pp. 747–758.

32. Nino Shervashidze. "Scalable graph kernels". PhD thesis. Universität Tübingen, 2012.

33. Nino Shervashidze et al. "Efficient graphlet kernels for large graph comparison". In: *AISTATS*. Vol. 5. 2009, pp. 488–495.

34. Nino Shervashidze et al. "Weisfeiler-lehman graph kernels". In: *Journal of Machine Learning Research* 12.Sep (2011), pp. 2539–2561.

35. Jean Talairach and Pierre Tournoux. *Co-planar stereotaxic atlas of the human brain. 3-Dimensional proportional system: an approach to cerebral imaging*. Thieme, 1988.

36. Victor Veitch and Daniel M Roy. "The class of random graphs arising from exchangeable random measures". In: *arXiv preprint arXiv:1512.03099* (2015).

37. Cédric Villani. *Optimal transport: old and new*. Vol. 338. Springer Science & Business Media, 2008.

38. S Vichy N Vishwanathan et al. "Graph kernels". In: *Journal of Machine Learning Research* 11.Apr (2010), pp. 1201–1242.

39. SVN Vishwanathan, Karsten M Borgwardt, Nicol N Schraudolph, et al. "Fast computation of graph kernels". In: *NIPS*. Vol. 19. 2006, pp. 131–138.

40. Patrick J Wolfe and Sofia C Olhede. "Nonparametric graphon estimation". In: *arXiv preprint arXiv:1309.5936* (2013).

41. Pinar Yanardag and SVN Vishwanathan. "Deep graph kernels". In: *Proceedings of the 21th ACM SIGKDD International Conference on Knowledge Discovery and Data Mining*. ACM. 2015, pp. 1365–1374.

Gröbner Bases of Convex Neural Code Ideals (Research)

Kaitlyn Phillipson, Elena S. Dimitrova, Molly Honecker, Jingzhen Hu, and Qingzhong Liang

1 Introduction and Background

Humans and animals perceive their surroundings based on previous encounters. Their brains have to store information about those encounters to be accessed in the future, and the way this information is stored and processed is the subject of active research in neuroscience. Great strides have also been made towards a mathematical understanding of the brain. For example, the theory of neural codes studies how the brain represents external stimulation. These codes are extracted from stereotyped stimulus-response maps, associating to each neuron a convex receptive field. An important problem confronted by the brain is to infer properties of a represented stimulus space without knowledge of the receptive fields, using only the intrinsic structure of the neural code. To understand how the brain does this, one must first determine what stimulus space features can be extracted from neural codes.

In this paper, we study neural codes through an algebraic object called a *neural ideal* which was introduced in [5] to better understand the combinatorial

K. Phillipson (✉)
Department of Mathematics, St. Edward's University, Austin, TX, USA
e-mail: kphillip@stedwards.edu

E. S. Dimitrova
Department of Mathematics, California Polytechnic State University, San Luis Obispo, CA, USA
e-mail: edimitro@calpoly.edu

M. Honecker
School of Mathematical and Statistical Clemson University, Clemson University, Clemson, SC, USA
e-mail: mhoneck@clemson.edu

J. Hu · Q. Liang
Department of Mathematics, Duke University, Durham, NC, USA
e-mail: jingzhen.hu@duke.edu; qingzhong.liang@duke.edu

© The Author(s) and the Association for Women in Mathematics 2020
B. Acu et al. (eds.), *Advances in Mathematical Sciences*, Association for Women in Mathematics Series 21, https://doi.org/10.1007/978-3-030-42687-3_8

127

structure of neural codes. More specifically, we focus on convex neural codes (and their corresponding ideals) since they have been observed experimentally in brain activity. In Sect. 2 we begin with a survey on what is known so far about convex neural codes. In Sect. 3 we discuss the structure of neural ideals and their Gröbner bases. We then introduce results on the connection between the canonical form of a neural ideal and its reduced Gröbner basis, suggesting that neural ideals which have a unique reduced Gröbner bases are of particular interest. Thus, in Sect. 4, we introduce a method for identifying neural codes with unique Gröbner bases. These results suggest a conjecture, stated in Sect. 5, that provides a characterization of convex neural codes based on their Gröbner bases.

We first review some terminology and results here (see [5]). Given a *neural code* C written as a set of binary strings of length n (alternatively, it can be written as subsets of $[n]$), we can construct the ideal of polynomials that vanish on C:

$$I_C := \{p \in \mathbb{F}_2[x_1, \ldots, x_n] : p(c) = 0 \text{ for all } c \in C\}, \tag{1}$$

where $\mathbb{F}_2$ is the finite field of two elements (0 and 1), and $\mathbb{F}_2[x_1, \ldots, x_n]$ is the polynomial ring in n variables with coefficients in $\mathbb{F}_2$. Note that since $0^2 = 0$ and $1^2 = 1$, I_C always contains the set of Boolean relations $\mathcal{B} = \langle x_i^2 - x_i : i \in [n] \rangle$.

We can construct a generating set for the rest of the elements of I_C, via indicator functions: Given a codeword $v \in \mathbb{F}^2$, define

$$\rho_v := \prod_{i:v_i=1} x_i \prod_{j:v_j=0} (1 + x_j).$$

Note that $\rho_v(v) = 1$ and $\rho_v(c) = 0$ for $c \neq v$. From these functions, we can build the *neural ideal* of C:

$$J_C := \langle \rho_v : v \in \mathbb{F}_2^n \setminus C \rangle$$

Note that $I_C = \mathcal{B} + J_C$ [5]. The functions ρ_v that generate J_C are examples of *pseudo-monomials*: these are polynomials $f \in \mathbb{F}_2[x_1, \ldots, x_n]$ of the form

$$f = x_\sigma \prod_{j \in \tau} (1 + x_j),$$

where $x_\sigma := \prod_{i \in \sigma} x_i$ and $\sigma, \tau \subseteq [n]$ with $\sigma \cap \tau = \emptyset$.

Given an ideal $J \subset \mathbb{F}_2[x_1, \ldots, x_n]$, a pseudo-monomial $f \in J$ is *minimal* if there does not exist another pseudo-monomial $g \in J$ with $\deg(g) < \deg(f)$ and $f = hg$ for some $h \in \mathbb{F}_2[x_1, \ldots, x_n]$. We define the *canonical form* of J_C to be the set of all minimal pseudo-monomials of J_C, denoted $CF(J_C)$. For any neural code C, the set $CF(J_C)$ is a generating set for the neural ideal J_C. The canonical form $CF(J_C)$ can be constructed algorithmically from the code C (see [5, 13]).

Example 1 Given the code $C = \{000, 100, 110, 101, 001, 111\}$, there are two elements in $\mathbb{F}_2^3$ that are not in C: 010 and 011. From these, we construct the neural ideal:

$$J_C = \langle x_2(1 + x_1)(1 + x_3), x_2 x_3(1 + x_1) \rangle$$

The canonical form is $CF(J_C) = \{x_2(1 + x_1)\}$. Observe that if a codeword c satisfies $x_2(1 + x_1) = 0$, then whenever neuron 2 is firing ($x_2 = 1$), we must have neuron 1 firing, as well ($x_1 = 1$).

2 Convexity of Neural Codes

We will now investigate combinatorial codes arising from covers of a stimulus space. Let X be a topological space. A collection of non-empty open sets $\mathcal{U} = \{U_1, U_2, \ldots, U_n\}$, $U_i \subset X$, is called an *open cover*. Given an open cover $\mathcal{U}$, the *code of the cover* is the neural code defined as:

$$C(\mathcal{U}) = \{\sigma \subseteq [n] : \bigcap_{i \in \sigma} U_i \setminus \bigcup_{j \in [n] \setminus \sigma} U_j \neq \emptyset\}.$$

Given a combinatorial code C, we say that C is realized by an open cover $\mathcal{U}$ if $C = C(\mathcal{U})$. If C can be realized by $\mathcal{U}$, where $\mathcal{U} = \{U_1, \ldots, U_n\}$ with each U_i a convex subset of $\mathbb{R}^d$, then C is a *convex code* with geometric realization $\mathcal{U}$.

Not all combinatorial codes are convex. For example, the code $C = \{\emptyset, 1, 2, 13, 23\}$ cannot be realized with convex sets, as the set U_3 is the disjoint union of open sets $U_1 \cap U_3$ and $U_2 \cap U_3$, forcing it to be disconnected (and thus, non-convex). A complete condition for convexity is still unknown; we summarize here the known results.

Note that in the previous example the relationship in the receptive fields forced the non-convexity of one of the sets, and the presence of the single codeword 3 would eliminate this topological inconsistency. This is an example of a local obstruction to convexity, instrinsic to the combinatorial structure of the code itself.

Definition 1 ([3]) Let $C = C(\mathcal{U})$ be a code on n neurons, with $\mathcal{U} = \{U_1, \ldots, U_n\}$ a realization of C. Let $U_\sigma = \bigcap_{i \in \sigma} U_i$. A *receptive field relationship* (*RF relationship*) of C is a pair (σ, τ) corresponding to the set containment

$$U_\sigma \subseteq \bigcup_{i \in \tau} U_i,$$

where $\sigma \neq \emptyset$, $\sigma \cap \tau = \emptyset$, and $U_\sigma \cap U_i \neq \emptyset$ for all $i \in \tau$. A receptive field relationship is *minimal* if no single neuron from σ or τ can be removed without destroying the containment.

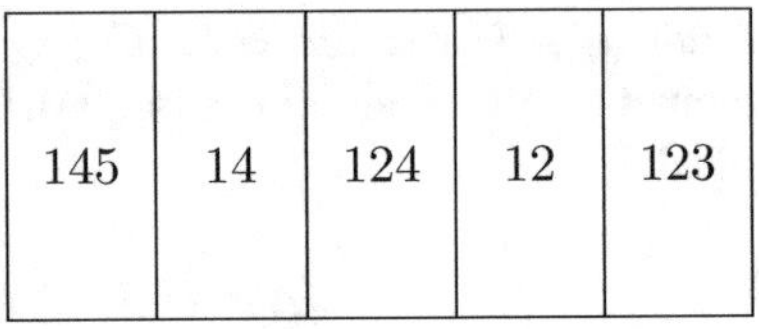

Fig. 1 Convex realization of
$C1$

145	14	124	12	123

In general, we can detect local obstructions via the simplicial complex of a code. Given a code C, its *simplicial complex* is $\Delta(C) := \{\sigma \subseteq [n] : \sigma \subseteq c$ for some $c \in C\}$. For a simplicial complex Δ, the *restriction* of Δ to σ is the simplicial complex $\Delta|_\sigma := \{\omega \in \Delta : \omega \subset \sigma\}$. For any $\sigma \in \Delta$, the *link* of σ in Δ is $Lk_\sigma(\Delta) = \{w \in \Delta : \sigma \cap w = \emptyset, \sigma \cup w \in \Delta\}$.

Definition 2 ([3]) Let (σ, τ) be a receptive field relationship, and let $\Delta = \Delta(C)$. We say that (σ, τ) is a *local obstruction* of C if $\tau \neq \emptyset$ and $Lk_\sigma(\Delta|_{\sigma \cup \tau})$ is not contractible.

Note that in $C = \{\emptyset, 1, 2, 13, 23\}$, $(\sigma, \tau) = (\{3\}, \{1, 2\})$ is a receptive field relationship $(U_3 \subseteq U_1 \cup U_2)$, and $Lk_3(\Delta|_{123}) = \{1, 2\}$, which is disconnected (and thus, not contractible).

Notice that the simplicial complex of a code C is defined by its maximal codewords. A *maximal codeword* σ of a code C is maximal under inclusion in C. A code is *max intersection-complete* if it is closed under taking all intersections of its maximal codewords.

We can now state necessary and sufficient conditions for convexity:

Proposition 1 *For a neural code C:*

1. *If C is max intersection-complete, then C is convex.*
2. *If C is convex, then C has no local obstructions.*

Part 1 of Proposition 1 is due to [2], while Part 2 is due to [3] as a consequence of the Nerve Lemma.

The converses of Part 1 and Part 2 of Proposition 1 hold for $n \leq 4$ (see [3]); however, these statements fail for $n \geq 5$. An example of a convex code which is not max intersection-complete can be seen via $C1 = \{123, 124, 145, 14, 12\}$ in Fig. 1. An example of a non-convex code which has no local obstructions was found in [12], which is code $C4 = \{2345, 123, 134, 145, 13, 14, 23, 34, 45, 3, 4, \emptyset\}$. The case for $n = 5$ neurons has also been fully classified; see [9].

3 Structure of the Neural Ideal

We now turn to a discussion relating convexity to the structure of the neural ideal. As we saw in Example 1 in Sect. 1, the canonical form encodes minimal descriptions of the relationships between the sets U_i. The following lemma given in [5] generalizes this observation:

Lemma 1 *Let $C = C(\mathcal{U})$ be a neural code on n neurons with neural ideal J_C. For $\sigma, \tau \in [n]$ with $\sigma \cap \tau = \emptyset$, $x_\sigma \prod_{j \in \tau}(1 + x_j) \in J_C$ if and only if (σ, τ) is an RF relationship (i.e., $U_\sigma \subseteq \bigcup_{j \in \tau} U_j$.).*

Moreover, $x_\sigma \prod_{j \in \tau}(1 + x_j) \in CF(J_C)$ if and only if (σ, τ) is a minimal RF relationship.

From Example 1, the minimal pseudo-monomial $x_2(1 + x_1)$ gives us the minimal relationship $U_2 \subseteq U_1$.

3.1 Gröbner Basis of a Neural Ideal

The canonical form $CF(J_C)$ is a particular generating set for J_C that gives information about the structure of the sets U_i. Another well-known generating set for a polynomial ideal is a Gröbner basis.

Given an ideal in a polynomial ring $R = k[x_1, \ldots, x_n]$ and a monomial ordering $<$ on R, we can let $LT_<(I)$ denote the ideal generated by the leading terms of elements in I. If G is a finite subset of I whose leading terms generate $LT_<(I)$, then G is a *Gröbner basis* for I. A Gröbner basis for I is always a generating set for the ideal I. A Gröbner basis G is *reduced* if, given any element $f \in G$, f has leading coefficient 1 and no term of f is divisible by the leading term of any $g \in G$ with $g \neq f$. We often also talk about *marked* reduced Gröbner bases to emphasize that the leading term of each polynomial in a Gröbner basis is distinguished. For a given monomial order $<$, the marked reduced Gröbner basis exists and is unique.

A *universal Gröbner basis* for an ideal I is a Gröbner basis that is a Gröbner basis with respect to any monomial order. *The universal Gröbner basis $\widehat{G}$ of an ideal I is the union of all reduced Gröbner bases of I*. Since the set of all reduced Gröbner bases is finite, the universal Gröbner basis always exists and is unique.

If a set is a Gröbner basis, it is not necessarily a reduced Gröbner basis nor a universal Gröbner basis. However, it was shown in [10] that if the canonical form is a Gröbner basis, then it is in fact the universal Gröbner basis for J_C. This result leads to the following proposition:

Proposition 2 ([10]) *Let C be a neural code with neural ideal J_C. The following are equivalent:*

1. *The canonical form of J_C is a Gröbner basis of J_C.*
2. *The canonical form of J_C is the universal Gröbner basis of J_C.*
3. *The universal Gröbner basis of J_C consists of pseudo-monomials.*

In particular, this gives a way to certify that the canonical form is not a Gröbner basis: If, for a given term order, the reduced Gröbner basis contains polynomials which are not pseudo-monomials, this implies that the canonical form is not a Gröbner basis.

The following proposition refines Proposition 2 by replacing its second statement with *"The canonical form of J_C has a unique marked reduced Gröbner basis."*

Proposition 3 *Let C be a code and J_C its neural ideal. $CF(J_C)$ is a Gröbner basis if and only if J_C has a unique marked reduced Gröbner basis.*

Proof In [7], it is shown that an ideal has a unique marked reduced Gröbner basis if and only if all marked reduced Gröbner basis generators are factor-closed, i.e., the non-leading terms of each polynomial divide its leading term. Furthermore, in [10] the authors prove that if the universal Gröbner basis of J_C consists solely of pseudo-monomials, then its canonical form is a Gröbner basis. Since over $\mathbb{F}_2$ all polynomials that are factor-closed and square-free are pseudo-monomials, the result follows.   □

Notice that by Proposition 3, the goal of classifying codes whose neural ideals have canonical forms that are Gröbner bases becomes identical to classifying codes whose ideals of points (or neural ideals) have unique marked reduced Gröbner basis. In Sect. 4 we present an efficient algorithm for testing whether a code has a neural ideal with a unique marked reduced Gröbner basis.

Lemma 2 *If there is a pseudo-monomial $f \in CF(J_C)$ whose leading term is divisible by any term of another pseudo-monomial $g \in CF(J_C)$, then the canonical form is not a Gröbner basis for J_C for any monomial order.*

Proof If $f \in CF(J_C)$ has leading term that is divisible by a term of another pseudo-monomial $g \in CF(J_C)$, then the canonical form cannot be a reduced Gröbner basis, which by Proposition 2 implies that it is not a Gröbner basis.   □

We will utilize this fact in the next subsection.

3.2   Canonical Form and Gröbner Bases of J_C

Recall from Sect. 2 that if a code has a local obstruction, then it is not convex. Since the canonical form $CF(J_C)$ encodes information about the minimal relationships between the sets U_i, the canonical form can be used to detect certain local obstructions in the code. The following definition was introduced in [4].

Definition 3 A local obstruction (σ, τ) is CF-*detectable* if there exists a local obstruction (σ', τ') with $\sigma' \subset \sigma$ and $\tau' \subset \tau$ such that (σ', τ') is a minimal RF relationship.

The next proposition connects the convexity of C to the Gröbner basis of J_C.

Proposition 4 *Given a code C, if C has a CF-detectable local obstruction, then the canonical form of J_C is not a Gröbner basis.*

Proof By Theorem 5.4 in [4], if C has a CF-detectable local obstruction, then there exist $\sigma, \tau \subset [n], \tau \neq \emptyset$ with $x_\sigma \prod_{i \in \tau}(1 + x_i) \in CF(J_C)$ and $x_\sigma x_\tau \in J_C$. Since $x_\sigma x_\tau$ is a pseudo-monomial in J_C and $CF(J_C)$ is a generating set for J_C, there exists $x_\alpha \in CF(J_C)$ with $\alpha \subset \sigma \cup \tau$, so the canonical form is not a Gröbner basis by Proposition 5. $\square$

Thus, if a code C has a CF-detectable local obstruction, C is both not convex and its canonical form is not a Gröbner basis for J_C.

Proposition 5 *Let C be a neural code with neural ideal J_C and canonical form $CF(J_C)$. If there exist two distinct pseudo-monomials $f = x_\sigma \prod_{i \in \tau}(1 + x_i)$ and $g = x_\alpha \prod_{j \in \beta}(1 + x_j) \in CF(J_C)$ with $\alpha \cup \beta \subseteq \sigma \cup \tau$, then the canonical form $CF(J_C)$ is not a Gröbner basis of J_C.*

Proof For any monomial order, the leading term of f is $x_\sigma x_\tau$ while the leading term of g is $x_\alpha x_\beta$. Since $\alpha \cup \beta \subseteq \sigma \cup \tau$ implies that $x_\alpha x_\beta$ divides $x_\sigma x_\tau$, by Lemma 2 we have that the canonical form is not a Gröbner basis. $\square$

Unfortunately, the converse of Proposition 5 fails as the following example shows.

Example 2 The code
$C = \{\emptyset, 1, 2, 3, 4, 5, 134, 1234, 234, 1235, 125, 13, 15, 23, 25, 14, 24, 235, 135, 1245, 35, 123, 12345\}$ has canonical form $CF(J_C) = \{x_3 x_4(1 + x_1)(1 + x_2), x_1 x_2(1 + x_3)(1 + x_5), x_4 x_5(1 + x_1), x_4 x_5(1 + x_2)\}$, with leading terms $x_1 x_2 x_3 x_4, x_1 x_2 x_3 x_5, x_1 x_4 x_5, x_2 x_4 x_5$, none of which are divisible by the others. However, the universal Gröbner basis of J_C has the polynomial $x_4(x_1 x_2 + x_1 x_3 + x_2 x_3 + x_3 x_4 + x_3 + x_5)$, which is not a pseudo-monomial. Thus, by Proposition 2, the canonical form of this code is not a Gröbner basis.

We do have the following partial converse to Proposition 5:

Proposition 6 *Let C be a neural code with canonical form $CF(J_C)$. If, for all minimal pseudomonomials $x_\sigma \prod_{i \in \tau}(1 + x_i)$ and $x_\alpha \prod_{j \in \beta}(1 + x_j) \in CF(J_C)$, we have $(\sigma \cup \tau) \cap (\alpha \cup \beta) = \emptyset$, then $CF(J_C)$ is a Gröbner basis for J_C.*

Proof Let $g = x_\sigma \prod_{i \in \tau}(1 + x_i)$ and $h = x_\alpha \prod_{j \in \beta}(1 + x_j) \in CF(J_C)$. Since the leading terms of g and h are $x_\sigma x_\tau$ and $x_\alpha x_\beta$ respectively, if $(\sigma \cup \tau) \cap (\alpha \cup \beta) = \emptyset$, then the leading terms of g and h are relatively prime. By Proposition 4 in [6], this guarantees that the S-polynomial of g and h has standard representation. Since this is true for any pair of pseudo-monomials, this shows that $CF(J_C)$ is a Gröbner basis for J_C by Theorem 3 in [6]. $\square$

Note that the hypothesis of Proposition 6 is not a necessary condition for the canonical form to be a Gröbner basis, as will be seen in Examples 3 and 4. We now give several examples of convex and non-convex codes with their canonical forms and universal Gröbner bases $\widehat{G}$. The labeling of the codes follow the classification given in [9].

Example 3 The code $C4 = \{2345, 123, 134, 145, 13, 14, 23, 34, 45, 3, 4, \emptyset\}$ is non-convex, non-max intersection complete, with no local obstructions (see [12]). It has canonical form $CF(C4) = \{x_5(1 + x_4), x_1 x_2 x_4, x_2 x_4(1 + x_5), x_2(1 + x_3), x_1 x_2 x_5, x_1 x_3 x_5, x_3 x_5(1 + x_2), x_1(1 + x_3)(1 + x_4)\}$. The universal Gröbner basis is $\widehat{G}(C4) = \{x_1x_2x_5, x_1x_2x_4, x_5(x_2 + x_3), x_5(1 + x_4), x_2(1 + x_3), x_1(1 + x_3)(1 + x_4), x_1x_3x_5, x_2x_4 + x_3x_5, x_2(x_4 + x_5)\}$.

It was shown in [12] that adding either the codeword 1 or the codewords 234 and 345 to $C4$ would make it convex. Upon adding 1, the universal Gröbner basis and the canonical form lose pseudo-monomials, but $\widehat{G}$ still does not equal the canonical form. Adding the codewords 234 and 345 instead makes the canonical form equal to the Gröbner basis: $CF = \{x_1x_2x_5, x_1x_2x_4, x_1x_3x_5, x_5(1 + x_4), x_2(1 + x_3), x_1(1 + x_3)(1 + x_4)\}$. Note that it is still not max-intersection complete.

Example 4 The code $C22 = \{145, 124, 135, 235, 125, 123, 234, 35, 1, 23, 15, 25, 5, 13, 2, 24, 3, 14, 12\}$ is convex with geometric realization in $\mathbb{R}^3$ and not max-intersection complete (see [9]). The universal Gröbner basis and the canonical form are the same: $CF(C22) = \{x_2x_4x_5, x_1x_2x_3x_5, x_3x_4(1 + x_2), x_3x_4x_5, x_4x_5(1 + x_1), x_4(1 + x_1)(1 + x_2)\}$.

4   Identifying Neural Codes with Unique Marked Reduced Gröbner Bases

Based on Proposition 3, the goal of classifying codes whose neural ideals have canonical forms that are Gröbner bases becomes identical to classifying codes whose ideals of points have unique marked reduced Gröbner basis. In this section we outline a method for testing whether a neural ideal has a unique marked reduced Gröbner basis. We begin with two relevant definitions from [1].

Definition 4 A *staircase* is a set $\lambda \subseteq \mathbb{N}^d$ of nonnegative integer vectors such that $u \leq v \in \lambda$ (coordinatewise) implies $u \in \lambda$. The staircase of exponent vectors of standard monomials of an ideal I is called an *initial* staircase.

Definition 5 A staircase λ is *basic* for an ideal I if the congruence classes modulo I of the monomials x^v with $v \in \lambda$ form a vector space basis for $\mathbb{Z}_p[x_1, \ldots, x_n]/I$.

As we will see in Proposition 7, if we want to find out whether $I(V)$ has a unique marked reduced Gröbner basis, we just need to check whether $I(V)$ has a unique basic staircase.

Definition 6 Given a staircase S on n variables and number of points m, let $\alpha_S = (\alpha_S^1, \cdots, \alpha_S^n)$ be an n-dimensional vector, where $\alpha_S^i = 0$ if S has zeros for all points in its ith direction. Otherwise $\alpha_S^i = 1$. We use $\sum \alpha_S$ to denote the summation of all entries in α_S, and call it the *dimension* of S.

Example 5 The following two examples illustrate the concept of staircase dimension which is needed for the algorithm at the end of this section.

1. Let $S = \{(0, 0), (0, 1), (0, 2), (0, 3)\}$. Then $\alpha_S = (0, 1)$ and $\sum \alpha_S = 1$.
2. If $S = \{(0, 0, 0), (0, 0, 1), (0, 1, 0), (1, 0, 0)\}$, then $\alpha_S = (1, 1, 1)$ and $\sum \alpha_S = 3$.

We now construct the following matrix. Let $\lambda = \{u^1, \ldots, u^r\}$ be an r-subset of $\mathbb{Z}_p^n$ and let $V = \{v^1, \ldots, v^s\}$ be an s-subset of $\mathbb{Z}_p^n$. The *evaluation matrix* $\mathbb{X}(x^\lambda, V)$ is the s-by-r matrix whose element in position (i, j) is $x^{u^j}(v^i)$, the evaluation of x^{u^j} at v^i.

Example 6 Let $\lambda_1 = \{(0, 0), (1, 0)\}$, $\lambda_2 = \{(0, 0), (0, 1)\}$, and $V = \{(2, 0), (0, 1)\}$ be subsets of $\mathbb{Z}_3^2$. Then $\mathbb{X}(x^{\lambda_1}, V) = \begin{bmatrix} 1 & 2 \\ 1 & 0 \end{bmatrix}$ and $\mathbb{X}(x^{\lambda_2}, V) = \begin{bmatrix} 1 & 0 \\ 1 & 1 \end{bmatrix}$.

Theorem 1 ([1]) *Let λ and V be subsets of $\mathbb{Z}_p^n$. Then λ is basic for $I(V)$ if and only if $\mathbb{X}(x^\lambda, V)$ is invertible.*

An initial staircase must be basic, while a basic staircase might not be initial; however, if $I(V)$ has a unique initial staircase (and thus a unique reduced Gröbner basis), then $I(V)$ has a unique basic staircase. The following lemma is found in [8] without proof.

Lemma 3 *Let x^α, x^β be monomials with $x^\alpha \nmid x^\beta$. There exists a weight vector γ and monomial order $\prec_\gamma$ such that $x^\beta \prec_\gamma x^\alpha$.*

Proof Let $x^\alpha \nmid x^\beta$. As $x^\alpha \nmid x^\beta$, $\alpha_j > \beta_j$ for some coordinate j. Take γ to be a vector in $\mathbb{R}^n$ with a sufficiently large rational value in entry j and square roots of distinct prime numbers elsewhere such that $\gamma \cdot \alpha > \gamma \cdot \beta$. Then the entries of γ are linearly independent over $\mathbb{Q}$ and so γ defines a weight order. Define $\prec_\gamma$ to be the monomial order weighted by γ. It follows that $x^\beta \prec_\gamma x^\alpha$. □

Proposition 7 ([8]) *An ideal $I(V)$ has a unique initial staircase if and only if $I(V)$ has a unique basic staircase.*

Proof Follows directly from Proposition 2.2 in [1] and Lemma 3. □

Based on Proposition 7, if we want to find out whether $I(V)$ has a unique marked reduced Gröbner basis, we just need to check if there exist a unique staircase $\lambda \subseteq \mathbb{Z}_p^n$ such that $\mathbb{X}(x^\lambda, V)$ is invertible.

The above paragraph is the basis of the following method we propose for identifying if a set of points has an ideal with a unique marked reduced Gröbner basis: Given a set of points V, the algorithm goes over all possible staircases with $|V|$ elements and checks if the corresponding evaluation matrix is invertible. Notice that no Gröbner basis computation is required. Unfortunately, finding all staircases is equivalent to the NP-complete integer partitioning problem [11] but there are pseudo-polynomial time dynamic programming solutions. For example, one can use the Sherman-Morrison formula [14]: Given an invertible matrix $A \in \mathbb{R}^{n \times n}$, and two column vectors $u, v \in \mathbb{R}^n$, $A + uv^T$ is invertible if and only if $1 + v^T A^{-1} u \neq 0$.

The following algorithm is based on the theory summarized in this section. Its goal is to identify data sets $V \subseteq \mathbb{Z}_p^n$ of fixed size, dimension, and finite field cardinality having an ideal with a unique marked reduced Gröbner basis. Before we present it, we need one last definition.

Definition 7 ([8]) For $V_1, V_2 \subset \mathbb{Z}_p^n$ with $|V_1| = |V_2|$, we say V_1 is a *linear shift* of V_2, if there exists $\phi = (\phi_1, \cdots, \phi_n) : \mathbb{Z}_p^n \to \mathbb{Z}_p^n$ such that $\phi(V_1) = V_2$ and for each $i \in \{1, \cdots, n\}$, $\phi_i(x_i) = a_i x_i + b_i : \mathbb{Z}_p \to \mathbb{Z}_p$ with $a_i \in (\mathbb{Z}_p \backslash \{0\})$ and $b_i \in \mathbb{Z}_p$.

The linear shift is a bijection between two data sets, defining an equivalence relation. We note that by a "good" representative of an equivalence class E we mean one of the data sets with smallest total Euclidean distance to the origin among all data sets in E.

4.1 Data Preparation

Input: n (dimension), p (characteristic of finite field), m (number of points in the data set)
Purpose: Prepare the data for use in the main iterations
Steps:

1. Generate all staircases $\{S\}$ and their corresponding dimensions $\{\alpha_S\}$.
2. For each S, calculate all evaluation matrices $\{\mathbb{X}(x^S, S)\}$ and their inverses $\{\mathbb{X}(x^S, S)^{-1}\}$.
   Note: Since $\{\mathbb{X}(x^S, S)\}$ is a square Vandermonde matrix and S is a set of distinct points, $\{\mathbb{X}(x^S, S)\}$ is invertible.
3. Find "good" representatives $\{E_\ell\}$, for all the equivalence classes.

Note: The number of staircases has an upper bound of $O(m(\log m)^{n-1})$ [1].

4.2 Main Iterations

Input: $\{S\}, \{\alpha_S\}, \{\mathbb{X}(x^S, S)\}, \{\mathbb{X}(x^S, S)^{-1}\}, \{E_\ell\}$.
Output: Good representatives of equivalence classes in which an ideal of the data sets have unique reduced Gröbner bases.

Create a list called storage to store all the previous results
for $\ell \in \{E_\ell\}$ **do**
   create an empty vector called flag = [ ]
   for $S \in \{S\}$ **do**
      if ℓ and S are only different in one point **then**
         compute $D = \mathbb{X}(x^S, \ell) - \mathbb{X}(x^S, S)$
         decompose $D = uv^T$, where $u, v \in \mathbb{F}_p^m$ are two column vectors
         if $1 + v^T \mathbb{X}(x^S, S)^{-1} u = 0 \in \mathbb{Z}_p$ **then**
            flag.append(False)
         else
            flag.append(True)
         end if

else if $\sum \alpha_S < n$ and storage has the result of ℓ' such that ℓ' have exactly
the same value of ℓ at non-zero entries in α_S **then**
    flag.append(the previous result)
else if $\det(\mathbb{X}(x^S, \ell)) \neq 0 \in \mathbb{Z}_p$ **then**
    flag.append(True)
else
    flag.append(False)
end if
if there are two Trues in flag **then**
    use storage to store flags
    break the inside loop
end if
end for
use storage to store flags
if flag has only one T **then**
    print ℓ
end if
end for

5 Discussion and Future Work

We explored convex neural codes by considering the canonical forms and Gröbner bases of their ideals. While we still do not have a complete algebraic characterization of convex codes, the results we presented lead us to believe that there is a strong connection between convexity of a code and the number of the marked reduced Gröbner bases of its ideal. In particular, it would seem that the relations among the U_i from Definition 1 cannot be too "contradictory" for the canonical form of a neural ideal to be a Gröbner basis. From the comparisons and computations of canonical forms and Gröbner bases for convex and non-convex codes thus far, the authors make the following conjecture to strengthen Proposition 4:

Conjecture 1 Given a neural code C with neural ideal J_C, if the canonical form $CF(J_C)$ is a Gröbner basis, then the code C is convex.

Notice that in light of Proposition 3, the above conjecture can also be stated as "Given a neural code C with neural ideal J_C, if J_C has a unique marked reduced Gröbner basis, then the code C is convex."

In addition, Section 4 of [10] gives three examples of families of codes whose canonical forms are Gröbner bases, which we can verify will always be convex codes, thus further suggesting that Conjecture 1 is worth future work:

1. C is a simplicial complex: then C is intersection complete, so C is convex.
2. C is the singleton $C = \{(c_1, \ldots, c_n)\}$. Then $U_i = X$ for $c_i = 1$, and $U_j = \emptyset$ for $c_j = 0$. If X is chosen to be convex, then the code will be convex.

3. C is missing one codeword from $[n]$. If $11 \cdots 1 \in C$, then C is convex (see [3]).
   If $C = \{0, 1\}^n \setminus \{11 \cdots 1\}$, then C is a simplicial complex, which is convex by
   (1).

In [8] we characterize geometrically a family of codes whose ideals have a unique marked reduced Gröbner basis and the codes above are in that family. By Proposition 3, the above conjecture would imply that all codes in the family are convex which remains to be verified. Furthermore, in [7], we show that if the neural ideal of a code has a unique marked reduced Gröbner basis, so does the neural ideal of its complement. It remains to be verified if convex codes whose neural ideals have unique marked reduced Gröbner bases always have convex complements.

Acknowledgements This work was supported by the National Science Foundation under Awards DMS-1419038 and DMS-1419023. The authors thank the anonymous reviewers for the thoughtful comments. E.S. Dimitrova thanks Brandilyn Stigler for many productive discussions.

References

1. E. Babson, S. Onn, and R. Thomas. The Hilbert zonotope and a polynomial time algorithm for universal Gröbner bases. *Advances in Applied Mathematics*, 30(3):529–544, 2003.
2. J. Cruz, C. Giusti, V. Itskov, and Bill Kronholm. On open and closed convex codes. *Discrete & Computational Geometry*, 61(2):247–270, 2019.
3. C. Curto, E. Gross, Jack Jeffries, K. Morrison, M. Omar, Z. Rosen, A. Shiu, and N. Youngs. What makes a neural code convex? *SIAM Journal on Applied Algebra and Geometry*, 1:222–238, 2017.
4. C. Curto, E. Gross, J. Jeffries, K. Morrison, Z. Rosen, A. Shiu, and N. Youngs. Algebraic signatures of convex and non-convex codes. *Journal of Pure and Applied Algebra*, 223(9):3919–3940, 2019.
5. C. Curto, V. Itskov, A. Veliz-Cuba, and N. Youngs. The neural ring: an algebraic tool for analyzing the intrinsic structure of neural codes. *Bulletin of Mathematical Biology*, 75:1571–1611, 2013.
6. D. A. Cox, J. Little, and D. O'Shea. *Ideals, Varieties, and Algorithms*. Springer, New York City, 4 edition, 2015.
7. E. S. Dimitrova, Q. He, L. Robbiano, and B. Stigler. Small Gröbner fans of ideals of points. *Journal of Algebra and Its Applications*, 2019.
8. E. S. Dimitrova, Q. He, B. Stigler, and A. Zhang. Geometric characterization of data sets with unique reduced Gröbner bases. *Bulletin of Mathematical Biology*, 2019.
9. S. A. Goldrup and K. Phillipson. Classification of open and closed convex codes on five neurons, 2019.
10. R. Garcia, L. D. García Puente, R. Kruse, J. Liu, D. Miyata, E. Petersen, K. Phillipson, and A. Shiu. Gröbner bases of neural ideals. *International Journal of Algebra and Computation*, 28(4):553–571, 2018.
11. B. Hayes. Computing science: The easiest hard problem. *American Scientist*, 90(2):113–117, 2002.
12. C. Lienkaemper, A. Shiu, and Z. Woodstock. Obstructions to convexity in neural codes. *Advances in Applied Mathematics*, 85:31–59, 2017.
13. E. Petersen, N. Youngs, R. Kruse, D. Miyata, R. Garcia, and L. D. García Puente. Neural ideals in sagemath, 2016.
14. J. Sherman and W. J. Morrison. Adjustment of an inverse matrix corresponding to a change in one element of a given matrix. *Ann. Math. Statist.*, 21(1):124–127, 03 1950.

The Number of Gröbner Bases in Finite Fields (Research)

Brandilyn Stigler and Anyu Zhang

1 Introduction

Polynomial systems are ubiquitous across the sciences. While linear approximations are often desired for computational and analytic feasibility, certain problems may not permit such reductions. In 1965 Bruno Buchberger introduced Gröbner bases, which are multivariate nonlinear generalizations of echelon forms [3, 5]. Since this landmark thesis, the adoption of Gröbner bases has expanded into diverse fields, such as geometry [24], image processing [18], oil production [23], quantum field theory [20], and systems biology [17].

While working with a Gröbner basis (GB) of a system of polynomial equations is just as natural as working with a triangularization of a linear system, their complexity can make them cumbersome with which to work: for a general system, the complexity of Buchberger's Algorithm is doubly exponential in the number of variables [4]. The complexity improves in certain settings, such as systems with finitely many real-valued solutions ([6] is a classic example, whereas [12] is a more contemporary example), or solutions over finite fields [15]. Indeed much research has been devoted to improving Buchberger's Algorithm and analyzing the complexity and memory usage in more specialized settings (for example, [11, 19]), and even going beyond traditional ways of working with Gröbner bases [16]; however most results are for characteristic-0 fields, such $\mathbb{R}$ or $\mathbb{Q}$.

The goal of our work is to consider the *number* of Gröbner bases for a system of polynomial equations over a finite field (which has positive characteristic and consequently all systems have finitely many solutions). The motivation comes from the work of [17], in which the authors presented an algorithm to reverse

B. Stigler (✉) · A. Zhang
Southern Methodist University, Dallas, TX, USA
e-mail: bstigler@smu.edu; anyuz@smu.edu

© The Author(s) and the Association for Women in Mathematics 2020
B. Acu et al. (eds.), *Advances in Mathematical Sciences*, Association for
Women in Mathematics Series 21, https://doi.org/10.1007/978-3-030-42687-3_9

engineer a model for a biological network from discretized experimental data and made a connection between the number of distinct *reduced* GBs and the number of (possibly) distinct *minimal* polynomial models. The number of reduced GBs associated to a data set gives a quantitative measure for how "underdetermined" the problem of reverse engineering a model for the underlying biological system is.

The Gröbner fan geometrically encapsulates all reduced Gröbner bases [21]. In [13] the authors provided an algorithm to compute all reduced GBs. When their number is too large for enumeration, the method in [9] allows one to sample from the fan. Finally in [22], the authors provide an upper bound for systems with finitely many solutions; however this bound is much too large for data over a finite field. To our knowledge, there is no closed form for the number of reduced Gröbner bases, in particular for systems over finite fields with finitely many solutions.

In this paper we make the following contributions:

1. a formula and some upper bounds of the number of reduced Gröbner bases for data sets over finite fields
2. geometric characterization of data associated with different numbers of reduced Gröbner bases.

In Sect. 2, we provide the relevant background, definitions, and results. In Sect. 3, we discuss the connection between the number of distinct reduced Gröbner bases for ideals of two points and the geometry of the points; furthermore, we provide a formula to two-point data sets. We provide upper bounds for data sets of three points in Sect. 4 and geometric observations for larger sets in Sect. 5. Then in Sect. 6, we consider the general setting of any fixed number of points over any finite field and provide an upper bound. We close with a discussion of possible future directions. We have verified all of the computations referenced in this work, provided illustrative examples throughout the text, and listed data tables in the Appendix.

2 Background

2.1 Algebraic Geometry Preliminaries

Let K be a field and let $R = K[x_1, \ldots, x_n]$ be a polynomial ring over K. Most definitions and known results in this section can be found in [8].

A *monomial order* $\prec$ is a total order on the set of all monomials in R that is closed with respect to multiplication and is a well-ordering. The *leading term* of a polynomial $g \in R$ is thus the largest monomial for the chosen monomial ordering, denoted as $LT_\prec(g)$. Also we call $LT_\prec(I) = \langle LT_\prec(g) : g \in I \rangle$ the *leading term ideal* for an ideal I.

Definition 1 Let $\prec$ be a monomial order on R and let I be an ideal in R. Then $G \subset I$ is a *Gröbner basis* for I with respect to $\prec$ if for all $f \in I$ there exists $g \in G$ such that the leading term $LT_\prec(g)$ divides $LT_\prec(f)$.

It is well known that Gröbner bases exist for every $\prec$ and make multivariate polynomial division well defined in that remainders are unique; for example, see [8]. While there are infinitely many orders, there are only finitely many reduced GBs for a given ideal, that is monic polynomials whose leading terms do not divide other terms. This results in an equivalence relation where the leading terms of the representative of each equivalence class can be distinguished (underlined) [21]. In fact there is a one-to-one correspondence between *marked* reduced Gröbner bases and leading term ideals [7].

In this work all Gröbner bases are reduced.

Definition 2 The monomials which do not lie in $LT_\prec(I)$ are *standard* with respect to $\prec$; the set of standard monomials for an ideal I is denoted by $SM_\prec(I)$.

A set of standard monomials $SM_\prec(I)$ for a given monomial order forms a basis for R/I as a vector space over K. Given their construction, it follows that the sets of standard monomials associated to an ideal I are in bijection with the leading term ideals of I.

It is straightforward to check that standard monomials satisfy the following divisibility property: if $x^\alpha \in SM_\prec(I)$ and x^β divides x^α, then $x^\beta \in SM_\prec(I)$. This divisibility property on monomials is equivalent to the following geometric condition on lattice points.

Definition 3 A set $\lambda \subset \mathbb{N}^n$ is a *staircase* if for all $u \in \lambda$, $v \in \mathbb{N}^n$ and $v_i \leq u_i$ for $1 \leq i \leq n$ imply $v \in \lambda$.

Let $\binom{\mathbb{N}^n}{m}$ denote the collection of all sets of m points in $\mathbb{N}^n$. Then for $\lambda = \{\lambda_1, \ldots, \lambda_m\} \in \binom{\mathbb{N}^n}{m}$, let $\sum \lambda$ denote the vector sum $\sum_{i=1}^m \lambda_i \in \mathbb{N}^n$. Let Λ denote the set of all staircases in $\binom{\mathbb{N}^n}{m}$. The *staircase polytope* of Λ is the convex hull of all points $\sum \lambda$ where $\lambda \in \Lambda$ (see [2, 22] for more details). For an ideal I, we call $\mathcal{P}$ the *staircase polytope of I* if $\mathcal{P}$ is the staircase polytope of the exponent vectors of the standard monomial sets associated to I for any monomial order.

For $S \subseteq K^n$, we call the set $I(S) := \{h \in R \mid h(s) = 0 \,\forall s \in S\}$ of polynomials that vanish on S an *ideal of points*. An ideal is *zero dimensional* if $\dim_K R/I < \infty$; when K is algebraically closed and $|S| = m < \infty$, then $m = \dim_K R/I(S)$. The number of reduced Gröbner bases for an ideal is in bijection with the number of vertices of the staircase polytope, which was proved for ideals of points in [22] and for all other zero-dimensional ideals in [2].

The following results provide an upper bound for the number of reduced Gröbner bases for an ideal over any field.

Lemma 1 ([1]) *The number of vertices of a lattice polytope $P \subset \mathbb{R}^n$ is $\#vert(P) = O\left(vol(P)^{(n-1)/(n+1)}\right)$.*

Theorem 1 ([2, 22]) *Let I be an ideal such that $\dim_K R/I = m$. Let $\Lambda(I)$ be the set of standard monomial sets for I over all monomial orders. Then the number of* *distinct reduced Gröbner bases of I is in bijection with the number of vertices of the staircase polytope of I; that is,* $\#GBs = O\left(m^{2n\frac{n-1}{n+1}}\right)$.

Example 1 Let $S = \{(1, 1), (2, 3), (3, 5), (4, 6)\} \subset \mathbb{R}^2$. So $\dim_{\mathbb{R}} \mathbb{R}[x, y]/I(S) = 4$. Also $\Lambda(I(S)) = \{(1, x, x^2, x^3), (1, x, x^2, y), (1, x, y, y^2), (1, y, y^2, y^3)\}$. So the number of reduced Gröbner bases for $I(S)$ is four. Note that there are five staircases in $\binom{\mathbb{N}^2}{4}$, namely $\Lambda = \{\{(0, 0), (1, 0), (2, 0), (3, 0)\}, \{(0, 0), (1, 0), (2, 0), (0, 1)\}, \{(0, 0), (1, 0), (0, 1), (1, 1)\}, \{(0, 0), (1, 0), (0, 1), (0, 2)\}, \{(0, 0), (0, 1), (0, 2), (0, 3)\}\}$. The staircase polytope of Λ is the convex hull of the vector sums $\{(6,0), (3,1), (2,2), (1,3), (0,6)\}$, which has vertices $(6,0)$, $(3,1)$, $(1,3)$, and $(0,6)$, corresponding to the four standard monomial sets of $I(S)$.

Now we summarize the bijective correspondences for the number of reduced Gröbner bases for an ideal of points.

Theorem 2 *Let I be an ideal. There is a one-to-one correspondence among the following:*

1. *distinct marked reduced Gröbner bases of I*
2. *leading term ideals of I*
3. *sets of standard monomials for I*
4. *vertices of the staircase polytope of I.*

Proof Equivalence $1 \iff 2$ is a result in [7]; $2 \iff 3$ is by construction of standard monomials; and $1 \iff 4$ was proved in [22] for ideals of points and in [2] for other zero-dimensional ideals.   □

2.2 Ideals Over Finite Fields

In this section and following, we will work over a finite base field. Let F be a finite field of characteristic $p > 0$. We will typically consider the finite field $\mathbb{Z}_p = \{0, 1, \ldots, p - 1\}$, that is the field of remainders of integers upon division by p with modulo-p addition and multiplication. Let $R = F[x_1, \ldots, x_n]$ be a polynomial ring over F. Finally let m denote the number of points in a subset of F^n.

A *polynomial dynamical system* (PDS) over F is a function $f = (f_1, \ldots, f_n) : F^n \to F^n$ where each component f_i is a polynomial in R. Below is an algorithm, first introduced in [17], to compute a PDS from a given set of data written using the ideal of the input points. This algorithm motivates the leading question in this work.

The general strategy is given input-output data $V = \{(s_1, t_1), \ldots, (s_m, t_m)\} \subset F^n \times F^n$, find all PDSs that fit V and select a minimal PDS with respect to polynomial division. This is done as follows. For each x_j, compute one interpolating function $f_j \in R$ such that $f_j(s_i) = t_{ij}$; note that $s_i \in F^n$ while $t_{ij} \in F$. Then compute the ideal $I := I(\{s_1, \ldots, s_m\})$ of the inputs in V. The *model space* for V is the set

$$f + I := \{(f_1 + h_1, \dots, f_n + h_n) : h_i \in I\}$$

of all PDSs which fit the data in V and where $f = (f_1, \dots, f_n)$ is as computed above. A PDS can be selected from $f + I$ by choosing a monomial order $\prec$, computing a Gröbner basis G for I, and then computing the remainder (*normal form*) $\overline{f}^G$ of each f_i by dividing f_i by the polynomials in G. We call

$$(\overline{f_1}^G, \overline{f_2}^G, \dots, \overline{f_n}^G)$$

the *minimal* PDS with respect to $\prec$, where G is a Gröbner basis for I with respect to $\prec$.

Changing the monomial order may change the resulting minimal PDS. While it is possible for two reduced Gröbner bases to give rise to the same normal form (see [17]), it is still the case that in general a set of data points may have *many* GBs associated to it. In this way, the number of distinct reduced GBs of I gives an upper bound for the number of different minimal PDSs. Therefore, we aim to find the number of distinct reduced Gröbner bases for a given data set.

Example 2 Consider two inputs $S = \{(0, 0), (1, 1)\} \subset (\mathbb{Z}_2)^2$. The corresponding ideal I of the points in S has 2 distinct reduced Gröbner bases, namely

$$G_1 = \{\underline{x_1} - x_2, \underline{x_2^2} - x_2\}, G_2 = \{\underline{x_2} - x_1, \underline{x_1^2} - x_1\}$$

Here, '_' marks the leading terms of the polynomials in a Gröbner basis. There are two resulting minimal models: any minimal PDS with respect to G_1 will be in terms of x_2 only as all x_1's are divided out, while any minimal PDS with respect to G_2 will be in terms of x_1 only as all x_2's are divided out. Instead if the inputs are $\{(0, 0), (0, 1)\}$, then I has a unique GB, $\{x_2^2 - x_2, \underline{x_1}\}$, resulting in a unique minimal PDS.

It is the polynomial $g = x_1 - x_2$ that has different leading terms for different monomial orders. In fact, for monomial orders with $x_1 \succ x_2$, the leading term of g is x_1, while for orders with $x_2 \succ x_1$ the opposite will be true. We say that g has *ambiguous* leading terms. We will mark only ambiguous leading terms.

As the elements of the quotient ring R/I are equivalence classes of functions defined over the inputs $S = \{s_1, \dots s_m\}$ in V and since a set of standard monomials is a basis for R/I, it follows that each reduced polynomial $\overline{f}^G$ is written in terms of standard monomials. When working over a finite field, extensions of classic results in algebraic geometry state that when the number m of input points is finite, then m coincides with the dimension of the vector space $R/I(S)$ over F [14], which is stated below for convenience.

Theorem 3 ([14]) *Let $S \subseteq \mathbb{F}^n$ and $I(S)$ be the ideal of the points in S. Then $|S| = \dim_F R/I(S)$.*

Next we state a result about data sets and their complements.

Theorem 4 ([10]) *Let I be the ideal of input points S, and let I^c be ideal of the complement $F^n \setminus S$ of S. Then we have $SM_{\prec}(I) = SM_{\prec}(I^c)$ and $LT_{\prec}(I) = LT_{\prec}(I^c)$ for a given monomial order $\prec$. Hence, we have $\#GB(S) = \#GB(F^n \setminus S)$.*

We say that a polynomial $f \in R$ is *factor closed* if every monomial $m \in supp(f)$ is divisible by all monomials in $supp(f)$ smaller than m with respect to an order $\prec$. The following result gives an algebraic description of ideals with unique reduced Gröbner bases for any monomial order.

Theorem 5 ([10]) *A reduced Gröbner basis G with factor-closed generators is reduced for every monomial order; that is, G is the unique reduced Gröbner basis for its corresponding ideal.*

We end this section with a discussion on the number of distinct reduced Gröbner bases for extreme cases. The set $\mathbb{Z}_p^n$ contains p^n points. For $n = 1$, all ideals have a unique reduced GB since all polynomials are single-variate and as such are factor closed. We consider cases for $n > 1$. For empty sets or singletons in $\mathbb{Z}_p^n$, it is straightforward to show that the ideal of points has a unique reduced GB for any monomial order; that is, for a point $s = (s_1, \ldots, s_n)$, the ideal of s is $I = \langle x_1 - s_1, \ldots, x_n - s_n \rangle$ whose generators form a Gröbner basis and hence is unique (via Theorem 5). According to Theorem 4, the same applies to $p^n - 1$ points. In the rest of this work, we consider the number of reduced Gröbner bases for an increasing number of points.

Note that over a finite field, the relation $x^p - x$ always holds.

3   Data Sets with $m = 2$ Points

In this section we consider bounds for the number of Gröbner bases for ideals of two points and relate the geometry of the points to these numbers.

Define $NGB(p, n, m)$ to be the number of reduced Gröbner bases for ideals of m points in $\mathbb{Z}_p^n$. The following theorem provides a formula for sets with $m = 2$ points in any number of coordinates and over any finite field $\mathbb{Z}_p$.

Theorem 6 *Let $P = (p_1, \ldots, p_n), Q = (q_1, \ldots, q_n) \in \mathbb{Z}_p^n$ where $P \neq Q$, and let $I \subset \mathbb{Z}_p[x_1, \ldots, x_n]$ be the ideal of the points P, Q. The number of distinct reduced Gröbner bases for I is given by*

$$NGB(p, n, 2) = \sum_{\substack{p_i \neq q_i \\ i=1,\ldots,n}} 1.$$

Proof Let $S = \{P, Q\} \subset \mathbb{Z}_p^n$ with $P = (p_1, \ldots, p_n), Q = (q_1, \ldots, q_n)$. Let $I \subset \mathbb{Z}_p[x_1, \ldots, x_n]$ be the ideal of the points in S. By Theorem 3, the number of elements of any set of standard monomials for I is $|S| = 2$. Since sets of standard monomials must be closed under division, the only option for such a set is $\{1, x_i\}$

for some $i = 1, \ldots, n$. So the possible associated minimally generated leading term ideals are of the form $\langle x_1, \ldots, x_{i-1}, x_i^2, x_{i+1}, \ldots, x_n \rangle$. We consider the number of leading terms ideals in regards to the number of coordinate changes between the points.

If P and Q have one different coordinate, say $p_1 \neq q_1$, then the only possible minimal generating set for the leading term ideal of I is $\{x_1^2, x_2, \ldots, x_n\}$. If P, Q have two different coordinates, say $p_i \neq q_i$ for $i = 1, 2$, then the possible minimal generating sets for the leading term ideal of I are $\{x_1^2, x_2, \ldots, x_n\}$ when $x_1 \prec x_2$ and $\{x_1, x_2^2, x_3, \ldots, x_n\}$ when $x_2 \prec x_1$. Increasing the number of coordinate changes will add another leading term ideal. In general, if $p_i \neq q_i$ for $i = 1, \ldots, k$ where $k \leq n$, then the possible minimal generating sets for the leading term ideal of I are as follows:

1. $\{x_1^2, x_2, \ldots, x_n\}$ when x_1 is the smallest variable in the monomial order among $x_1, \ldots, x_k$
2. $\{x_1, x_2^2, x_3, \ldots, x_n\}$ when x_2 is smallest among $x_1, \ldots, x_k$

   $\vdots$

k. $\{x_1, \ldots, x_{k-1}, x_k^2, x_{k+1}, \ldots, x_n\}$ when x_k is smallest among $x_1, \ldots, x_k$.

$\square$

Corollary 1 *The maximum number of distinct reduced Gröbner bases for an ideal of two points in $\mathbb{Z}_p^n$ is $NGB(p, n, 2) \leq n$.*

With different choices of smallest coordinate, there are up to n different sets of standard monomials, each corresponding to a distinct reduced Gröbner basis. So, there are up to n reduced Gröbner bases, with the maximum achieved by two points with no coordinates in common.

In applications, modeling is often driven by data. So geometric descriptions of data sets can reveal essential features in the underlying network. We illustrate the above results by considering different configurations of points. We begin with Boolean data.

Example 3 Consider two points in $\mathbb{Z}_2^2$. The left graph in Fig. 1 is the plot of all points in $\mathbb{Z}_2^2$. By decomposing the 2-square on which they lie, we find that pairs of points that lie along horizontal lines have unique reduced Gröbner bases for any monomial order; see Fig. 2. For example, $\{(0, 0), (0, 1)\}$ has ideal of points $\langle x_1, x_2^2 - x_2 \rangle$. By Theorem 5 we see that the generators of I form a unique reduced GB. Similarly

Fig. 1 The lattice of points in $\mathbb{Z}_2^2$ (left), in $\mathbb{Z}_2^3$ (center), and in $\mathbb{Z}_3^2$ (right)

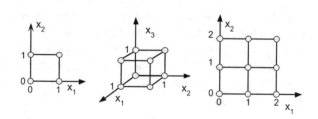

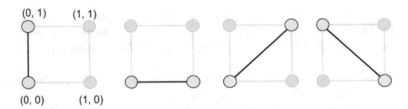

Fig. 2 Four configurations of pairs of points in $\mathbb{Z}_2^2$. From left to right: $\{(1,0),(0,1)\}$ and $\{(0,0),(1,0)\}$ each have 1 GB, while $\{(0,0),(1,1)\}$ and $\{(1,0),(0,1)\}$ have 2 distinct GBs

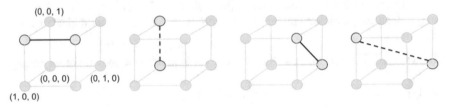

Fig. 3 Four configurations of pairs of points in $\mathbb{Z}_2^3$. From left to right: $\{(1,0,1),(1,1,1)\}$ and $\{(0,0,0),(0,0,1)\}$ have 1 GB; $\{(1,1,1),(0,1,0)\}$ has 2 GBs; and $\{(1,0,1),(0,1,0)\}$ has 3 GBs

$\{(1,0),(1,1)\}$ has ideal of points $\langle x_1 - 1, x_2^2 - x_2 \rangle$, which also has a unique reduced GB. Note that while they have different GBs, they have the same leading term ideal, namely, $\langle x_1, x_2^2 \rangle$. In the same way, pairs of points that lie along vertical lines have unique reduced GBs: sets $\{(0,0),(1,0)\}$ and $\{(0,1),(1,1)\}$ have the unique leading term ideal $\langle x_1^2, x_2 \rangle$. In each case, these sets have points with one coordinate change.

On the other hand, pairs of points that lie on diagonals have 2 distinct reduced Gröbner bases as such points have two coordinate changes. For example, the set of points $\{(0,0),(1,1)\}$ has GBs $\{\underline{x_1} - x_2, x_2^2 - x_2\}$ and $\{x_1^2 - x_1, \underline{x_2} - x_1\}$ with leading term ideals $\langle x_1, x_2^2 \rangle$ and $\langle x_1^2, x_2 \rangle$ respectively. Similarly the set $\{(0,1),(1,0)\}$ has $\{\underline{x_1} - x_2 - 1, x_2^2 - x_2\}$ and $\{x_1^2 - x_1, \underline{x_2} - x_1 - 1\}$ as Gröbner bases with leading term ideals $\langle x_1, x_2^2 \rangle$ and $\langle x_1^2, x_2 \rangle$ respectively.

Example 4 Now consider two points in $\mathbb{Z}_2^3$. The center graph in Fig. 1 is the plot of all points in $\mathbb{Z}_2^3$. In Fig. 3, pairs of points that lie on edges of the 3-cube have 1 reduced Gröbner basis, as the points have one coordinate change: for example the set $\{(1,0,1),(1,1,1)\}$ (first from the left in Fig. 3) has the unique reduced GB $\{x_1 - 1, x_2^2 - x_2, x_3 - 1\}$ and $\{(0,0,0),(0,0,1)\}$ (second) has the unique GB $\{x_1, x_2, x_3^2 - x_3\}$. Points that lie on faces of the 3-cube have 2 GBs as they have 2 coordinate changes: the third set $\{(1,1,1),(0,1,0)\}$ in Fig. 3 has GBs $\{\underline{x_1} - x_3, x_2 - 1, x_3^2 - x_3\}$ and $\{x_1^2 - x_1, x_2 - 1, \underline{x_3} - x_1\}$. Finally points that lie on lines through the interior have 3 GBs as they have 3 coordinate changes: the fourth set $\{(1,0,1),(0,1,0)\}$ has GBs $\{\underline{x_1} - x_3, \underline{x_2} - x_3 - 1, x_3^2 - x_3\}$, $\{\underline{x_1} - x_2 - 1, x_2^2 - x_2, \underline{x_3} - x_2 - 1\}$, and $\{x_1^2 - x_1, \underline{x_2} - x_1 - 1, \underline{x_3} + x_1\}$.

Fig. 4 Three configurations of points in $\mathbb{Z}_3^2$. From left to right: $\{(0, 0), (0, 2)\}$ has 1 GB, while $\{(1, 2), (2, 1)\}$ and $\{(0, 2), (1, 0)\}$ each have 2 distinct GBs

Next we consider data over the field $\mathbb{Z}_3$.

Example 5 Let $p = 3$ and $n = 2$. The right graph in Fig. 1 is the plot of all points in $\mathbb{Z}_3^2$. Similar to the Boolean case in Fig. 2, pairs of points that lie on horizontal or vertical lines have one associated reduced Gröbner basis for any monomial order, while pairs of points that lie on any skew line have two distinct GBs. For example, the set $\{(0, 0), (0, 2)\}$ in Fig. 4 has ideal of points $\langle x_1, x_2^2 + x_2 \rangle$, which has a unique reduced Gröbner basis via Theorem 5. On the other hand, the set of points $\{(1, 2), (2, 1)\}$ has two GBs, namely $\{\underline{x_1} + x_2, x_2^2 + 1\}$ and $\{x_1^2 - 1, \underline{x_2} + x_1\}$ with leading term ideals $\langle x_1, x_2^2 \rangle$ and $\langle x_1^2, x_2 \rangle$ respectively.

In the case of $m = 2$ points, we see that data that lie on horizontal or vertical edges have ideals of points with unique Gröbner bases, that is unique models, while data whose coordinates change simultaneously have multiple models associated with them. Though the number n of coordinates impacts the number of resulting models, the field cardinality p does not.

4 Data Sets with $m = 3$ Points

Theorem 7 *The number of distinct reduced Gröbner bases for ideals of three points in $\mathbb{Z}_p^n$ is*

$$NGB(p, n, 3) \leq \begin{cases} \frac{n(n-1)}{2} & \text{for } p = 2 \\ \frac{n(n+1)}{2} & \text{for } p \geq 3. \end{cases}$$

Proof We begin by considering the Boolean base field. By Theorem 3, the form of a set of standard monomials for an ideal of three points is $\{1, x_i, x_j\}$ for $x_i \neq x_j$. Considering the choice of x_i and x_j, there are up to $\frac{n(n-1)}{2}$ different standard monomial sets, each corresponding to a distinct reduced Gröbner basis by Theorem 2.

For a base field with $p > 2$, the two possible forms of standard monomial sets are $\{1, x_i, x_j\}$ for $x_i \neq x_j$, and $\{1, x_i, x_i^2\}$. As we showed above, there are up to $\frac{n(n-1)}{2}$ distinct reduced Gröbner bases corresponding to $\{1, x_i, x_j\}$. Further, the maximum number for the standard monomial form $\{1, x_i, x_i^2\}$ is n. As the two standard monomial forms can both be associated to the same data set, the upper bound for a non-Boolean field is $\frac{n(n-1)}{2} + n = \frac{n(n+1)}{2}$.   □

Example 6 Let $p = 2$ and $n = 2$. Then $NGB(2, 2, 3) \leq 1$; that is, all ideals of three points in $\mathbb{Z}_2^2$ have a unique reduced Gröbner basis, which is corroborated by Theorem 4 and the fact that ideals of a single point have only one distinct Gröbner basis for any monomial order.

Unlike the bound for two points, there are sets of three points for which the upper bound is not sharp. For example when $n = 4$, the upper bound is $NGB(2, 4, 3) \leq 6$; however the maximum number is 5, which we tested exhaustively (data not shown).

Next we connect configurations of three points to the number of associated Gröbner bases. We start with Boolean data.

Example 7 Let $p = 2$ and $n = 3$. In this case, $NGB(2, 3, 3) \leq 3$. Consider the configurations of points in $\mathbb{Z}_2^3$ in Fig. 5. The data set corresponding to the green triangle on the top "lid" of the leftmost 3-cube is $S_1 = \{(0, 0, 1), (0, 1, 1), (1, 0, 1)\}$ and its ideal of points has a unique Gröbner basis, namely $\{x_2^2 + x_2, x_3 + 1, x_1 x_2, x_1^2 + x_1\}$. The data set corresponding to the pink triangle in the center 3-cube is $S_2 = \{(0, 0, 1), (0, 1, 1), (1, 1, 0)\}$ and has two distinct associated GBs, with ambiguous leading terms distinguished:

$$\{x_3^2 + x_3, x_2 x_3 + x_2 + x_3 + 1, x_2^2 + x_2, \underline{x_1} + x_3 + 1\}, \{x_1 + \underline{x_3} + 1, x_2^2 + x_2, x_1 x_2 + x_1, x_1^2 + x_1\}.$$

Finally the data set corresponding to the red triangle in the rightmost 3-cube is $S_3 = \{(1, 0, 0), (0, 1, 0), (1, 1, 1)\}$ and has three GBs:

$$\{x_3^2 + x_3, x_2 x_3 + x_3, x_2^2 + x_2, \underline{x_1} + x_2 + x_3 + 1\},$$

$$\{x_3^2 + x_3, x_1 + \underline{x_2} + x_3 + 1, x_1 x_3 + x_3, x_1^2 + x_1\},$$

$$\{x_1 + x_2 + \underline{x_3} + 1, x_2^2 + x_2, x_1 x_2 + x_1 + x_2 + 1, x_1^2 + x_1\}.$$

The example illustrates that points that lie on faces of the 3-cube have 1 Gröbner basis; points forming a triangle which lies in the interior with 2 collinear vertices have 2 distinct GBs, and points in other configurations have 3 GBs.

Now we consider data in $\mathbb{Z}_3$.

Example 8 Let $p = 3$ and $n = 2$. By Theorem 7, we have that $NGB(3, 2, 3) \leq 3$. Consider the point configurations in Fig. 6. The data set corresponding to the green triangle (left) is $S_1 = \{(0, 0), (0, 1), (1, 1)\}$ and has a unique reduced Gröbner basis: $\{x_2^2 - x_2, x_1 x_2 - x_1, x_1^2 - x_1\}$. The data set corresponding to the pink triangle (center) is $S_2 = \{(0, 1), (1, 2), (2, 0)\}$ and has two distinct associated reduced GBs:

$$\{x_2^3 - x_2, \underline{x_1} - x_2 + 1\}, \{-x_1 + \underline{x_2} - 1, x_1^3 - x_1\}.$$

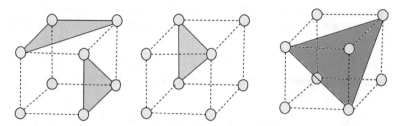

Fig. 5 Configurations of sets of three points in $\mathbb{Z}_2^3$ corresponding to different numbers of GBs. Points that are in configurations similar to the green triangles (left) have a unique reduced Gröbner basis for any monomial order; the pink triangle (center) has two distinct GBs; and the red triangle (right) has three distinct GBs

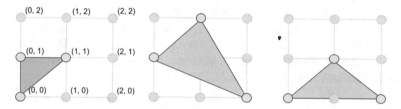

Fig. 6 Configurations of sets of three points in $\mathbb{Z}_3^2$ corresponding to unique and non-unique Gröbner bases. Points that are in configurations similar to the green triangle (left) have a unique reduced Gröbner basis for any monomial order; the pink triangles (center and right) have two distinct GBs

The data set corresponding to the pink triangle (right) is $S_3 = \{(0, 1), (1, 2), (2, 0)\}$ and also has two GBs:

$$\{x_2^3-x_2, x_1x_2^2-x_1x_2+x_2^2-x_2, x_1^2-x_1x_2+x_1-x_2\}, \{x_2^3-x_2, -x_1^2+x_1x_2-x_1+x_2, x_1^3-x_1\}.$$

Using Fig. 6, we see that three points that are collinear or have two adjacent collinear points have unique Gröbner bases, while other configurations result in 2 distinct ones. There are no data sets of three points in $\mathbb{Z}_3^2$ that have 3 associated Gröbner bases which we verified exhaustively (data not shown). Therefore the upper bound in Theorem 7 is not sharp for $p = 3, n = 2$.

5 Geometric Observations for Larger Sets

In this section, we offer empirical observations for the number r of distinct reduced Gröbner bases for data sets of m points, where $2 \le m \le 6$. Furthermore, we state a conjecture for decreasing r by adding points in so-called linked positions, using the geometric insights from $m = 2, 3$ points.

To generalize the observations from small data sets to larger data sets, we start with configurations of two points, and then consider changes in r as points are added.

Definition 4 Given a set S of points, we say that a point q is in a *linked* position with respect to the points in S if q is adjacent to a point in S and has minimal sum of distances to the points in S.

Figure 7 shows the changes in the number of Gröbner bases when points are added at either linked or non-linked positions.

Example 9 Consider the set $S = \{(0, 1), (1, 2)\}$, which has $r = 2$ Gröbner bases associated to it. We aim to add a point so that the augmented set has $r = 1$. There are four points adjacent to the points in S, namely $(0, 0)$, $(0, 2)$, $(1, 1)$ and $(2, 2)$; see the green points in the top panel of Fig. 7. The sum of the distances between $(0, 0)$ and the points in S is $\sqrt{5} + 1$; similarly for $(2, 2)$. On the other hand, $(0, 2)$ and $(1, 1)$ both have a distance sum of 2. So $(0, 2)$ and $(1, 1)$ are in linked positions with respect to S. Note that inclusion of either $(0, 2)$ or $(1, 1)$ to S reduces r to 1, while inclusion of either of $(0, 0)$ or $(2, 2)$ keeps $r = 2$.

Example 10 Consider the set $S = \{(0, 1), (1, 1)\}$, which has a unique Gröbner basis. There are five points adjacent to S, namely $(0, 0)$, $(0, 2)$, $(1, 0)$, $(1, 2)$, and $(2, 1)$; see the green points in the bottom panel of Fig. 7. The first four points have a distance sum of $\sqrt{2} + 1$, while the last point $(2, 1)$ has a distance sum of 3. So these four points are in linked positions with respect to S and inclusion of any one of them keeps $r = 1$. On the other hand, $(2, 1)$ is not in linked position; nevertheless adding it to S results in a unique Gröbner basis due to it being collinear to the points in S.

Adding a red point in Fig. 7, which is not in a linked position with respect to the starting data set, will not reduce the number of Gröbner bases as its inclusion does not aid in removing ambiguous leading terms. In fact, the pink triangles in the last column in Fig. 7 give instances in which r increases.

For $p = 3$ and $n = 2$, we computed the number of Gröbner bases for data sets up to six points; see Fig. 8. The points at the vertices of the green polygons have $r = 1$. The uniqueness can be maintained by adding points in linked positions; however the points at the vertices of the pink polygons have non-unique Gröbner bases.

Based on the geometric observations from Figs. 7 and 8, we provide heuristic rules to aid in decreasing the number of candidate models as enumerated by the number of Gröbner bases:

1. *For two points*, fewer changing coordinates in the data points will lead to fewer Gröbner bases. In the simplest case, if only one coordinate changes, a unique model will be generated.
2. *For three points*, more points lying on horizontal or vertical edges will reduce the number of Gröbner bases. A unique Gröbner basis arises when the data lie on a horizontal line, a vertical line or form a right triangle.

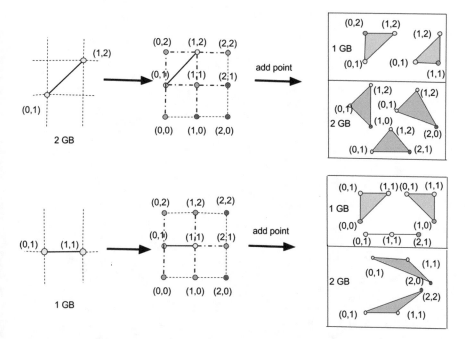

Fig. 7 The green points are adjacent to the blue points. Green triangles are associated with a unique GB, while pink triangles are associated with non-unique GBs

3. *In the process of adding points*, to decrease or maintain the number of minimal models, add points in linked positions with respect to an existing data set: this guarantees more points lying on horizontal or vertical edges.

By adding points in linked positions, data sets with multiple Gröbner bases can be transformed to data sets with unique one, as the following example suggests.

Example 11 Consider data sets in $\mathbb{Z}_2^4$. Let S_{max} be a data set whose ideal of points has the maximum number of Gröbner bases. Define $S_{unique} = S_{max} \cup S_{add}$ where S_{add} is a collection of points such that the augmented data set S_{unique} has an ideal of points with a unique GB. The table summarizes for different sized sets how many points must be added to guarantee a unique Gröbner basis from a data set associated with the maximum number of Gröbner bases.

$\max(\#GBs)$	4	5	6	13	12	13	9	13	12	13	6	5	4
$|S_{max}|$	2	3	4	5	6	7	8	9	10	11	12	13	14
$|S_{unique}|$	5	5	8	11	11	11	11	12	15	15	15	15	15
$|S_{add}|$	3	2	4	6	5	4	3	3	5	4	3	2	1

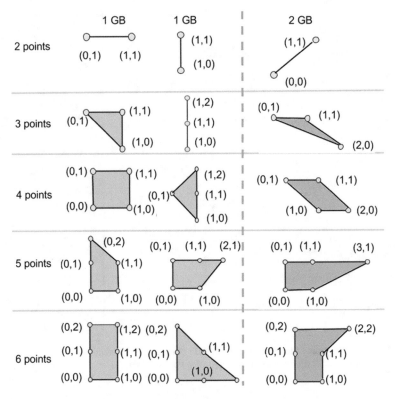

Fig. 8 Point configurations based on the number of Gröbner bases for $2 \leq m \leq 6$. The left two columns contain points that form green polygons and correspond to a unique Gröbner basis. The right column contains the pink polygons corresponding to non-unique GBs

We end this discussion with a conjecture about points in linked positions.

Conjecture 1 Let S be a set of points, $q \notin S$, and $T = S \cup \{q\}$. If q is in a linked position and the convex hull of the points in T does not contain "holes" (i.e., lattice points not in T), then $\#GB(T) \leq \#GB(S)$.

6 Upper Bound for the Number of Gröbner Bases

We now focus on the general setting of subsets of any size m in $\mathbb{Z}_p^n$, for any p and any n.

In Theorem 1, the stated upper bound for the number of Gröbner bases for an ideal I of m points in K^n is $m^{2n\frac{n-1}{n+1}}$, where K is any field; furthermore the number of Gröbner bases coincides with the number of vertices of the staircase polytope of I. When the base field is finite, however, this bound becomes unnecessarily

Fig. 9 The staircase $\lambda \subset \mathbb{R}^2$ (left) has $\sum \lambda = (0, 6)$ while the staircase $\lambda \subset \mathbb{Z}_3^2$ (right) has $\sum \lambda = (1, 3)$

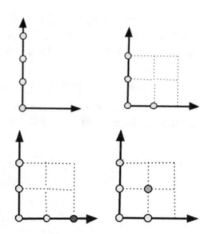

Fig. 10 The staircase $\lambda \subset \mathbb{Z}_3^2$ with red point (left) has $\sum \lambda = (3, 3)$ while the staircase $\lambda \subset \mathbb{Z}_3^2$ with green point (right) has $\sum \lambda = (2, 4)$

large for even small m. Unlike in characteristic-0 fields, all coordinates in positive-characteristic fields are bounded above by p; for example see Fig. 9. We will use the fact that staircases in a finite field are contained in a hypercube of volume p^n to modify the bound. The only part of the construction of the staircase polytope that is affected by the field characteristic is the maximum value of any vertex. As a vertex is a vector sum $\sum \lambda$ of points in a staircase λ, the modification comes from placing staircase points aimed to maximize the sum.

Consider any staircase λ of 5 elements. In the following discussion, we will consider the placement of points so that the vector sum is maximized. We proceed in a "greedy" manner by maximizing a fixed coordinate. Suppose four (blue) points have already been placed so as to maximize the value of the second coordinate of $\sum \lambda$; see Fig. 10. Placing the green point $(1, 1)$ contributes 1 to the running sum, that is, $\sum_{j=1}^{m} \lambda_{j2} = 4$ while placing the red point $(2, 0)$ keeps the sum of the coordinate unchanged. In fact, to maximize the sum of second coordinate, choose any point whose second coordinate is largest among the available positions, that is so that the configuration continues to be a staircase.

Theorem 8 *The number of distinct reduced Gröbner bases for an ideal of m points in $\mathbb{Z}_p^n$ is*

$$
NGB(p, n, m) = \begin{cases} O\left(\left(p^2 \lfloor m/p \rfloor + (m \ (\mathrm{mod} \ p))^2\right)^{n\frac{n-1}{n+1}}\right) & : 0 < m \leq \lfloor p^n/2 \rfloor \\ O\left(\left(p^2 \lfloor (p^n - m)/p \rfloor + (-m \ (\mathrm{mod} \ p))^2\right)^{n\frac{n-1}{n+1}}\right) & : \lfloor p^n/2 \rfloor \leq m < p^n \\ 1 & : m = 0, p^n. \end{cases}
$$

Proof Let I be an ideal of m points in $\mathbb{Z}_p^n$. Recall that the number of Gröbner bases of I is bijective with the number of vertices of the staircase polytope $\mathcal{P}$ of I by Theorem 2. The cases $m = 0, p^n$ are trivial. So we proceed with $0 < m \leq \lfloor p^n/2 \rfloor$.

As $\mathcal{P}$ is the convex hull of the points $\sum \lambda$ where λ is a staircase corresponding to the exponent vectors of the standard monomial sets of I, we will show that the staircase polytope of I is contained in a larger convex body whose volume can be computed easily.

Let $\lambda = \{\lambda_1, \ldots, \lambda_m\}$. Then $\sum \lambda = \sum_{i=1}^{m} \lambda_i = \sum_{i=1}^{m} \left(\sum_{j=1}^{m} \lambda_{ji} \right) e_i$ where λ_{ji} denotes the i-th coordinate of the j-th point and e_i is the standard basis vector. Note that the maximum sum of the i-th coordinate is

$$
M := \max \sum_{j=1}^{m} \lambda_{ji} = \underbrace{(1 + \ldots + p - 1)\lfloor m/p \rfloor}_{p\lfloor m/p \rfloor \text{ points}} + \underbrace{(1 + \ldots + m \pmod{p} - 1)}_{\text{remaining } m \pmod{p} \text{ points}}
$$
$$
= \frac{p(p-1)}{2} \lfloor m/p \rfloor + \frac{(m \pmod{p})(m \pmod{p} - 1)}{2}.
$$

So the staircase polytope $\mathcal{P} \subset \mathbb{R}^n$ is contained in the hypercube $[0, M]^n$, which has volume M^n. Therefore $vol(\mathcal{P}) \leq M^n$. By Lemma 1 and Theorem 1, we have that

$$
NGB(p, n, m) = O\left(vol(\mathcal{P})^{(n-1)/(n+1)} \right)
$$
$$
= O\left((M^n)^{\frac{n-1}{n+1}} \right)
$$
$$
= O\left(\left(p^2 \lfloor m/p \rfloor + (m \pmod{p})^2 \right)^{n \frac{n-1}{n+1}} \right). \tag{1}
$$

For the final case when $m \geq \lfloor p^n/2 \rfloor$, the number of Gröbner bases can be computed by plugging $p^n - m$ into the second argument of the above bound, according to Theorem 4.   □

It is straightforward to show that our bound grows much slower than the bound $O\left(m^{2n \frac{n-1}{n+1}} \right)$ reported in [22], which we have also verified computationally. In the Appendix Tables 1, 2, 3, and 4 contain numerical results of the new upper bound in comparison to the values of the original upper bound in [22]. Figure 11 provides a comparison for selected cases among $p = 2, 3$ and $n = 2, 3, 4$.

Not only are the values from Theorem 8 closer to the actual number of GBs, including an application of Theorem 4 in our bound retains the symmetric nature of the maximum number of Gröbner bases for ideals of points in $\mathbb{Z}_p^n$. For example, for $p = 2$, $n = 4$, and $m = 5$ in Fig. 11, the original bound is over 2000, while the modified bound is in the same order of magnitude as the actual maximum number of GBs.

The significance of this result is that Theorem 8 provides a more accurate representation of the maximum number of models associated to a data set, which may aid in experimental design.

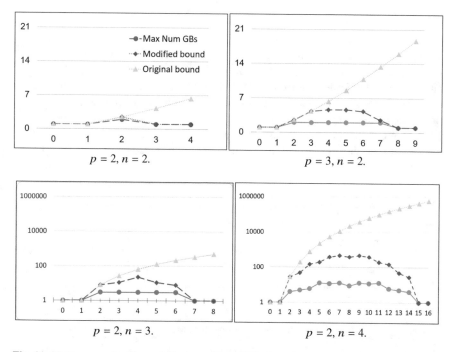

Fig. 11 Plots comparing the maximum number of Gröbner bases. The caption in each plot indicates the values of p and n for $\mathbb{Z}_p^n$. In each case, all subsets of size m are computed, where m ranges from 0 to p^n and listed on the horizontal axis. The vertical axis is the maximum number of GBs for a set of size m. The blue solid line with dots shows the actual maximum number of GBs. The yellow dotted line with triangles is the original upper bound given by Theorem 1, where the red dashed line with squares is the modified upper bound given by Theorem 8. The data for the four plots is available in Tables 1, 2, 3, and 4 in the Appendix

7   Discussion

This work relates the geometric configuration of data points with the number of associated Gröbner bases. In particular we provide some insights into which configurations lead to uniqueness. We give an upper bound for the number of Gröbner bases for any set over a finite field. We also provide a heuristic for decreasing the number by adding points in so-called linked positions. An implication of this work is a more computationally accurate way to predict the number of distinct minimal models which may aid modelers in estimating the computational cost before running physical experiments.

Increasing p, n or m inflates the difference between the estimated number of Gröbner bases and the actual number. The performance of the bound in Theorem 8 works well for large p and m. Though the bound is tighter than the original bound in [22], it still has large differences from the actual values for $n > 4$; see Table 5 in the Appendix. Hence, improving this bound further or finding a closed form for the number of Gröbner bases remains an important direction for future work.

Appendix

We provide tables comparing of the maximum number of distinct reduced Gröbner bases to the predictions made by the original bound (third column) in Theorem 1 and the modified bound (last column) in Theorem 8. In Tables 1, 2, 3, and 4, the second column shows the actual maximum number as computed for all sets in $\mathbb{Z}_p^n$ of size given in the first column. In Table 5, the maximum number of Gröbner bases is compared to the predictions made by the two bounds with regards to an increasing number of coordinates (first column). All values are rounded up to 2 decimal places.

Table 1 $p = 2, n = 2$

# of points	Max # of GBs	Original bound	Modified bound
1	1	1	1
2	2	2.52	2.52
3	1	4.33	1
4	1	6.35	1

Table 2 $p = 2, n = 3$

# of points	Max # of GBs	Original bound	Modified bound
1	1	1	1
2	3	8	8
3	3	27	11.18
4	3	64	22.63
5	3	125	11.18
6	3	216	8
7	1	343	1
8	1	512	1

Table 3 $p = 2, n = 4$

# of points	Max # of GBs	Original bound	Modified bound
1	1	1	1
2	4	27.86	27.86
3	5	195.07	47.59
4	6	776.05	147.03
5	13	2264.94	195.07
6	12	5434.08	389.08
7	13	11,388.61	471.48
8	9	21,618.82	389.08

Half of the table is listed due to space constraints

Table 4 $p = 3, n = 2$

# of points	Max # of GBs	Original bound	Modified bound
1	1	1	1
2	2	2.52	2.52
3	2	4.33	4.33
4	2	6.35	4.64
5	2	8.55	4.64
6	2	10.90	4.33
7	2	13.39	2.52
8	1	16	1
9	1	18.72	1

Table 5 $p = 2$ and $m = 4$

# of coordinates	Max # of GBs	Original bound	Modified bound
2	1	6.35	1
3	3	64	22.63
4	6	776.05	147.03
5	8	10,321.27	1024

Here we show how the number of Gröbner bases changes as the number of coordinates changes.

References

1. G. Andrews. A lower bound for the volume of strictly convex bodies with many boundary lattice points. *Transactions of the American Mathematical Society*, 106(2):270–279, February 1963.
2. E. Babson, S. Onn, and R. Thomas. The Hilbert zonotope and a polynomial time algorithm for universal Gröbner bases. *Advances in Applied Mathematics*, 30(3):529–544, 2003.
3. B. Buchberger. *Ein Algorithmus zum Auffinden der Basiselemente des Restklassenringes nach einem nulldimensionalen Polynomideal*. PhD thesis, Universität Innsbruck, 1965.
4. B. Buchberger. A note on the complexity of constructing Groebner-Bases. In J. von Hulzen, editor, *Computer Algebra: Proceedings of EUROCAL 83*, volume 162 of *Lecture Notes in Computer Science*, pages 137–145. Springer Berlin, 1983.
5. B. Buchberger. Bruno Buchberger's PhD thesis 1965: An algorithm for finding the basis elements of the residue class ring of a zero dimensional polynomial ideal. *Journal of Symbolic Computation*, 41(3–4):475–511, March-April 2006.
6. B. Buchberger and M. Möller. The construction of multivariate polynomials with preassigned zeroes. In J. Calmet, editor, *Computer Algebra: EUROCAM '82*, volume 144 of *Lecture Notes in Computer Science*, pages 24–31. Springer Berlin, 1982.
7. D. Cox, J. Little, and D. O'Shea. *Using Algebraic Geometry*, volume 185. Springer Science & Business Media, 2006.
8. D. Cox, J. Little, and D. O'Shea. *Ideals, Varieties, and Algorithms*. Springer, 2007.
9. E.S. Dimitrova. Estimating the relative volumes of the cones in a Gröbner fan. *Special issue of Mathematics in Computer Science: Advances in Combinatorial Algorithms II*, 3(4):457–466, 2010.
10. E.S. Dimitrova, Q. He, L. Robbiano, and B. Stigler. Small Gröbner fans of ideals of points. *Journal of Algebra and Its Applications*, 2019.

11. C. Eder and J.-C. Faugére. A survey on signature-based Gröbner basis computations. *Journal of Symbolic Computation*, 80(3):719–784, May 2014.
12. J. Farr and S. Gao. Computing Gröbner bases for vanishing ideals of finite sets of points. In M Fossorier, H Imai, S Lin, and et al., editors, *Applied Algebra, Algebraic Algorithms and Error-Correcting Codes*, pages 118–127. Ann N Y Acad Sci., Springer, Berlin, 2006.
13. F. Fukuda, A. Jensen, and R. Thomas. Computing Gröbner fans. *Mathematics of Computation*, 76(260):2189–2212, 2007.
14. S. Gao, A. Platzer, and E. Clarke. Quantifier elimination over finite fields using Gröbner bases. In Franz Winkler, editor, *Algebraic Informatics*, pages 140–157, Berlin, Heidelberg, 2011. Springer Berlin Heidelberg.
15. W. Just and B. Stigler. Computing Gröbner bases of ideals of few points in high dimensions. *ACM Communications in Computer Algebra*, 40(3/4):67–78, September/December 2006.
16. V. Larsson, M. Oskarsson, K. Astrom, A. Wallis, T. Pajdla, and Z. Kukelova. Beyond Gröbner bases: Basis selection for minimal solvers. In *2018 IEEE/CVF Conference on Computer Vision and Pattern Recognition*, pages 3945–3954. IEEE, June 2018.
17. R. Laubenbacher and B. Stigler. A computational algebra approach to the reverse engineering of gene regulatory networks. *J. Theor. Biol.*, 229(4):523–537, 2004.
18. Z. Lin, L. Xu, and Q. Wu. Applications of Gröbner bases to signal and image processing: A survey. *Linear Algebra and its Applications*, 391:169–202, 2004.
19. R. Makarim and M. Stevens. M4GB: An efficient Gröbner-basis algorithm. In *Proceedings of the 2017 ACM on International Symposium on Symbolic and Algebraic Computation*, ISSAC 17, pages 293–300. ACM, July 2017.
20. M. Maniatis, A. von Manteuffel, and O. Nachtmann. Determining the global minimum of Higgs potentials via Groebner bases - applied to the NMSSM. *The European Physical Journal C*, 49(4):1067–1076, 2007.
21. T. Mora and L. Robbiano. The Gröbner fan of an ideal. *Journal of Symbolic Computation*, 6(2–3):183–208, 1988.
22. S. Onn and B. Sturmfels. Cutting corners. *Advances in Applied Mathematics*, 23(1):29–48, 1999.
23. M. Torrente. *Applications of Algebra in the Oil Industry*. PhD thesis, Scuola Normale Superiore di Pisa, 2009.
24. Y.-L. Tsai. Estimating the number of tetrahedra determined by volume, circumradius and four face areas using Groebner basis. *Journal of Symbolic Computation*, 77:162–174, 2016.

Part IV
Commutative Algebra

Depth of Powers of Squarefree Monomial Ideals (Research)

Louiza Fouli, Huy Tài Hà, and Susan Morey

Mathematics Subject Classfication (2010): 13C15, 13D05, 13F55, 05E40

1 Introduction

During the past two decades, many papers have appeared with various approaches to computing lower bounds for the depth, or equivalently upper bounds for the projective dimension, of R/I for a *squarefree* monomial ideal I (cf. [7, 8, 22, 25, 26, 31]). The general idea has been to associate to the ideal I a graph or hypergraph G and use *dominating* or *packing* invariants of G to bound the depth of R/I.

In general, given an ideal $I \subseteq R$, it is not just the depth of R/I that attracts significant attention; rather, it is the entire *depth function* depth R/I^s, for $s \in \mathbb{N}$. A result by Burch, which was later improved by Brodmann, states that $\lim_{s \to \infty}$ depth $R/I^s \leq \dim R - \ell(I)$, where $\ell(I)$ is the analytic spread of I [3, 5]. Moreover, Eisenbud and Huneke [9] showed that if, in addition, the associated graded ring, $\mathrm{gr}_I(R)$, of I is Cohen-Macaulay, then the above inequality becomes an equality. Therefore, one can say that the limiting behavior of the depth R/I^s is quite well understood. It is then natural to consider the initial behavior of the depth function (cf. [1, 11, 16, 17, 19, 21, 23, 24, 27–30, 33]).

Examples have been exhibited to show that the initial behavior of depth R/I^s can be wild, see [1]. In fact, it was conjectured by Herzog and Hibi [19] that for

L. Fouli (✉)
Department of Mathematical Sciences, New Mexico State University, Las Cruces, NM, USA
e-mail: lfouli@nmsu.edu

H. T. Hà
Department of Mathematics, Tulane University, New Orleans, LA, USA
e-mail: tha@tulane.edu

S. Morey
Department of Mathematics, Texas State University, San Marcos, TX, USA
e-mail: morey@txstate.edu

© The Author(s) and the Association for Women in Mathematics 2020
B. Acu et al. (eds.), *Advances in Mathematical Sciences*, Association for
Women in Mathematics Series 21, https://doi.org/10.1007/978-3-030-42687-3_10

any numerical function $f : \mathbb{N} \to \mathbb{Z}_{\geq 0}$ that is asymptotically constant, there exists an ideal I in a polynomial ring R such that $f(s) = \operatorname{depth} R/I^s$ for all $s \geq 1$. This conjecture has recently been resolved affirmatively in [15]. It was proven in [15] that the depth function of a monomial ideal can be any numerical function that is asymptotically constant. Yet, it is still not clear what depth functions are possible for squarefree monomial ideals.

Unlike the case for depth R/I, few lower bounds for depth R/I^s, $s \in \mathbb{N}$, are known (cf. [11, 29, 33]). One reason for this is that powers of squarefree monomial ideals are not squarefree and so many of the known bounds for R/I do not apply to R/I^s. To address this situation, we adapt a proof technique from [2] to generalize bounds for depth R/I that were given by Dao and Schweig [8] in terms of the *edgewise domination number*, and by the authors [12] in terms of the length of an *initially regular sequence*. We provide lower bounds for the depth function depth R/I^s, $s \in \mathbb{N}$, when I is a squarefree monomial ideal corresponding to a hyperforest or a forest, Theorems 1 and 2.

Our results, Theorems 1 and 2, predict correctly the general behavior, as computation indicates for random hyperforests and forests, that the depth function depth $R/I(G)^s$ decreases incrementally as s increases. For specific examples, our bound in Theorem 1 could be far from the actual values of the depth function—and this is because the starting bound for depth R/I in terms of the edgewise domination number is not always optimal. For forests, Theorem 2 could provide a more accurate starting bound for depth R/I using initially regular sequences and, thus, be closer to the depth function.

The common underlying idea behind Theorems 1 and 2 is that if $\alpha(G)$ is an invariant associated to a hyperforest G that gives depth $R/I(G) \geq \alpha(G)$ and satisfies a certain inequality when restricted to subhypergraphs then one should have

$$\operatorname{depth} R/I(G)^s \geq \max\{\alpha(G) - s + 1, 1\}.$$

Our work in this paper, thus, could be interpreted as the starting point of a research program in finding such combinatorial invariants $\alpha(G)$ to best describe the depth function of squarefree monomial ideals, which we hope to continue to pursue in future works.

2 Background

For unexplained terminology, we refer the reader to [4] and [18]. Throughout the paper, $R = k[x_1, \ldots, x_n]$ is a polynomial ring over an arbitrary field k and all hypergraphs will be assumed to be *simple*, that is, there are no containments among the edges. For a hypergraph $G = (V_G, E_G)$ over the vertex set $V_G = \{x_1, \ldots, x_n\}$, the *edge ideal* of G is defined to be

$$I(G) = \left\langle \prod_{x \in e} x \mid e \in E_G \right\rangle \subseteq R.$$

This construction gives a one-to-one correspondence between squarefree monomial ideals in $R = k[x_1, \ldots, x_n]$ and (simple) hypergraphs on the vertex set $V = \{x_1, \ldots, x_n\}$.

For a vertex x in a graph or hypergraph G, we say y is a *neighbor* of x if there exists an edge $E \in E_G$ such that $x, y \in E$. The *neighborhood* of x in G is $N_G(x) = \{y \in V_G \mid y$ is a neighbor of $x\}$. The *closed neighborhood* of x in G is $N_G[x] = N_G(x) \cup \{x\}$. Note that the G in the notation will be suppressed when it is clear from the context.

Simplicial forests were defined by Faridi in [10], where it was shown that the edge ideals of these hypergraphs are always sequentially Cohen-Macaulay. They have also been used in the study of standard graded (symbolic) Rees algebras of squarefree monomial ideals [20]. We first recall the definition of a simplicial forest (or a hyperforest for short).

Definition 1 Let $G = (V, E)$ be a simple hypergraph.

1. An edge $e \in E$ is called a *leaf* if either e is the only edge in G or there exists $e \neq g \in E$ such that for any $e \neq h \in E$, $e \cap h \subseteq e \cap g$.
2. A leaf e in G is called a *good leaf* if the set $\{e \cap h \mid h \in E\}$ is totally ordered with respect to inclusion.
3. G is called a *simplicial forest* (or simply, a *hyperforest*) if every subhypergraph of G contains a leaf. A *simplicial tree* (or simply, a *hypertree*) is a connected hyperforest.

It follows from [20, Corollary 3.4] that every hyperforest contains good leaves. It is also immediate that every graph that is a forest is also a hyperforest.

Example 1 For the hypergraphs depicted below, the first one is not a hypertree while the second one is, see also [10, Examples 1.4, 3.6].

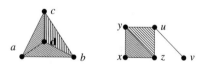

In this paper, we will focus on two invariants that are known to bound the depth of R/I when I is the edge ideal of an arbitrary hypergraph. When G is a simplicial forest, we will provide a linearly decreasing lower bound for the depths of the powers of I using each of these invariants. The first of these bounds for the depth function of a squarefree monomial ideal is the edgewise domination number introduced in [8]. Recall that for a hypergraph $G = (V, E)$, a subset $F \subseteq E$ is called *edgewise dominant* if for every vertex $v \in V$ either $\{v\} \in E$ or v is adjacent to a vertex contained in an edge of F.

Definition 2 ([8]) The *edgewise domination number* of G is defined to be

$$\epsilon(G) = \min\{|F| \mid F \subseteq E \text{ is edgewise dominant}\}.$$

The second invariant used in this paper will be a variation on the depth bound for monomial ideals introduced in [12]. For an arbitrary vertex b_0 in a hypergraph G, define a *star* on b_0 to be a linear sum $b_0+b_1+\cdots+b_t$ such that for each edge E_i of G, if $b_0 \in E_i$, then there exists a $j > 0$ such that $b_j \in E_i$. It was shown in [12, Theorem 3.11] that a set of vertex-disjoint stars that can be embedded in a hypergraph G forms an initially regular sequence and, thus, gives a lower bound for the depth of $R/I(G)$. While much of [12] focuses on strengthening this bound by weakening the disjoint requirement and allowing for additional types of linear sums, in this article we will apply the bound to graphs, where the situation is more restricted. Notice that for a graph G, a star on b_0 is the sum of all vertices in the closed neighborhood of b_0, while for a hypergraph, a subset of the closed neighborhood can suffice. A *star packing* is a collection S of vertex-disjoint stars in G such that if $x \in V_G$ then $N_G[x] \cap \mathrm{Supp}(S) \neq \emptyset$. In other words, S is maximal in the sense that no additional disjoint stars exist. This leads to the following definition, whose notation reflects its relationship to a 2-packing of closed neighborhoods in graph theory.

Definition 3 The *star packing number* α_2 of a hypergraph G is given by

$$\alpha_2(G) = \max\{|S| \mid S \text{ is a star packing of } G\}.$$

Remark 1 If $x_1, \ldots, x_k \in R$ are variables in R that do not appear in any edge of G, then $x_1, \ldots, x_k$ is a regular sequence on $R/I(G)$ and depth $R/I(G) = k + $ depth $R/(x_1, \ldots, x_k, I(G))$.

Note that if S is any set of disjoint stars in a hypergraph G, then $\alpha_2(G) \geq |S|$ since S can be extended to a full star packing. Note also that for the special case when G is a graph, a star packing is equivalent to a closed neighborhood packing and, by focusing on the centers of the stars, to a maximal set of vertices such that the distance between any two is at least 3.

3 Depth of Powers of Squarefree Monomial Ideals

In this section, we use a technique introduced in [2] to give a general lower bound for the depth function of a squarefree monomial ideal when the underlying hypergraph is a hyperforest. In the case of a forest, we extend the result to show that an additional, often stronger, bound holds. For simplicity of notation, we write V_G and E_G to denote the vertex and edge sets of a hypergraph G.

Theorem 1 *Let G be a hyperforest with at least one edge of cardinality at least 2, and let $I = I(G)$. Then for all $s \geq 1$,*

$$\operatorname{depth} R/I^s \geq \max\{\epsilon(G) - s + 1, 1\}.$$

Proof It follows from [20, Corollary 3.3] (see also [13]) that the symbolic Rees algebra of I is standard graded. That is, $I^{(s)} = I^s$ for all $s \geq 1$. In particular, this implies that I^s has no embedded primes for all $s \geq 1$. Thus, $\operatorname{depth} R/I^s \geq 1$ for all $s \geq 1$.

It remains to show that $\operatorname{depth} R/I^s \geq \epsilon(G) - s + 1$. Indeed, this statement and, hence, Theorem 1 follows from the following slightly more general result. □

Proposition 1 *Let G be a hyperforest. Let H and T be subhypergraphs of G such that*

$$E_H \cup E_T = E_G \text{ and } E_H \cap E_T = \emptyset.$$

Then we have

$$\operatorname{depth} R/[I(H) + I(T)^s] \geq \max\{\epsilon(G) - s + 1, 0\}.$$

Proof It suffices to show that $\operatorname{depth} R/[I(H) + I(T)^s] \geq \epsilon(G) - s + 1$. We shall use induction on $|E_T|$ and s. If $|E_T| = 0$ then the statement follows from [7, Theorem 3.2]. If $s = 1$ then the statement also follows from [7, Theorem 3.2]. Suppose that $|E_T| \geq 1$ and $s \geq 2$.

Let e be a good leaf of T. Then by the proof of [6, Theorem 5.1], we have $I(T)^s : e = I(T)^{s-1}$. This implies that

$$(I(H) + I(T)^s) : e = (I(H) : e) + I(T)^{s-1}.$$

Moreover,

$$I(H) + I(T)^s + (e) = I(H + e) + I(T \setminus e)^s.$$

Thus, we have the exact sequence

$$0 \to R/[(I(H) : e) + I(T)^{s-1}] \to R/[I(H) + I(T)^s] \to R/[I(H+e) + I(T \setminus e)^s] \to 0$$

which, in turns, gives

$$\operatorname{depth} R/[I(H) + I(T)^s] \geq$$
$$\min\{\operatorname{depth} R/[(I(H) : e) + I(T)^{s-1}], \operatorname{depth} R/[I(H + e) + I(T \setminus e)^s]\}. \quad (1)$$

Observe that $G = (H + e) + (T \setminus e)$ and $E_{H+e} \cap E_{T \setminus e} = \emptyset$. Thus, by induction on $|E_T|$, we have

$$\text{depth } R/[I(H+e)+I(T \setminus e)^s] \geq \epsilon(G)-s+1.$$

On the other hand, let $Z = \{z \in V_H \mid \exists h \in E_H \text{ such that } \{z\} = h \setminus e\}$. Let H' be the hypergraph obtained from $I(H):e$ by deleting the vertices in Z and any vertex in H that does not belong to any edge. Let T' be the hypergraph whose edges are obtained from edges of T after deleting all those that contain any vertex in $V_T \cap Z$. Then

$$I(H):e = I(H')+(z \mid z \in Z).$$

Let $G' = H'+T'$, let $R' = k[V_{H'} \cup V_{T'}]$, and let $W = V_G \setminus (V_{G'} \cup Z)$. It follows by induction on s that

$$\begin{aligned}
\text{depth } R/[(I(H):e)+I(T)^{s-1}] &= \text{depth } R/[I(H')+I(T')^{s-1}+(z \mid z \in Z)] \\
&= \text{depth } R'/[I(H')+I(T')^{s-1}]+|W| \\
&\geq \epsilon(G')-(s-1)+1+|W| \\
&= (\epsilon(G')+1+|W|)-s+1.
\end{aligned}$$

Now, let $F' \subseteq E_{G'}$ be an edgewise dominant set in G'. By the construction of H', for each $f' \in F' \cap E_{H'}$, there is an edge $f \in E_H$ such that $f' = f \setminus e$. Let F be the set obtained from F' by replacing each $f' \in F' \cap E_{H'}$ by such an f. Observe that for any vertex $v \in V_G$, either $v \in W$, or $v \in Z$, or $v \in V_{G'}$. If $v \in Z$, then v is dominated by e. If $v \in V_{G'}$, then v is dominated by some edge in F'. Thus, $F \cup \{e\}$ together with one edge for each vertex in W will form an edgewise dominant set in G. This implies that

$$\epsilon(G')+1+|W| \geq \epsilon(G).$$

Therefore,

$$\text{depth } R/[(I(H):e)+I(T)^{s-1}] \geq \epsilon(G)-s+1.$$

Hence, by (1), we have

$$\text{depth } R/[I(H)+I(T)^s] \geq \epsilon(G)-s+1,$$

which concludes the proof. $\square$

A close examination of the proof of Proposition 1 shows that we can replace $\epsilon(G)$ by any invariant $\alpha(G)$, for which depth $R/I(G) \geq \alpha(G)$ and $\alpha(G')+1+|W| \geq \alpha(G)$, where G' and W are defined as in the proof of Proposition 1.

Corollary 1 *If $\alpha(G)$ is any invariant of a hyperforest G for which* depth $R/I(G) \geq \alpha(G)$ *and* $\alpha(G')+1+|W| \geq \alpha(G)$, *then*

$$\text{depth } R/I^s \geq \max\{\alpha(G) - s + 1, 0\}.$$

For a random hypertree G, computations indicate that the depth function depth $R/I(G)^s$ decreases incrementally as s increases as predicted by Theorem 1. However, for low powers of I, the ϵ-bound is often less than optimal, as can be seen by comparing the results to the bounds on depth $R/I(G)$ obtained from [12]. For hypertrees G for which depth $R/I(G) = \epsilon$, the depth function depth $R/I(G)^s$ usually does not initially decrease incrementally as s increases. These statements are illustrated by the following pair of examples.

Example 2 Let $I = (x_1x_2, x_2x_3, x_3x_4, x_3x_5, x_3x_6, x_6x_7, x_6x_8, x_8x_9, x_8x_{10}, x_8x_{11}, x_8x_{12})$ in $R = \mathbb{Q}[x_1, \ldots, x_{12}]$ be the edge ideal of the graph G depicted below.

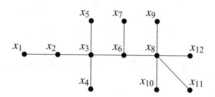

Computation in Macaulay 2 [14] shows that the depth function of I is $4, 3, 2, 1, 1, \ldots$. Thus, Theorem 1 predicts correctly how the depth function behaves. However, in this example, $\epsilon(G) = 2$ does not give the right value for depth R/I.

Example 3 Let $I = (x_1x_2, x_1x_3, x_1x_4, x_4x_5, x_5x_6, x_5x_7, x_4x_8, x_8x_9, x_8x_{10}, x_8x_{11}, x_8x_{12})$ in $R = \mathbb{Q}[x_1, \ldots, x_{12}]$ be the edge ideal of the graph G depicted below.

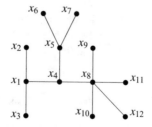

Then $\epsilon(G) = 3$. Computation in Macaulay 2 [14] shows that the depth function of I is $3, 3, 3, 1, 1, \ldots$. The bound in Theorem 1 gives the depth function of I to be at least $3, 2, 1, 1, 1, \ldots$. In this example, while $\epsilon(G)$ gives the right value for depth R/I, Theorem 1 does not predict correctly how the depth function of I behaves.

Examples 2 and 3 show that to get a sharp bound for the depth function of random hypertrees, we may want to start with invariants other than $\epsilon(G)$ which give stronger bounds for depth $R/I(G)$. In order to do so, one often needs to assume additional structure on G. For example, if G is a forest, the invariant from Definition 3 can be used.

Proposition 2 *Let G be a forest with connected components $G_1, \ldots, G_t$. Let H and T be subforests of G such that $E_H \cup E_T = E_G$, $E_H \cap E_T = \emptyset$, and $T \cap G_i$ is connected for each i. Then*

$$\text{depth } R/[I(H) + I(T)^s] \geq \max\{\alpha_2(G) - s + 1, 0\}.$$

Proof The proof follows the outline of that of Proposition 1 with special care toward the end. If $|E_T| = 0$ or $s = 1$, then the statement follows from [12, Theorem 3.11], so we assume $|E_T| \geq 1$ and $s \geq 2$.

Consider an edge $\{x, y\}$ of H. Then, $\{x, y\} \in G_i$ for some i. Since $T \cap G_i$ is connected, if $x, y \in V_T$, then there is a path in T from x to y. This path, together with $\{x, y\}$, forms a cycle in G, which is a contradiction. Thus, no edge of H can have both endpoints in V_T.

Let e be a leaf of T. Since T is a forest, e is a good leaf of T. Thus, as in the proof of Proposition 1, we have

$$\text{depth } R/[I(H) + I(T)^s] \geq$$
$$\min\{\text{depth } R/[(I(H) : e) + I(T)^{s-1}], \text{depth } R/[I(H + e) + I(T \setminus e)^s]\}. \quad (2)$$

Observe further that $G = (H + e) + (T \setminus e)$, $E_{H+e} \cap E_{T \setminus e} = \emptyset$, and $(T \setminus e) \cap G_i$ is connected for each i. Thus, by induction on $|E_T|$, we have

$$\text{depth } R/[I(H + e) + I(T \setminus e)^s] \geq \alpha_2(G) - s + 1.$$

On the other hand, let $Z = \{z \in V_H \mid \exists h \in E_H \text{ such that } \{z\} = h \setminus e\}$. Let H' be the graph obtained from $I(H) : e$ by deleting the vertices in Z and any vertex of H that does not belong to any edge. Note that $V_T \cap Z = \emptyset$, since otherwise there would be an edge of H having both endpoints in V_T (one in Z and the other in e). Then

$$I(H) : e = I(H') + (z \mid z \in Z).$$

Let $G' = H' + T$, let $R' = k[V_{H'} \cup V_T]$, and let $W = V_G \setminus (V_{G'} \cup Z)$. It follows by induction on s that

$$\text{depth } R/[(I(H) : e) + I(T)^{s-1}] = \text{depth } R/[I(H') + I(T)^{s-1} + (z \mid z \in Z)]$$
$$= \text{depth } R'/[I(H') + I(T)^{s-1}] + |W|$$
$$\geq \alpha_2(G') - (s - 1) + 1 + |W|$$
$$= (\alpha_2(G') + 1 + |W|) - s + 1.$$

We will show that $\alpha_2(G') + 1 + |W| \geq \alpha_2(G)$. Fix a set of disjoint stars of G of cardinality $\alpha_2(G)$ and let $S = \{x_1, \ldots, x_{\alpha_2(G)}\}$ denote the set of the centers of these stars.

Let $S' = \{x_i \mid x_i \in R'\}$ and notice that the set of stars in G' centered at x_i for each $x_i \in R'$ is a set of disjoint stars and thus $\alpha_2(G') \geq |S'|$. If $x_i \notin R'$, then $x_i \in Z \cup W$. Since the stars with centers in S are disjoint, there can be at most two elements in $Z \cap S$. If $|Z \cap S| \leq 1$, then $|S'| \geq |S| - 1 - |W| = \alpha_2(G) - 1 - |W|$, and so $\alpha_2(G') + 1 + |W| \geq \alpha_2(G)$.

Suppose that $|Z \cap S| = 2$. Write $e = ab$ and notice that if either a or b is in S, then $Z \cap S = \emptyset$. Hence, we may assume that $a, b \notin S$. We will construct a new set of stars in G' of cardinality at least $\alpha_2(G) - 1 - |W|$ and, thus, also give $\alpha_2(G') + 1 + |W| \geq \alpha_2(G)$ in this case.

Indeed, let $\{z_1, z_2\} = Z \cap S$. Then, $z_1, z_2 \in N_G(a) \cup N_G(b)$ and, without loss of generality, we may assume that $z_1 \in N_G(a)$ and $z_2 \in N_G(b)$. Since e is a leaf in T, we may also assume that b is a leaf vertex in T; that is, $N_T(b) = a$. Then, $N_G(b) \setminus \{a\} \subseteq Z$. Let $\widehat{S'} = S' \cup \{b\}$. We claim that the stars in G' centered on the elements of $\widehat{S'}$ are disjoint. Any two stars centered at elements of S' are already disjoint. Consider then a star centered at an element $x_i \in S'$ and the star centered at b in G'. Since $x_i \neq z_1$, and the stars in G centered at x_i and z_1 are disjoint, we have $a \notin N_{G'}(x_i)$. Thus, $N_{G'}[x_i] \cap N_{G'}[b] = \emptyset$. Clearly, $|\widehat{S'}| \geq |S| - 1 - |W| = \alpha_2(G) - 1 - |W|$.

Now, we have

$$\text{depth } R/[(I(H) : e) + I(T)^{s-1}] \geq \alpha_2(G) - s + 1,$$

and the assertion now follows from (2). $\qquad\square$

Using this result, we obtain the following bound which, while generally stronger than that of Theorem 1 when applicable, applies only to graphs that are trees or forests.

Theorem 2 *Let G be a forest with at least one nontrivial edge, and let $I = I(G)$. Then,*

$$\text{depth } R/I^s \geq \max\{\alpha_2(G) - s + 1, 1\}.$$

Proof It follows from [32, Theorem 5.9] that $I^{(s)} = I^s$ for all $s \geq 1$ and so depth $R/I^s \geq 1$ for all $s \geq 1$. By Proposition 2, depth $R/I^s \geq \alpha_2(G) - s + 1$ and the result follows. $\qquad\square$

Example 4 Let G be the graph in Example 2. Using x_1, x_5, x_7, x_9 as centers of stars, we have $\alpha_2(G) = 4$. Thus, Theorem 2 gives the correct depth function depth $R/I(G)^s$, for all $s \in \mathbb{N}$, for this graph.

On the other hand, let G be the graph as in Example 3. Then, $\alpha_2(G) = 3 = \epsilon(G)$, and so Theorem 2 gives the same bound as that of Theorem 1 for this graph.

It would be interesting to know whether the length of a more general initially regular sequence with respect to $I(G)$, or improved bounds for depth $R/I(G)$ obtained in [12, Section 4], could be used to get better bounds for the depth function than those given in Theorem 1 when G is a hyperforest.

Acknowledgements The first author was partially supported by a grant from the Simons Foundation (grant #244930). The second named author is partially supported by Louisiana Board of Regents (grant #LEQSF(2017-19)-ENH-TR-25). We also thank Seyed Amin Seyed Fakhari for pointing out an error in a previous version of the article.

References

1. S. Bandari, J. Herzog, and T. Hibi, Monomial ideals whose depth function has any given number of strict local maxima. Ark. Mat. 52 (2014), 11–19.
2. S. Beyarslan, H.T. Hà, and T.N. Trung, Regularity of powers of forests and cycles. Journal of Algebraic Combinatorics, 42 (2015), no. 4, 1077–1095.
3. M. Brodmann, The asymptotic nature of the analytic spread. Math. Proc. Cambridge Philos. Soc. 86 (1979), no. 1, 35–39.
4. W. Bruns and J. Herzog, Cohen-Macaulay rings. Cambridge Studies in Advanced Mathematics, 39. Cambridge University Press, Cambridge, 1993. xii+403 pp.
5. L. Burch, Codimension and Analytic Spread. Proc. Cambridge Philos. Soc. 72 (1972), 369–373.
6. G. Caviglia, H.T. Hà, J. Herzog, M. Kummini, N. Terai, and N.V. Trung, Depth and regularity modulo a principal ideal. J. Algebraic Combin. 49 (2019), no. 1, 1–20.
7. H. Dao and J. Schweig, Projective dimension, graph domination parameters, and independence complex homology. J. Combin. Theory Ser. A 120 (2013), 453–469.
8. H. Dao and J. Schweig, Bounding the projective dimension of a squarefree monomial ideal via domination in clutters. Proc. Amer. Math. Soc. 143 (2015), no. 2, 555–565.
9. D. Eisenbud and C. Huneke, Cohen-Macaulay Rees Algebras and their Specializations. J. Algebra 81 (1983) 202–224.
10. S. Faridi, Simplicial trees are sequentially Cohen-Macaulay. J. Pure Appl. Algebra 190 (2004), no. 1–3, 121–136.
11. L. Fouli and S. Morey, A lower bound for depths of powers of edge ideals. J. Algebraic Combin. 42 (2015), no. 3, 829–848.
12. L. Fouli, H.T. Hà, and S. Morey, Initially regular sequences and depth of ideals. J. Algebra 559 (2020), 33–57, https://arxiv.org/abs/1810.01512.
13. G. Isidoro, R. Enrique, and R.H. Villarreal, Blowup algebras of square-free monomial ideals and some links to combinatorial optimization problems. Rocky Mountain J. Math. 39 (2009), no. 1, 71–102.
14. D. R. Grayson and M. E. Stillman, Macaulay2, a software system for research in algebraic geometry, Available at https://faculty.math.illinois.edu/Macaulay2/
15. H.T. Hà, H.D. Nguyen, N.V. Trung, and T.N. Trung, Depth functions of powers of homogeneous ideals. Proc. Amer. Math. Soc. (2019), https://doi.org/10.1090/proc/15083.
16. H.T. Hà and M. Sun, Squarefree monomial ideals that fail the persistence property and non-increasing depth. Acta Math. Vietnam. 40 (2015), no. 1, 125–137.
17. N.T. Hang and T.N. Trung, The behavior of depth functions of cover ideals of unimodular hypergraphs. Ark. Mat. 55 (2017), no. 1, 89–104.
18. J. Herzog and T. Hibi, Monomial Ideals. Graduate Texts in Mathematics 260, Springer, 2011.
19. J. Herzog and T. Hibi, The depth of powers of an ideal. J. Algebra 291 (2005), no. 2, 534–550.

20. J. Herzog, T. Hibi, N.V. Trung, and X. Zheng, Standard graded vertex cover algebras, cycles and leaves. Trans. Amer. Math. Soc. 360 (2008), no. 12, 6231–6249.
21. J. Herzog and M. Vladoiu, Squarefree monomial ideals with constant depth function. J. Pure Appl. Algebra 217 (2013), no. 9, 1764–1772.
22. T. Hibi, A. Higashitani, K. Kimura, and A.B. O'Keefe, Depth of initial ideals of normal edge rings. Comm. Algebra 42 (2014), no. 7, 2908–2922.
23. L.T. Hoa, K. Kimura, N. Terai, and T.N. Trung, Stability of depths of symbolic powers of Stanley-Reisner ideals. J. Algebra 473 (2017), 307–323.
24. L.T. Hoa and T.N. Trung, Stability of depth and Cohen-Macaulayness of integral closures of powers of monomial ideals. Acta Math. Vietnam. 43 (2018), no. 1, 67–81.
25. K.-N. Lin and P. Mantero, Hypergraphs with high projective dimension and 1-dimensional hypergraphs. Internat. J. Algebra Comput. 27 (2017), no. 6, 591–617.
26. K.-N. Lin and P. Mantero, Projective dimension of string and cycle hypergraphs. Comm. Algebra 44 (2016), no. 4, 1671–1694.
27. K. Matsuda, T. Suzuki, and A. Tsuchiya, Nonincreasing depth function of monomial ideals. Glasg. Math. J. 60 (2018), no. 2, 505–511.
28. C.B. Miranda-Neto, Analytic spread and non-vanishing of asymptotic depth. Math. Proc. Cambridge Philos. Soc. 163 (2017), no. 2, 289–299.
29. S. Morey, Depths of powers of the edge ideal of a tree. Comm. Algebra 38 (2010), no. 11, 4042–4055.
30. S. Morey and R.H. Villarreal, Edge ideals: Algebraic and combinatorial properties. *Progress in Commutative Algebra 1*, 85–126, de Gruyter, Berlin, 2012.
31. D. Popescu, Graph and depth of a monomial squarefree ideal. Proc. Amer. Math. Soc. 140 (2012), no. 11, 3813–3822.
32. A. Simis, W.V. Vasconcelos, and R.H. Villarreal, On the ideal theory of graphs. J. Algebra 167 (1994), 389–416.
33. N. Terai and N.V. Trung, On the associated primes and the depth of the second power of squarefree monomial ideals. J. Pure Appl. Algebra 218 (2014), no. 6, 1117–1129.

A Note on the Uniqueness of Zero-Divisor Graphs with Loops (Research)

Aihua Li, Ryan Miller, and Ralph P. Tucci

1 Introduction

Let R be a finite commutative ring with $1 \neq 0$. Let $Z^*(R)$ denote the set of nonzero zero-divisors of R and $U(R)$ the set of units of R. The *zero-divisor graph* of R, denoted $\Gamma(R)$, is the undirected graph whose vertices are labeled by the elements of $Z^*(R)$. For two distinct vertices r and s in $Z^*(R)$, there is an edge in $\Gamma(R)$ between r and s if and only if $rs = 0$. In this case, we say that r and s are *adjacent*. As usual, the edge in between r and s is denoted as "$r - s$" which is a member of $E(\Gamma(R))$. This graph has no loop because the edges are only defined on distinct vertices. We generalize the definition to include loops. If $r \in Z^*(R)$ with $r^2 = 0$, then there is an edge from r to itself in a modified graph. Such an edge is called a *loop*. The new zero-divisor graph of R that allows loops, denoted $\Gamma_0(R)$, has the same vertex set as that of $\Gamma(R)$ and the edge set is given by

$$E(\Gamma_0(R)) = E(\Gamma(R)) \cup \{\text{all loops}\}.$$

For $r \in Z^*(R)$, the set of vertices adjacent to r, referred to as the *neighborhood* of r, is the set $N(r) = \{s \in R \mid s \neq 0, \ rs = 0\}$. The cardinality of $N(r)$ is called the *degree* of r in $\Gamma(R)$. By [7], we know that $\Gamma(R)$ is connected for any ring R. For a general background on graph theory, see [13]. In a zero-divisor graph including loops, the neighborhood of a vertex may have the vertex itself.

A. Li (✉)
Montclair State University, Montclair, NJ, USA
e-mail: lia@montclair.edu

R. Miller · R. P. Tucci
Loyola University New Orleans, New Orleans, LA, USA
e-mail: rlmiller@loyno.edu; tucci@loyno.edu

© The Author(s) and the Association for Women in Mathematics 2020
B. Acu et al. (eds.), *Advances in Mathematical Sciences*, Association for
Women in Mathematics Series 21, https://doi.org/10.1007/978-3-030-42687-3_11

Zero-divisor graphs (not allowing loops) were first defined for commutative rings by Beck ([10], 1988) when the coloring of graphs was studied. In the last 15 years, there has been a large number of papers on this topic [1, 4–7, 11, 12, 16]. Readers can refer to recent survey papers [2] (2011), [14] (2012), [8] (2017), and [9] (2017) for extensive bibliographies.

Redmond ([19, 20], 2002) introduced the concept of the zero-divisor graph for a noncommutative ring. It is a directed graph denoted as $\overrightarrow{\Gamma}(R)$. Let R be a finite noncommutative ring with identity. The vertices of $\overrightarrow{\Gamma}(R)$ are the nonzero zero-divisors of R. For two vertices r and s, if $rs = 0$, then we have a directed edge in the graph from r to s, denoted $r \to s$. We say that r is *adjacent to s* and s is *adjacent from r*. The directed graphs we consider also allow loops. As in the undirected graph case, we denote the zero-divisor graphs of R that may have loops by $\overrightarrow{\Gamma}_0(R)$.

Bŏzić and Petrović ([12], 2009) studied the zero-divisor graph of a ring of matrices over a commutative ring. Akbari and Mohammadian ([1], 2007) studied the problem of determining when the zero-divisor graphs of two rings are isomorphic, given that the zero-divisor graphs of their matrix rings are isomorphic. B. Li ([18], 2011) and A. Li and Tucci ([17], 2013) investigated the zero-divisor graphs of upper triangular matrix rings. Dolzan and Oblak ([15], 2012) have studied zero-divisor graphs of semirings and some other rings. Birch, Thibodeaux, and Tucci discovered some properties for the zero-divisor graphs of finite direct products of finite rings ([11], 2014).

It is well-known that non-isomorphic rings can have isomorphic zero-divisor graphs. For example, the zero-divisor graphs of $\mathbb{Z}_4$ and $\mathbb{Z}_2[x]/(x^2)$ are isomorphic, each consisting of a single vertex with a loop on it. However, we show that if two rings R and S have isomorphic zero-divisor graphs, then $R \cong S$ if both R and S are finite direct products of $\mathbb{Z}_n$'s and/or finite fields.

2 Direct Products of Rings of Integers Modulo Various n

Given two commutative rings R and S, we say that R and S are in the same zero-divisor class if $\Gamma_0(R) \cong \Gamma_0(S)$. A natural question is, "in what situations does $\Gamma_0(R) \cong \Gamma_0(S)$ imply $R \cong S$?" The following theorem provides the first step in answering this question. Note that $\Gamma(R)$ always stands for the zero-divisor graph of R not including loops and $\Gamma_0(R)$ for the zero-divisor graph of R by adding loops in $\Gamma(R)$.

Theorem 1 (Theorem 2.1, [3]) *Let* $\{R_i\}_{i \in I}$ *and* $\{S_j\}_{j \in J}$ *be two families of integral domains and let* $R = \prod_{i \in I} R_i$ *and* $S = \prod_{j \in J} S_j$. *Then* $\Gamma(R) \cong \Gamma(S)$ *if and only if there is a bijection* $\phi : I \longrightarrow J$ *such that* $|R_i| = |S_{\phi(i)}|$ *for each* $i \in I$. *In particular, if* $\Gamma(R) \cong \Gamma(S)$ *and each* R_i *is a finite field, then each* S_j *is also a finite field and* $R_i \cong S_{\phi(i)}$ *for each* $i \in I$ *and thus* $R \cong S$.

Note that in a commutative (or noncommutative) ring R which has no nilpotent element, the zero-divisor graph $\Gamma_0(R)$ (or $\vec{\Gamma}_0(R)$) has no loop. In this case, $\Gamma(R) = \Gamma_0(R)$ (or $\vec{\Gamma}(R) = \vec{\Gamma}_0(R)$). In particular, $\Gamma(R) = \Gamma_0(R)$ for any integral domain R.

Redmond [21] provided the following useful result:

Theorem 2 ([21], Theorem 2.1) *Let G be a graph with four or more vertices that is the zero-divisor graph of a finite commutative ring with 1. Then G can be associated with a finite commutative ring $R \cong R_1 \times R_2 \times \cdots \times R_m$, where each R_i is a local ring, such that $\Gamma(R) \cong G$. Furthermore, this association is unique up to elements of the same zero-divisor class. More specifically, if $S \cong S_1 \times S_2 \times \cdots \times S_t$ with each S_i local and $\Gamma(S) \cong G$, then $m = t$ and, after possible reordering, either $R_i \cong S_i$ or R_i and S_i are in the same zero-divisor class for each $i = 1, 2, \ldots, m$.*

Let $R = \mathbb{Z}_{p_1}^{\alpha_1} \times \cdots \times \mathbb{Z}_{p_m}^{\alpha_m}$, where $m \geq 1$, and for each i, p_i is prime and α_i is a positive integer. It is straightforward to check that if the zero-divisor graph $\Gamma_0(R)$ (or $\Gamma(R)$) of R has 3 or fewer vertices, then R is isomorphic to one of the five rings: $\mathbb{Z}_2 \times \mathbb{Z}_2$, $\mathbb{Z}_4$, $\mathbb{Z}_6$, $\mathbb{Z}_8$, or $\mathbb{Z}_9$. All of these five rings are in different zero-divisor classes. We first give some basic graph theory properties for the zero-divisor graph $\Gamma_0(\mathbb{Z}_{p^\alpha})$, where p is prime and α is an integer at least 2. Please see Section 4 of [11] for related results.

Proposition 1 *Let p be any prime number and α be any integer > 1. Then the zero-divisor graph $G = \Gamma_0(\mathbb{Z}_{p^\alpha})$ has the following properties:*

1. *G has $p^{\alpha-1} - 1$ vertices;*
2. *The maximum degree of G is $\Delta(G) = p^{\alpha-1}$ (including loops) and there are $p-1$ such vertices. Each of these vertices is adjacent to all other vertices.*
3. *If $\alpha = 2$, the minimum degree of G is $\delta(G) = p = \Delta(G)$. If $\alpha > 2$, then the minimum degree of G is $\delta(G) = p - 1$. In either case, there are $p^{\alpha-2}(p - 1)$ vertices bearing the minimum degree.*

Proof

(1) Obviously, the nonzero zero-divisors of $\mathbb{Z}_{p^\alpha}$ are the nonzero elements not relatively prime with p^α. There are $p^\alpha - \phi(p^\alpha) - 1 = p^{\alpha-1} - 1$ such elements, where $\phi(n)$ is the Euler number of n.

(2) The nonzero zero-divisors of $\mathbb{Z}_{p^\alpha}$ are given by $ap^i \in \mathbb{Z}_{p^\alpha}$ with $1 \leq i \leq \alpha - 1$ and $\gcd(a, p^i) = 1$. Elements in the form of $ap^{\alpha-1}$ with $\gcd(a, p) = 1$ are exactly the elements which annihilate all the nonzero zero-divisors, including themselves. These are the $p - 1$ vertices having the maximum degree $p^{\alpha-1}$.

(3) The minimum degree of the graph is at least $p - 1$ because each vertex is adjacent to the $p - 1$ vertices with the maximum degree. Each element in the form of ap with $\gcd(a, p^{\alpha-1}) = 1$ is adjacent only to $bp^{\alpha-1}$ where $\gcd(b, p) = 1$. These are the vertices bearing the minimum degree and there are $p - 1$ of them. When $\alpha = 2$, there is a loop at ap because $(ap)(ap) = 0$. Thus, the minimum degree $\delta(G) = p = \Delta(G)$. When $\alpha > 2$, such a loop does not exist, thus $\delta(G) = p - 1$.

$\square$

Proposition 2 *Let $R \cong \mathbb{Z}_{p^\alpha}$ and $S \cong \mathbb{Z}_{q^\beta}$, where p, q are prime numbers and α, β are positive integers at least 2. If $\Gamma_0(R) \cong \Gamma_0(S)$ or $\Gamma(R) \cong \Gamma(S)$ then $R \cong S$.*

Proof Note that, for a ring R, $\Gamma_0(R)$ and $\Gamma(R)$ have the same number of vertices, which is the number of nonzero zero-divisors of R. By Proposition 1(1), $\Gamma_0(R)$ (or $\Gamma(R)$) has $p^{\alpha-1} - 1$ vertices and $\Gamma_0(S)$ (or $\Gamma(S)$) has $q^{\beta-1} - 1$ vertices. Since $\Gamma_0(R) \cong \Gamma_0(S)$, $p^{\alpha-1} - 1 = q^{\beta-1} - 1$ which implies that $p = q$ and $\alpha = \beta$. Therefore $R \cong S$. Similarly, $\Gamma(R) \cong \Gamma(S) \Longrightarrow R \cong S$. $\qquad\square$

Definition 1 For any finite ring R, we define $N(R)$ as the number of nilpotent elements of index 2 in R.

In the paper by Birch, Thibideaux, and Tucci [11], the number of nilpotent elements of index 2 in a finite product of certain finite rings is calculated. Let v be a nonzero zero-divisor of a ring R. There is a loop from v to itself if and only if $v^2 = 0$, that is, v is a nilpotent element of index 2. Thus, the number of nilpotent elements of index 2 is the same as the number of loops in the zero-divisor graph. We claim the following proposition by applying the results from [11].

Proposition 3 *If p is prime and $\beta > 1$, then the number of loops in $\Gamma_0(\mathbb{Z}_{p^\beta})$ is $p^{\lfloor \beta/2 \rfloor} - 1$. Let $R \cong R_1 \times \cdots \times R_m$, where each R_i is a finite commutative ring and R is not a field. Then the number of loops in $\Gamma_0(R)$ is $N(R) = -1 + \prod_{i=1}^{m} (N(R_i) + 1)$. In particular, assume for each i, $R_i \cong \mathbb{Z}_{p_i^{\alpha_i}}$ (p_i is prime and $\alpha_i > 0$), then the number of loops in $\Gamma_0(R)$ is given by*

$$N(R) = -1 + \prod_{i=1}^{m} p_i^{\lfloor \alpha_i/2 \rfloor}.$$

Proof By Corollary 4.3 in [11], $\mathbb{Z}_{p^\beta}$ has $p^{\lfloor \beta/2 \rfloor} - 1$ nilpotent elements of index 2. Thus, $\Gamma_0(\mathbb{Z}_{p^\beta})$ has $p^{\lfloor \beta/2 \rfloor} - 1$ loops. By Lemma 3.2 in [11], $N(R) = -1 + \prod_{i=1}^{m} N(R_i)$. Since $\Gamma_0(R)$ has exactly $N(R)$ loops, the result follows. $\qquad\square$

Example 1 We apply Proposition 3 to confirm the numbers of loops in the zero-divisor graphs $\Gamma_0(\mathbb{Z}_{36})$ and $\Gamma_0(\mathbb{Z}_{40})$ which have 5 loops and 1 loop respectively. Precisely, the ring $\mathbb{Z}_{36} \cong \mathbb{Z}_{2^2} \times \mathbb{Z}_{3^2}$ has $-1 + 2 \cdot 3 = 5$ nilpotent elements of index 2: $(0, 3), (0, 6), (2, 0), (2, 3), (2, 6)$. These are the five vertices producing the loops in $\Gamma_0(\mathbb{Z}_{36})$. The ring $\mathbb{Z}_{40} \cong \mathbb{Z}_5 \times \mathbb{Z}_{2^3}$ has $-1 + 1 \cdot 2 = 1$ nilpotent element of index 2: $(0, 4)$. So, there is only one loop in $\Gamma_0(\mathbb{Z}_{40})$: from $(0, 4)$ to itself.

Our main result is given below.

Theorem 3 *Let $R \cong R_1 \times R_2 \times \cdots \times R_m$, where either R_i is a finite field or $R_i \cong \mathbb{Z}_{p_i^{\alpha_i}}$ with p_i prime and $\alpha_i > 1$, $1 \le i \le m$. Likewise, let $S \cong S_1 \times S_2 \times \cdots \times S_t$, where either S_j is a finite field or $S_j \cong \mathbb{Z}_{q_j^{\beta_j}}$ with q_j prime and $\beta_j > 1$, $1 \le j \le t$. Suppose that either R or S has at least 4 nonzero zero-divisors. If $\Gamma_0(R) \cong \Gamma_0(S)$ then $R \cong S$.*

Proof The graph $\Gamma(R)$ (without loops) is the subgraph of $\Gamma_0(R)$ by removing the loops from $\Gamma_0(R)$. It is the same situation for $\Gamma(S)$ and $\Gamma_0(S)$. From $\Gamma_0(R) \cong \Gamma_0(S)$, we have $\Gamma(R) \cong \Gamma(S)$ (isomorphism of the two zero-divisor graphs after removing the loops respectively). By the hypothesis, each R_i or S_j is a finite local ring. Then by Theorem 2, $m = t$ and, after suitable rearrangement, either $R_i \cong S_i$ or $\Gamma(R_i) \cong \Gamma(S_i)$ for $i = 1, \ldots, m$.

Consider a fixed i such that $\Gamma(R_i) \cong \Gamma(S_i)$. Then both $\Gamma_0(R_i)$ and $\Gamma_0(S_i)$ have the same number of loops. Also, both R_i and S_i are not finite fields because a finite field has no nonzero zero-divisors. Thus $R_i \cong \mathbb{Z}_{p_i^{\alpha_i}}$ and $S_i = \mathbb{Z}_{q_i^{\beta_i}}$ with α_i and β_i at least 2. Then $\Gamma(R_i) \cong \Gamma(S_i) \Longrightarrow R_i \cong S_i$ by Proposition 2. Therefore, $R \cong S$. $\square$

Another approach to proving uniqueness is to count the number of edges and vertices in the appropriate zero-divisor graphs with loops. Unfortunately, this approach does not work. For example, $\Gamma_0(\mathbb{Z}_{36})$ and $\Gamma_0(\mathbb{Z}_{40})$ both have 23 vertices and 50 edges but are not isomorphic. However, $\Gamma_0(\mathbb{Z}_{36})$ has 5 loops, while $\Gamma_0(\mathbb{Z}_{40})$ has 1 loop. We therefore have the following conjecture.

Conjecture Consider two finite commutative rings R_1 and R_2. If $\Gamma_0(R_1)$ and $\Gamma_0(R_2)$ have the same number of vertices, edges, and loops, then $R_1 \cong R_2$.

2.1 Upper Triangular Matrix Rings Over Finite Fields

In this section, we focus on rings of upper triangular matrices over finite fields. The zero-divisor graphs are directed graphs allowing loops. A natural question is: "Does the isomorphism of the zero-divisor graphs of two upper triangular matrix rings imply the isomorphism of the two matrix rings?" Let n be a positive integer. We denote the $n \times n$ upper triangular matrix ring over a field F by $U_n(F)$.

Lemma 1 *Let n be a positive integer.*

1. *A matrix $M = [m_{ij}] \in U_n(F)$ is nilpotent iff $m_{jj} = 0$ for all $1 \leq j \leq n$.*
2. *Every nilpotent matrix $M \in U_n(F)$ is the sum of nilpotent matrices of index 2.*

Proof (1) is obvious. For (2), let E_{ij} be the elementary matrix with 1 in the (i, j) position and zeroes elsewhere. Then $M = [m_{ij}] = \Sigma_{1 \leq i \leq j \leq n} m_{ij} E_{ij}$. If M is nilpotent, then by (1), the sum is taken over all pairs (i, j) with $i < j$. We can easily check that $\left(m_{ij} E_{ij}\right)^2 = m_{ij}^2 E_{ij}^2 = 0$ if $i \neq j$. Thus for $i < j$, $m_{ij} E_{ij}$ is nilpotent of index 2. $\square$

Let X be the set of all nilpotent matrices in $U_n(F)$. Let $r(X) = \{M \in U_n(F) \mid XM = 0\}$ and let $l(X) = \{M \in U_n(F) \mid MX = 0\}$.

Lemma 2 *Let F be a finite field and $M = [m_{ij}] \in U_n(F)$.*

1. *If $M \in l(X)$, then the only nonzero entries of M occur in the nth column of M.*
2. *If $M \in r(X)$, then the only nonzero entries of M occur in the first row of M.*

3. *If $M \in l(X) \cap r(X)$, then $M = cE_{1n}$ with $c \in F$, meaning only the $(1, n)$-position entry of M may be nonzero.*

Proof Let $M \in l(X)$. Then $ME_{jk} = 0$ for every $E_{jk} \in X$ ($j < k$). It forces $m_{ij} = 0$ for all $i = 1, 2, \ldots, n$, that is, the jth column of M is 0. Since $j < k \leq n$, every column of M, except the nth column, consists of zeroes. Similarly, if $M \in r(X)$, then every row of M consists only zeros except for the first row. It follows that every matrix in $l(X) \cap r(X)$ is a scalar multiple of E_{1n}. $\qquad\square$

For a noncommutative ring R, let $L(R)$ be the set of vertices that produce loops in $\overrightarrow{\Gamma}_0(R)$. Let $\mathrm{adj}(L(R))$ be the set of vertices in $\overrightarrow{\Gamma}_0(R)$ which are both adjacent from and adjacent to each vertex in $L(R)$. The previous lemmas imply the following result:

Proposition 4 *Consider the matrix ring $U_n(F)$ over a field F. Then $\mathrm{adj}(L(R)) = l(X) \cap r(X)$.*

Theorem 4 *Let $R = U_n(F_1)$ and $S = U_m(F_2)$ be two upper triangular matrix rings over finite fields F_1 and F_2 respectively. If $\overrightarrow{\Gamma}_0(R) \cong \overrightarrow{\Gamma}_0(S)$, then $R \cong S$.*

Proof By Proposition 4 and Lemma 2,

$$\mathrm{adj}(L(R)) = \{aE_{1n} \mid 0 \neq a \in F_1\} \quad \text{and} \quad \mathrm{adj}(L(S)) = \{bE_{1m} \mid 0 \neq b \in F_2\}.$$

Because $\overrightarrow{\Gamma}_0(R) \cong \overrightarrow{\Gamma}_0(S)$, $|\mathrm{adj}(L(R)| = |\mathrm{adj}(L(S)| \implies |F_1| - 1 = |F_2| - 1 \implies |F_1| = |F_2|$. Thus $F_1 \cong F_2$. Now both R and S are upper triangular matrix rings over the same field. Since $\overrightarrow{\Gamma}_0(R) \cong \overrightarrow{\Gamma}_0(S)$, R and S have the same number of zero-divisors. This implies that R and S consist of matrices of the same dimension. Therefore $m = n$ and so $R \cong S$. $\qquad\square$

References

1. Akbari, S., Mohammadian, A.: On zero-divisor graphs of finite rings, J. Algebra 314 (2007), 168–184.
2. Anderson, D. F., Axtell, M., Stickles J.: Zero-divisor graphs in commutative rings, in Commutative algebra, Noetherian and non-Noetherian perspectives, Springer-Verlag, New York (2011).
3. Anderson, D. F., Levy, R., Shapiro, J.: Zero-divisor graphs, von Neumann regular rings, and Boolean algebras, J. Pure and Appl. Algebra 180, 221–241 (2003).
4. Anderson, D. D., Naseer, M.: Beck's coloring of a commutative ring, J. Algebra 159, 500–514 (1993).
5. Anderson, D. F., Badawi, A.: On the zero-divisor graph of a ring, Comm. Algebra 8, 3073–3092 (2008).
6. Anderson, D. F., Frazier, L., A. Livingston, A. P.: The zero-divisor graph of a commutative ring, II, In: Ideal Theoretic Methods in Commutative Algebra, Columbia, MO, (1999), Lecture Notes in Pure and Appl. Math., 220, Dekker, New York, 61–72 (2001).

7. Anderson, D. F. , Livingston, P. S.: The zero-divisor graph of a commutative ring, J. Algebra 217, 434–447 (1999).
8. Badawi, A.: Recent results on annihilator graph of a commutative ring: a survey, In: Nearrings, Nearfields, and Related Topics, edited by K. Prasad et al, New Jersey: World Scientific (2017).
9. Anderson, D.F., Badawi, A.: The zero-divisor graph of a commutative semigroup: A Survey, In Groups, Modules, and Model Theory-Surveys and Recent Developments, edited by Manfred Droste, László Fuchs, Brendan Goldsmith, Lutz Strüngmann, 23–39. Germany/NewYork: Springer (2017).
10. Beck, I.: Coloring of commutative rings, J. Algebra 118, 208–226 (1988).
11. Birch, L., Thibodeaux, J., Tucci, R.P.: Zero divisor graphs of finite direct products of finite rings, Comm. Algebra 42, 1–9 (2014).
12. Bŏzić, I., Petrović, Z.: Zero divisor graphs of matrices over commutative rings, Comm. Algebra 37, 1186–1192 (2009).
13. Chartrand, G., Lesniak, L., Chang, P.: Graphs and digraphs, 5th edition, ARC Press, Boca Raton, FL. (2011).
14. Coykendall, J., Sather–Wagstaff, S., Sheppardson, L., Spiroff, S.: On zero divisor graphs, Progress in Algebra 2, 241–299 (2012).
15. Dolzan, D., Oblak, P., The Zero-divisor Graphs of Rings and Semirings, Internat. J. Algebra Comput. 22, no. 4 , 1250033 (2012).
16. Lagrange, J. D.: On realizing zero-divisor graphs, Comm Algebra 36, 4509–4520 (2008).
17. Li, A., Tucci, R. P.: Zero-divisor graphs of upper triangular matrix rings, Comm Algebra 41, 4622–4636 (2013).
18. Li, B.: Zero-divisor graph of triangular matrix rings over commutative rings, International Journal of Algebra, Vol. 5(6), 255–260 (2011).
19. Redmond, S.: The zero-divisor graph of a non-commutative ring, In: Commutative Rings, Nova Sci. Publ., Hauppauge, NY., 39–47, (2002).
20. Redmond, S.: The zero-divisor graph of a non-commutative ring, Int. J. of Comm. Rings, Vol. 1(4), 203–221 (2002).
21. Redmond, S.: Recovering rings from zero-divisor graphs, J. Algebra Appl. Vol. 12, No.8, 1350047 (2013).

Some Combinatorial Cases of the Three Matrix Analog of Gerstenhaber's Theorem (Research)

Jenna Rajchgot, Matthew Satriano, and Wanchun Shen

1 Introduction

Let k be a field and let $M_d(k)$ be the space of $d \times d$ matrices with entries in k. In his 1961 paper [3], M. Gerstenhaber proved that the unital k-algebra generated by a pair of commuting matrices $X_1, X_2 \in M_d(k)$ has dimension at most d. Gerstenhaber's proof was algebro-geometric, and relied on the irreducibility of the scheme of pairs of $d \times d$ commuting matrices (a fact also proved in the earlier paper [7]). Linear algebraic proofs (see [2, 5]) and commutative algebraic proofs (see [1, 10]) of Gerstenhaber's theorem were later discovered.

The analog of Gerstenhaber's theorem for four or more pairwise commuting matrices is false. For example, if E_{ij} denotes the 4×4 matrix with a 1 in position (i, j) and 0s elsewhere, then the unital k-algebra generated by $E_{13}, E_{14}, E_{23}, E_{24}$ has k-vector space basis $I, E_{13}, E_{14}, E_{23}, E_{24}$ and thus has dimension $5 > 4$.

It is not known if the dimension of the unital k-algebra generated by three pairwise commuting matrices $X_1, X_2, X_3 \in M_d(k)$ can exceed d. Determining if this three matrix analog of Gerstenhaber's theorem is true is sometimes called the *Gerstenhaber problem*. For further details and history on Gerstenhaber's theorem and the Gerstenhaber problem, see [4, 9].

The Gerstenhaber problem can be viewed from a commutative-algebraic perspective, as in Proposition 1 below (see Proposition 2.4 and Corollary 2.9 of [8] for a proof). To state this result, we fix the following notation: given a set $X = \{X_1, \ldots, X_n\} \subseteq M_d(k)$ of pairwise commuting matrices, let $\mathcal{A}_X$ denote the

J. Rajchgot (✉)
Department of Mathematics and Statistics, University of Saskatchewan, Saskatoon, SK, Canada
e-mail: rajchgot@math.usask.ca

M. Satriano · W. Shen
Department of Pure Mathematics, University of Waterloo, Waterloo, ON, Canada
e-mail: msatrian@uwaterloo.ca; w35shen@uwaterloo.ca

© The Author(s) and the Association for Women in Mathematics 2020
B. Acu et al. (eds.), *Advances in Mathematical Sciences*, Association for
Women in Mathematics Series 21, https://doi.org/10.1007/978-3-030-42687-3_12

unital k-algebra generated by the matrices in X. Let $k[x_1, \ldots, x_n]$ be a polynomial ring in n variables over k and let (GP_n) denote the following statement (which is not always true):

(GP_n) Every $k[x_1, \ldots, x_n]$-module N which is finite dimensional over k and which has support $\operatorname{Supp} N = (x_1, \ldots, x_n)$ satisfies the inequality

$$\dim k[x_1, \ldots, x_n]/\operatorname{Ann}(N) \leq \dim N.$$

Proposition 1 *Fix a positive integer n. Statement (GP_n) is true if and only if for every positive integer d and every set of n pairwise commuting matrices $X = \{X_1, \ldots, X_n\} \subseteq M_d(k)$, the inequality $\dim A_X \leq d$ holds.*

Consequently, (GP_1) and (GP_2) are true, and (GP_n), $n \geq 4$, is false. Solving the Gerstenhaber problem is equivalent to determining if (GP_3) is true or false.

In this paper, we address (GP_3) in special cases. That is, we prove that the inequality

$$\dim k[x_1, x_2, x_3]/\operatorname{Ann} N \leq \dim N \tag{1}$$

holds for certain classes of modules. To motivate some of the classes that we treat, consider first the following example of a $k[x_1, x_2, x_3, x_4]$-module N for which $\dim k[x_1, x_2, x_3, x_4]/\operatorname{Ann} N > \dim N$. (This is the module associated to the standard counter-example, given above, to the four commuting matrix analog of Gerstenhaber's theorem. See [8, Example 1.7] for further explanation.)

Example 1 Let $S = k[x_1, x_2, x_3, x_4]$, let $\mathfrak{m} = (x_1, x_2, x_3, x_4)$, let $I = \mathfrak{m}^2 + (x_1, x_2)$, and let $J = \mathfrak{m}^2 + (x_3, x_4)$. Let $N = (S/I \times S/J)/\langle (x_3, -x_1), (x_4, -x_2) \rangle$. Note that N is 4-dimensional with basis $(1, 0), (x_3, 0), (x_4, 0), (0, 1)$. We have $\operatorname{Ann}(N) = \mathfrak{m}^2$ and so

$$5 = \dim S/\operatorname{Ann} N > \dim N = 4.$$

We make the following observations:

(i) N is an extension of a cyclic module by $S/\mathfrak{m}$. That is, it fits into the short exact sequence

$$0 \to S/I \xrightarrow{i} N \xrightarrow{\pi} S/\mathfrak{m} \to 0$$

such that $i(f) = (f, 0)$ and $\pi(f, g) = g$.

(ii) N is combinatorial in the sense that I and J are monomial ideals, the module N is obtained from S/I and S/J by identifying monomials in S/I with monomials in S/J, and $\operatorname{Ann} N$ is a monomial ideal.

   Equivalently, N can be described in terms of two 4-dimensional analogs of Young diagrams together with "gluing" data. Indeed, with respect to the usual correspondence between monomial ideals in $k[x_1, \ldots, x_n]$ and

n-dimensional analogs of Young diagrams (see Sect. 4.1), S/I corresponds to the 4-dimensional Young diagram λ drawn below and S/J corresponds to the 4-dimensional Young diagram μ drawn below; both λ and μ are supported in 2-dimensional coordinate spaces in this case. Boxes are labelled by their corresponding monomials. Grey boxes in λ are identified with grey boxes in μ, corresponding to the relations $(x_3, 0) = (0, x_1)$ and $(x_4, 0) = (0, x_2)$ in the module N.

$$\lambda = \begin{array}{|c|c|} \hline x_4 & \\ \hline 1 & x_3 \\ \hline \end{array} \qquad \mu = \begin{array}{|c|c|} \hline x_2 & \\ \hline 1 & x_1 \\ \hline \end{array} \tag{2}$$

Then, $\dim N$ is the total number of boxes in λ plus the number of unshaded boxes in μ, and $\operatorname{Ann} N$ is the monomial ideal associated to the diagram $\lambda \cup \mu$.

Motivated by Example 1 (i), the first two listed authors of the present paper proved in [8, Theorem 1.5] that inequality (1) holds whenever N is an extension of a finite dimensional cyclic module $k[x_1, x_2, x_3]/I$ by a simple module $k[x_1, x_2, x_3]/(x_1, x_2, x_3)$, i.e. $(GP3)$ holds for such N. In this paper, the cases we consider are motivated by Example 1 (ii), and by our result in [8]. We briefly discuss these cases now.

1.1 Towards Double Extensions of Cyclic Modules

In light of [8, Theorem 1.5], it is natural to ask if (GP_3) holds for finite dimensional modules which are double extensions by $S/(x_1, x_2, x_3)$ of a cyclic module. To be precise, let $S = k[x_1, \ldots, x_n]$, let $\mathfrak{m} = (x_1, \ldots, x_n)$, and define a *double extension module* to be an S-module N with the following properties:

- N is finite dimensional with support $\mathfrak{m}$
- there exists an ideal I and module N_1 which fits into the following short exact sequences

$$0 \to S/I \to N_1 \to S/\mathfrak{m} \to 0, \qquad 0 \to N_1 \to N \to S/\mathfrak{m} \to 0.$$

In Sect. 2, we begin our study of double extension modules by proving the following:

Proposition 2 *Let N be an S module with $\operatorname{Supp} N = \mathfrak{m}$. If N is a double extension module satisfying $\dim S/\operatorname{Ann} N > \dim N$, then there exists an ideal I' and a module map*

$$\beta : (x_1^2, x_2, \ldots, x_n) \to S/I'$$

satisfying $\dim I'/(I' \cap \ker \beta) > 2$.

Then, in Sect. 3 we prove the following:

Theorem 1 *Let $n = 3$ so that $S = k[x_1, x_2, x_3]$, and let $I \subseteq S$ be a monomial ideal with $\sqrt{I} = \mathfrak{m}$. If r is a positive integer and*

$$\beta : (x_1^r, x_2, x_3) \to S/I,$$

is a module map which maps monomials to monomials, then $\dim I/(I \cap \ker \beta) \leq r$.

In the proof of Proposition 2, we see that certain double extension modules N give rise to maps $\beta : (x_1^2, x_2, \ldots, x_n) \to S/I'$, $\sqrt{I'} = \mathfrak{m}$, and that

$$\dim S/\operatorname{Ann} N \leq \dim N \iff \dim I'/(I' \cap \ker(\beta)) \leq 2.$$

The following corollary is now immediate from Proposition 2 and the $r = 2$ case of Theorem 1:

Corollary 1 *Let N be a double extension $k[x_1, x_2, x_3]$-module which gives rise to a module map $\beta : (x_1^2, x_2, x_3) \to S/I'$ where I' is a finite colength monomial ideal and β maps monomials to monomials. Then N is not a counter-example to (GP_3).*

1.2 Other Combinatorial Classes

As pointed out in item (2) of Example 1, there are counter-examples to (GP_4) which can be described in terms of a pair of 4-dimensional Young diagrams λ and μ, together with gluing data. Indeed, in Example 1, λ and μ were glued to one another by identifying the two outer corners of λ with the two outer corners of μ. In Sect. 4, we investigate such 2-generated combinatorial modules. Our main result, which contrasts the four variable case, is the following (see Theorem 4 for a more precise statement):

Theorem 2 *(GP_3) holds for all modules obtained by gluing a subset of outer corners of one plane partition (a.k.a. 3-dimensional Young diagram) to another.*

Finally, in Sect. 5 we show that there are no counter-examples to (GP_n) of the form $N = J/I$ where $I, J \subseteq S$ are monomial ideals with $I \subseteq J$.[1]

Throughout the paper, we let $\mathbb{N}$ denote the set of non-negative integers.

---

[1] While we do not know of a reference where this is proved, we would not be surprised if this result is known, perhaps with a different proof.

2 A Reformulation of the Gerstenhaber Problem for Double Extensions of Cyclic Modules

By Proposition 1, the Gerstenhaber problem is true if and only if $\dim S/\operatorname{Ann} N \leq \dim N$ for all finite dimensional $k[x_1, x_2, x_3]$-modules with Supp $N = (x_1, x_2, x_3)$. Clearly, if N is a finite dimensional cyclic module, so that $N = S/I$ for an ideal $I \subseteq k[x_1, x_2, x_3]$, then $\dim S/\operatorname{Ann} N \leq \dim N$. Furthermore, it was proved in [8] that $\dim S/\operatorname{Ann} N \leq \dim N$ for all $k[x_1, x_2, x_3]$-modules N such that N has support (x_1, x_2, x_3), and N is an extension of a cyclic module by $S/(x_1, x_2, x_3)$. In this section, and the next, we consider modules obtained from such N by further extending by $S/(x_1, x_2, x_3)$. As discussed in Sect. 1, we call such modules *double extension modules*. In this section, we prove Proposition 2. The ideas in the proof are similar to those used in the proofs of [8, Propositions 1.10 and 2.2].

Proof of Proposition 2 Let $S = k[x_1, \ldots, x_n]$ and let $\mathfrak{m} = (x_1, \ldots, x_n)$. Let I be a finite-colength ideal in S, let N' be an extension of S/I by $S/\mathfrak{m}$, and let N be an extension of N' by $S/\mathfrak{m}$. Furthermore, assume Supp $N = \mathfrak{m}$.

The extension

$$0 \to N' \to N \to S/\mathfrak{m} \to 0$$

corresponds to a class $\alpha \in \operatorname{Ext}^1(S/\mathfrak{m}, N')$, and it was shown in the proof of [8, Proposition 2.2] that α lifts to a map $\alpha' : \mathfrak{m} \to N'$. It was furthermore shown in the same proof that $\dim S/\operatorname{Ann} N \leq \dim N$ if and only if

$$\dim S/\operatorname{Ann} N' + \dim \alpha'(\operatorname{Ann} N') \leq \dim N' + 1. \qquad (3)$$

We now consider two cases: α' factors through the submodule S/I of N', or it does not. In the first case, α' defines a map from $\mathfrak{m}$ to S/I, and so, by Rajchgot and Satriano [8, Theorem 3.1], we have $\dim \alpha'(I) \leq 1$. Furthermore, since N' is the extension of a cyclic module by $S/\mathfrak{m}$, [8, Theorem 1.5] implies that $\dim S/\operatorname{Ann} N' \leq \dim N'$. Adding these two inequalities together yields (3).

Consequently, if $\dim S/\operatorname{Ann} N > \dim N$, then α' does not factor through S/I. Let π be as in the bottom row of the diagram below. Then, since α' does not factor through $S/I \subseteq N'$, we see $\pi \circ \alpha'$ is surjective. We have a map of short exact sequences

$$
\begin{array}{ccccccccc}
0 & \longrightarrow & J & \longrightarrow & \mathfrak{m} & \longrightarrow & S/\mathfrak{m} & \longrightarrow & 0 \\
& & \downarrow{\scriptstyle \beta'} & & \downarrow{\scriptstyle \alpha'} & & \downarrow{\scriptstyle \simeq} & & \\
0 & \longrightarrow & S/I & \longrightarrow & N' & \overset{\pi}{\longrightarrow} & S/\mathfrak{m} & \longrightarrow & 0
\end{array}
$$

where $J = \ker(\pi \circ \alpha')$. The maps α' and β' define extensions

$$0 \to N' \to N \to S/\mathfrak{m} \to 0 \quad \text{and} \quad 0 \to S/I \to \widetilde{N} \to S/J \to 0,$$

respectively, and one checks that $\widetilde{N} \simeq N$.

Now, J has colength 2, so for some x_i, we have that $\{1, x_i\}$ is a basis of S/J. Without loss of generality, we may assume that $i = 1$. Consider the Lexicographic monomial order $x_n > x_{n-1} > \cdots > x_1$. Noting that $J \subseteq \mathfrak{m}$, it is easy to check that J has a Gröbner basis of the form

$$\{x_1^2 - a_1 x_1, x_2 - a_2 x_1, \ldots, x_n - a_n x_1\}, \quad a_j \in k.$$

Furthermore, the ideal generated by these terms has support at two distinct points unless $a_1 = 0$. Thus, $J = (x_1^2, x_2 - a_2 x_1, \ldots, x_n - a_n x_1)$. Let $x_i' = x_i - a_i x_1$, $2 \le i \le n$, and let $S' = k[x_1, x_2' \ldots, x_n']$. Let $\phi : S' \to S$ be the ring isomorphism given by $x_1 \mapsto x_1$, and $x_i' \mapsto x_i - a_i x_i$, $2 \le i \le n$. Then the short exact sequence of S-modules

$$0 \to S/I \to \widetilde{N} \to S/J \to 0$$

is also a short exact sequence of S'-modules via the ring map ϕ. Furthermore, the ring isomorphism ϕ induces an isomorphism between $S/\operatorname{Ann}_S(\widetilde{N})$ and $S'/\operatorname{Ann}_{S'}(\widetilde{N})$. Thus, $\dim S/\operatorname{Ann}_S(\widetilde{N}) \le \dim \widetilde{N}$ if and only if $\dim S'/\operatorname{Ann}_{S'}(\widetilde{N}) \le \dim \widetilde{N}$. Re-writing each module in our new coordinates $x_1, x_2' \ldots, x_n'$ yields a short exact sequence of S'-modules of the form

$$0 \to S'/I' \to M \to S'/(x_1^2, x_2', \ldots, x_n') \to 0 \tag{4}$$

and we have $\dim S'/\operatorname{Ann}_{S'} \widetilde{N} \le \dim \widetilde{N}$ if and only if $\dim S'/\operatorname{Ann}_{S'} M \le \dim M$. Consequently, $\dim S/\operatorname{Ann}_S N \le \dim N$ if and only if $\dim S'/\operatorname{Ann}_{S'} M \le \dim M$.

Let $\beta : (x_1^2, x_2', \ldots, x_n') \to S'/I'$ determine the extension in (4). Then, one can check that $\operatorname{Ann} M = I' \cap \ker \beta$. So,

$$\dim S'/\operatorname{Ann} M = \dim S'/(I' \cap \ker \beta)$$

$$= \dim S'/I' + \dim I'/(I' \cap \ker \beta).$$

Finally, since $\dim M = \dim S'/I' + 2$, the inequality $\dim S'/\operatorname{Ann} M \le \dim M$ holds if and only if $\dim I'/(I' \cap \ker \beta) \le 2$. $\qquad\square$

We end this section with an example of the usefulness of the main idea of the above proof (which was also a key idea in [8]), namely, the idea to translate the statement $\dim S/\operatorname{Ann} N \le \dim N$ into a statement about module maps.

Example 2 ($(GP)_n$ is true for extensions of S/I by S/I) Let $S = k[x_1, \ldots, x_n]$ and consider extensions of the form

$$0 \to S/I \to N \to S/I \to 0.$$

One may check (as in the proof of [8, Proposition 2.2]) that the corresponding class $\alpha \in \text{Ext}^1(S/I, S/I)$ is determined by a map $\beta : I \to S/I$ and that $\text{Ann } N = I \cap \ker \beta$. So,

$$\dim S/\text{Ann } N = \dim S/(I \cap \ker(\beta)) = \dim S/I + \dim I/(I \cap \ker(\beta))$$
$$= \dim S/I + \dim \beta(I).$$

Also, $\dim N = 2 \dim S/I$. Thus the inequality $\dim S/\text{Ann } N \leq \dim N$ is true if and only if the inequality $\dim \beta(I) \leq \dim S/I$ is true. This latter inequality obviously holds since the codomain of β is S/I.

3   Addressing the Gerstenhaber Problem for Double Extensions of Cyclic Modules in a Combinatorial Case

The purpose of this section is to prove Theorem 1. Throughout, let $S = k[x, y, z]$. Each ideal I will be assumed to be a finite colength monomial ideal. We say a module map $\beta : (x^r, y, z) \to S/I$ is a *monomial map* if β sends monomials to monomials. For each $\ell \geq 0$, we let

$$(S/I)_\ell := \{x^\ell y^i z^j \notin I \mid i, j \geq 0\}$$

and refer to this set of monomials as the x^ℓ-*slice of* S/I. We refer to any set of the form $(S/I)_\ell$ as an x-*slice of* S/I.

Notice that each x^ℓ-slice may be identified in a natural way with $k[y, z]/J_\ell$ where $J_\ell \subseteq k[y, z]$ is a monomial ideal. We say the x^ℓ-slice $(S/I)_\ell$ is *Gorenstein* if the socle $\text{Soc}(k[y, z]/J_\ell)$ is 1-dimensional as a k-vector space; recall the socle of a $k[y, z]$-module M is the subset of elements annihilated by (y, z).

3.1   The Case $I \subseteq (x^r, y, z)$

Here we assume that $I \subseteq (x^r, y, z)$ so that β restricts to a map $\beta|_I : I \to S/I$. Then $\ker(\beta|_I) = I \cap \ker(\beta)$, so

$$\beta(I) = I/(I \cap \ker(\beta)).$$

Consequently, $\dim I/(I \cap \ker(\beta)) \leq r$ if and only if $\dim \beta(I) \leq r$. We will show that this latter inequality holds. We begin by recording some properties that a map $\beta : (x^r, y, z) \to S/I$ would have to satisfy if it were a *counter-example*, that is, if $\dim \beta(I) > r$.

Our first goal is to identify those elements in I that could be mapped to nonzero elements in S/I by β. For this purpose, define

$$S_x := \{x^i \in I \mid r \le i\},$$

$$S_y := \{x^i y^j \in I \mid 0 \le i < r, j \ge 1\},$$

$$S_z := \{x^i z^l \in I \mid 0 \le i < r, l \ge 1\}.$$

Define a *border element* of S_y (respectively S_z) to be an $m \in S_y$ (respectively $m \in S_z$) such that $m/y \notin I$ (respectively $m/z \notin I$). Define a border element of S_x to be an $m \in S_x$ such that $m/x^r \notin I$. Let Ω_x, Ω_y and Ω_z be the set of border elements of S_x, S_y and S_z, respectively. Finally, define $\beta(\Omega_x)$, $\beta(\Omega_y)$, $\beta(\Omega_z)$ to be the submodule of S/I generated by the images, under β, of the monomials in Ω_x, Ω_y, Ω_z respectively.

Lemma 1 *If $\beta : (x^r, y, z) \to S/I$ is a monomial map then*

1. $\beta(I) \subseteq \beta(\Omega_x) + \beta(\Omega_y) + \beta(\Omega_z)$.
2. *If $x^r \mid \beta(x^r)$ (respectively $y \mid \beta(y)$, $z \mid \beta(z)$) then $\beta(\Omega_x) = 0$ (respectively $\beta(\Omega_y) = 0$, $\beta(\Omega_z) = 0$).*
3. $\dim \beta(\Omega_x) \le r$, $\dim \beta(\Omega_y) \le r$, and $\dim \beta(\Omega_z) \le r$.

Proof Suppose that $x^i y^j z^l \in I$. Then we must have $i \ge r$ or $j > 0$ or $l > 0$. Observe:

- If $i \ge r$ then $\beta(x^i y^j z^l) = \beta(x^r) x^{i-r} y^j z^l$, and so $y^j z^l$ divides $\beta(x^i y^j z^l)$.
- If $j > 0$ then $\beta(x^i y^j z^l) = x^i \beta(y) y^{j-1} z^l$, and so $x^i z^l$ divides $\beta(x^i y^j z^l)$.
- If $l > 0$ then $\beta(x^i y^j z^l) = x^i y^j \beta(z) z^{l-1}$, and so $x^i y^j$ divides $\beta(x^i y^j z^l)$.

Thus, if any two of the three conditions $i \ge r$, $j > 0$, $l > 0$ holds, we see that $x^i y^j z^l$ divides $\beta(x^i y^j z^l)$, and so $\beta(x^i y^j z^l) = 0$. This proves that every monomial in I which maps to a nonzero element is in one of S_x, S_y, or S_z, and so $\beta(I) \subseteq \beta(S_x) + \beta(S_y) + \beta(S_z)$.

Next, observe that the only elements of S_y that can map to non-zero elements of S/I are border elements in S_y: if $m \in S_y$ is not a border element, then $m = ym'$ for some $m' \in I$ and so $\beta(m) = \beta(y)m' = 0$. Similarly, the only elements of S_z and S_x which can map to non-zero elements of S/I are border elements. This proves (1).

Item (2) follows by noting that all elements of Ω_x, Ω_y, Ω_z are in I.

For item (3), one can check that there are only r distinct monomials in each of Ω_x, Ω_y, Ω_z. For example, for each $0 \le i < r$, there is a unique j such that $x^i y^j \in \Omega_y$. □

Lemma 2 *If $\beta : (x^r, y, z) \to S/I$ is a monomial map which is a counter-example, then $x^r y, x^r z, yz \in \ker(\beta)$.*

Proof *We only prove $\beta(x^r y) = 0$, the other two statements being similar. Proceed by contradiction and assume that $\beta(x^r y) \ne 0$. We have the following equality of nonzero monomials*

$$x^r \beta(y) = \beta(x^r) y,$$

which implies that y divides $\beta(y)$ *and* x^r *divides* $\beta(x^r)$. *Thus,* $\beta(\Omega_x) = \beta(\Omega_y) = 0$ *by (2) of Lemma 1. It follows that* $\beta(I) \subseteq \beta(\Omega_z)$ *and so* $\dim \beta(I) \leq r$ *by (1) and (3) Lemma 1. This contradicts the fact that* β *is a counter-example.* $\qquad\square$

Lemma 3 *Suppose* $\beta : (x^r, y, z) \to S/I$ *is a monomial map which is a counter-example. Then every element of* $\beta(I)$ *is contained in the socle of an x-slice. Moreover, if* $\ell \geq 0$ *and* $(S/I)_\ell$ *contains a non-zero element* $\omega \in \beta(\Omega_y)$, *then* $(S/I)_\ell$ *is Gorenstein. Similarly for* $\beta(\Omega_z)$.

Proof Since $\beta(x^r)$ is killed by y and z by Lemma 2, it is clear that $\beta(\Omega_x)$ is always mapped to a socle of an x-slice.

Next, say $\omega \in \Omega_y$ maps to the x^ℓ-slice. We show $\beta(\omega)$ is in the socle of this slice. It is clear, again by Lemma 2, that $\beta(\omega)$ is in the annihilator of z. To see that $\beta(\omega)$ is also killed by y, notice that since $\omega \in \Omega_y$ we have $j > 0$. So,

$$y\beta(\omega) = \omega\beta(y) = 0$$

as $\omega \in I$. Therefore, $\beta(\omega)$ is in the socle of the x^ℓ-slice.

By symmetry in y and z, every element of $\beta(\Omega_z)$ also maps to the socle of an x-slice. Therefore, every element of $\beta(I)$ maps to the socle of an x-slice by (1) of Lemma 1.

It remains to prove that if $\omega = x^i y^j \in \Omega_y$ and $0 \neq \beta(\omega)$ is in the x^ℓ-slice, then the slice is Gorenstein. Since $\beta(\Omega_y) \neq 0$, we have that $y \nmid \beta(y)$ by (2) of Lemma 1. Thus, we may assume $\beta(y) = x^u z^v$ for some $u, v \in \mathbb{N}$. Then $\beta(\omega) = x^{i+u} y^{j-1} z^v$, so $\ell = i + u$.

Now, $\beta(x^i y) = x^\ell z^v$ is a nonzero element in the x^ℓ-slice that is killed by z, so there are no monomials in the socle of the x^ℓ-slice which have strictly smaller y-coordinate (and strictly larger z-coordinate) than $\beta(\omega)$. On the other hand, there are also no monomials in the socle of the x^ℓ-slice with strictly larger y-coordinate (and strictly smaller z-coordinate) than $\beta(\omega)$ because any monomial in the x^ℓ-slice with y-coordinate strictly larger than $\beta(\omega) = x^{i+u} y^{j-1} z^v$ would be a multiple of $\omega = x^i y^j$, which is in I. $\qquad\square$

Corollary 2 *Let* $\beta: (x^r, y, z) \to S/I$ *be a monomial map with* $I \subseteq (x^r, y, z)$. *If* β *is a counter-example, then each x-slice contains at most one non-zero element of* $\beta(I)$.

Proof Fix an x-slice and suppose that two different monomials in I map to nonzero elements of this slice. Then, without loss of generality, our x-slice contains an element of $\beta(\Omega_y)$ as well as an element of $\beta(\Omega_x)$ or $\beta(\Omega_z)$. Since our x-slice contains an element of $\beta(\Omega_y)$, Lemma 3 tells us that our slice is Gorenstein and the $\beta(\Omega_y)$ element is in the socle. Since Lemma 3 also tells us that every element of $\beta(I)$ is in the socle of a slice, necessarily the other element of $\beta(\Omega_x)$ or $\beta(\Omega_z)$ maps to the same (unique) element of the socle. $\qquad\square$

Lemma 4 *Let* $\beta: (x^r, y, z) \to S/I$ *be a monomial map with* $I \subseteq (x^r, y, z)$ *and* $\beta(x^r) = 0$. *Then* β *is not a counter-example.*

Proof We proceed by induction on r. The base case $r = 1$ is a corollary of the main theorem of [8]; in fact the main theorem implies the $r = 1$ case *without* the assumption that $\beta(x) = 0$.

Now suppose $r \geq 2$. Let x^d, y^e, z^f be among the minimal generators of I. Then, $r \leq d$ as $I \subseteq (x^r, y, z)$.

We claim $\beta(y^e)$ and $\beta(z^f)$ are linearly independent in S/I, otherwise we may remove the x^0-slice to get a smaller counter-example. To make this precise, first notice that y^e and z^f are the only border elements in the x^0-slice, so by Lemma 1 (1), $\beta((S/I)_0) \subseteq (\beta(y^e), \beta(z^f))$ is at most 1-dimensional if $\beta(y^e)$ and $\beta(z^f)$ are linearly dependent. Define a map

$$\gamma : K = (x^{r-1}, y, z) \to (x)/(I \cap (x)) \simeq S/(I : x)$$

by $\gamma(f) = \beta(xf)$. Since $x(I : x) = I \cap (x)$,

$$\dim \gamma((I : x)) = \dim \beta(x(I : x)) = \dim \beta(I \cap (x)).$$

Note $\gamma(x^{r-1}) = \beta(x^r) = 0$ and $(I : x) \subseteq (x^{r-1}, y, z)$. Thus by the induction hypothesis, γ is not a counter-example, so that $\dim \beta(I \cap (x)) \leq r - 1$. It follows that

$$\dim \beta(I) = \dim \beta((S/I)_0) + \dim \beta(I \cap (x)) \leq r$$

and so β is not a counter-example.

In particular, $y \nmid \beta(y)$, for otherwise $\beta(y^e) = y^{e-1}\beta(y)$ would be divisible by $y^e \in I$, hence is zero in S/I, contradicting linear independence of $\beta(y^e)$ and $\beta(z^f)$. Similarly, $z \nmid \beta(z)$.

Therefore, we may assume

$$\beta(y) = x^u z^v, \quad \beta(z) = x^s y^t, \quad \beta(x^r) = 0.$$

for some $u, v, s, t \in \mathbb{N}$. Without loss of generality, we may assume $u \leq s$.

Now we show $v = f - 1$. Since $\beta(z^f) = x^s y^t z^{f-1} \notin I$ and $\beta(yz) = x^u z^{v+1} \in I$ by Lemma 2, we must have $f - 1 < v + 1$. On the other hand, $z^f \in I$ and $\beta(y^e) = x^u y^{e-1} z^v \notin I$, so $v < f$. Thus $v = f - 1$, $\beta(y^e) = x^u y^{e-1} z^{f-1}$.

Let w be the smallest integer such that $x^{u+w} z^{f-1} = x^w \beta(y) \in I$; note that $w \leq r$ by Lemma 2. Then all nonzero elements in $\beta(\Omega_y)$ must be contained in an x^l-slice with $u \leq l < u + w$. Indeed, for $x^i y^{e_i} \in \Omega_y$ with $i \geq w$, we have $\beta(x^i y^{e_i}) = x^{i+u} y^{e_i-1} z^{f-1} = 0$ as $x^{u+w} z^{f-1} \in I$.

Next, we consider the possible contributions from Ω_z. Let $h = x^i z^{f_i} \in \Omega_z$ be a border element, so $i \leq r$ and $f_i \leq f$.

For $i \leq u + w - 1$, we claim $\beta(h)$ is either zero or contained in an x^l-slice with $u \leq l < u + w$. By minimality of w, we see $x^{u+w-1} z^{f-1} \notin I$. Since $x^i z^{f_i} \in I$, we must have $f - 1 < f_i$; hence, $f = f_i$ and $\beta(h) = x^{s+i} y^t z^{f-1}$. Recall $x^{u+w} z^{f-1} \in$

I, so if $s + i \geq u + w$, then $\beta(h) = 0$; otherwise $s + i < u + w$, i.e. $\beta(h)$ is in some x^l-slice with $u \leq l < u + w$.

Hence, only $x^i z^{f_i} \in \Omega_z$ with $u + w - 1 < i < r$ can be mapped to nonzero elements in x^l-slices with $l \geq u + w$. The number of such elements is at most $\max\{r - u - w, 0\}$.

In other words, nonzero monomials in $\beta(I) \subseteq \beta(\Omega_y) + \beta(\Omega_z)$ are either contained in an x^l-slice with $u \leq l < u + w$, or of the form $\beta(h)$ with $h = x^i z^{f_i} \in \Omega_z$ and $u + w - 1 < i < r$. It then follows from Corollary 2 that

$$\dim \beta(I) \leq w + \max\{r - u - w, 0\} \leq \max\{r - u, w\} \leq r.$$

Thus, β is not a counter-example.   □

Lemma 5 *Let $\beta: (x^r, y, z) \to S/I$ be a monomial map with $I \subseteq (x^r, y, z)$ and*

$$\beta(\Omega_x) \subseteq \beta(\Omega_y) + \beta(\Omega_z).$$

Then β is not a counter-example.

Proof Consider the map $\gamma: (x^r, y, z) \to S/I$ with $\gamma(x^r) = 0$, $\gamma(y) = \beta(y)$, $\gamma(z) = \beta(z)$. Then $\gamma(I) = \beta(I)$, so it suffices to prove the result when $\beta(x^r) = 0$. This follows directly from Lemma 5.   □

Theorem 3 *If $\beta: (x^r, y, z) \to S/I$ is a monomial map with $I \subseteq (x^r, y, z)$, then β is not a counter-example.*

Proof For contradiction suppose β is a counter-example. By Lemma 1 (2) and Lemma 5, x^r does not divide $\beta(x^r)$, so $\beta(x^r) = x^a y^n z^m$ for some nonnegative integers a, n, m with $a \leq r$. Let d be minimal such that $x^d \in I$. Then $\beta(\Omega_x)$ is contained in the x^i-slices with $d - (r - a) \leq i < d$.

If $\beta(I)$ intersects an x^i-slice with $a \leq i < d - (r - a)$, then we claim that $\beta(\Omega_x) \subseteq \beta(\Omega_y) + \beta(\Omega_z)$, and so we are done by Lemma 5. To see this, we may assume without loss of generality that $\beta(\Omega_y)$ intersects the x^i-slice. Then the slice is Gorenstein and $\beta(x^{r+i-a})$ is in this slice. Moreover, $\beta(x^{r+i-a})$ is killed by y and z, i.e. it is the unique monomial in the socle of the slice, hence is contained in $\beta(\Omega_y)$. Since $\beta(\Omega_y)$ is closed under multiplication by x, it follows that $\beta(\Omega_x) \subseteq \beta(\Omega_y)$.

So, we may assume $\beta(I)$ does not intersect any x^i-slice with $a \leq i < d - (r-a)$. So, $\beta(I)$ is contained within the x^i slices for $i \in [0, a) \cup [d - (r - a), d)$. There are r such slices, so applying Corollary 2, we find $\dim \beta(I) \leq r$.   □

3.2   The Case $I \not\subseteq (x^r, y, z)$, and Finishing the Proof of Theorem 1

As above, let I be a finite colength monomial ideal and let $\beta: (x^r, y, z) \to S/I$, which sends monomials to monomials. Assume that $I \not\subseteq (x^r, y, z)$. Since I is a monomial ideal, there exists some minimal integer $m < r$ such that $x^i \in I$ for all $i \geq m$. Furthermore, our choice of m ensures that $I \subseteq (x^m, y, z)$.

Lemma 6 *In the above situation, we have* $\beta(x^m y) = 0$, *and* $\beta(x^m z) = 0$.

Proof This is clear since $x^m \in I$.   □

Define a map $\beta' : \langle x^m, y, z \rangle \rightarrow S/I$ by $\beta'(x^m) = \beta(x^r)$, $\beta'(y) = \beta(y)$, and $\beta'(z) = \beta(z)$.

Lemma 7 β' *is a module map.*

Proof We first observe that $y\beta'(x^m) - x^m \beta'(y) = 0$. This is true since

$$y\beta'(x^m) - x^m \beta'(y) = y\beta(x^r) - x^m \beta(y) = 0 - 0 = 0.$$

Similarly, $z\beta'(x^m) - x^m \beta'(z) = 0$. Finally, $y\beta'(z) - z\beta'(y) = 0$ since $\beta'(z) = \beta(z)$ and $\beta'(y) = \beta(y)$.   □

We are now ready to prove the main result of this section.

Proof of Theorem 1 If $I \subseteq (x^r, y, z)$, then we are done by Theorem 3. So, assume that $I \nsubseteq (x^r, y, z)$ and choose m to be the minimal integer $m < r$ such that $x^i \in I$ for all $i \geq m$. Let $\beta' : (x^m, y, z) \rightarrow S/I$ be as above. Then $I \subseteq (x^m, y, z)$ and by Theorem 3, we have that $\dim \beta'(I) \leq m$. Thus, by construction of β', we have

$$m \geq \dim \beta'(I) = \dim \beta(I \cap (x^r, y, z)) = \dim \frac{I \cap (x^r, y, z)}{(I \cap (x^r, y, z)) \cap \ker(\beta)}$$

$$= \dim \frac{I \cap (x^r, y, z)}{I \cap \ker(\beta)}.$$

Now I is a monomial ideal, and the only monomials in I which are not in $I \cap (x^r, y, z)$ are $x^m, x^{m+1}, \ldots, x^{r-1}$. Thus, $\dim I/(I \cap (x^r, y, z)) = r - m$. This, together with the above inequality shows

$$\dim \frac{I}{I \cap \ker(\beta)} = \dim \frac{I}{I \cap (x^r, y, z)} + \dim \frac{I \cap (x^r, y, z)}{I \cap \ker(\beta)} \leq (r - m) + m = r,$$

yielding the desired result.   □

4 Gluing Plane Partitions: Addressing the Gerstenhaber Problem for Some Two-Generated Combinatorial Modules

4.1 Young Diagrams and Skew-Diagrams

Let $S = k[x_1, \ldots, x_n]$ and let I be a finite colength monomial ideal in S. Associate to S/I the set of lattice points $\mathbf{c} := (c_1, \ldots, c_n) \in \mathbb{N}^n$ such that the monomial

$\mathbf{x^c} := x_1^{c_1} \cdots x_n^{c_n} \in S \setminus I$. This set of lattice points is naturally identified with an n-dimensional *Young diagram* (a.k.a. *standard set* or *staircase diagram*). See [6, Ch. 3] for details. If K is a finite colength monomial ideal with $I \subseteq K$, associate to K/I the set of lattice points $\mathbf{c} \in \mathbb{N}^n$ such that $\mathbf{x^c} \in K \setminus I$. This set of lattice points is naturally identified with $\nu := \lambda \setminus \lambda'$ where λ and λ' are the n-dimensional Young diagrams associated to S/I and S/K respectively. Observe that ν can be decomposed uniquely into $\nu_1 \cup \cdots \cup \nu_r$ such that the following hold:

1. For each ν_j and each pair of boxes $\mathbf{b}_1, \mathbf{b}_2 \in \nu_j$, there exists a sequence of moves of the form "move over one box in direction $\pm \vec{e}_i$" so that by starting at $\mathbf{b}_1$ and applying these moves, we end at $\mathbf{b}_2$, and we never leave ν_j in the process. Note that $\vec{e}_i = (0, \ldots, 0, 1, 0, \ldots, 0) \in \mathbb{N}^n$ denotes the ith standard basis vector.
2. $r \in \mathbb{N}$ is minimal such that 1. holds.

We call each ν_j a *skew-diagram*, and the union $\nu = \nu_1 \cup \cdots \cup \nu_r$ the *decomposition of ν into skew-diagrams*. Note that each n-dimensional Young diagram is a skew-diagram.

Using the correspondence between monomials $\mathbf{x^c}$ and their exponent vectors $\mathbf{c} \in \mathbb{N}^n$, we sometimes label a box in a skew-diagram by its coordinate $\mathbf{c} \in \mathbb{N}^n$, and sometimes by its associated monomial $\mathbf{x^c}$. Along these lines, we say that $\mathbf{c} \in \nu$ is a *socle* of ν if $\mathbf{x^c} \in \mathrm{Soc}(K/I)$. We let $\mathrm{Soc}(\nu)$ denote the set of socles of ν.

4.2 Background on Gluing

Now, let I, J, K, and L be finite colength monomial ideals in S such that $I \subseteq K$, $J \subseteq L$, and there is an S-module isomorphism $\phi : K/I \rightarrow L/J$ mapping monomials to monomials. In this section, we consider modules of the form

$$N = (S/I \times S/J)/\langle (k, -\phi(k)) \mid k \in K/I \rangle. \tag{5}$$

Modules N from (5) have a combinatorial description, which extends the correspondence between monomial ideals in S and n-dimensional analogs of *Young diagrams*. Indeed, let λ and μ denote the n-dimensional Young diagrams associated to S/I and S/J respectively. Let ν_λ and ν_μ denote the unions of skew-diagrams associated to K/I and L/J respectively. The isomorphism $\phi : K/I \rightarrow L/J$ is a partial gluing of λ to μ by identifying the skew-diagrams in ν_λ with those in ν_μ. Note that the shapes of the skew diagrams in ν_λ agree with the shapes of those in ν_μ, otherwise ϕ would fail to be an isomorphism.

Example 3 The module N from Example 1 is of the form of (5). Here $S = k[x_1, x_2, x_3, x_4]$, $I = \mathfrak{m}^2 + (x_1, x_2)$, $J = \mathfrak{m}^2 + (x_3, x_4)$, $K = L = \mathfrak{m}$ and $\phi : K/I \rightarrow K/J$ is defined by $\phi(x_3) = x_1$ and $\phi(x_4) = x_2$. With this presentation, N corresponds to the gluing of λ and μ along the grey boxes, as depicted in (2).

Example 4 Let $S = k[x, y]$. Consider the following Young diagrams:

Observe that $I = (x^5, x^4y, x^2y^3, xy^4, y^5)$ is the monomial ideal corresponding to λ and $J = (x^6, x^4y^2, x^3y^3, x^2y^4, y^5)$ is the monomial ideal corresponding to μ. Let $\nu_\lambda \subseteq \lambda$ consist of the two shaded skew-diagrams in λ (one with the boxes labelled 1, 2, 3 and the other with the boxes labelled 4, 5, 6), so that ν_λ corresponds to $K/I = ((x^3y, x^2y^2, y^3) + I)/I$. The union of two grey skew-diagrams $\nu_\mu \subseteq \mu$ corresponds to $L/J = ((x^5, x^4y, xy^3) + J)/J$. If $\phi : K/I \to L/J$ is the map which identifies box i in ν_λ with box i in ν_μ, then we obtain a module $N = (S/I \times S/J)/\langle (k, -\phi(k)) \mid k \in K/I \rangle$.

We next we translate the inequality $\dim S/\operatorname{Ann} N \le \dim N$ for the modules in (5) into a purely combinatorial one in terms of n-dimensional Young diagrams. This translation uses the following lemma.

Lemma 8 *Let $S = k[x_1, \ldots, x_n]$ and let $N = (S/I \times S/J)/\langle (k, -\phi(k)) \mid k \in K \rangle$ be as in (5). Then $\operatorname{Ann} N = I \cap J$.*

Proof Clearly $I \cap J \subseteq \operatorname{Ann} N$. On the other hand, suppose that $r \in \operatorname{Ann} N$. Then $r \cdot (1, 0) = (r, 0)$ is 0 in N and so $(r, 0)$ must be an element of the submodule $\langle (k, -\phi(k)) \mid k \in K/I \rangle \subseteq S/I \times S/J$. Since ϕ is an isomorphism, we have that $r = 0$ in S/I and thus $r \in I$. A similar argument shows that $r \in J$. □

If λ is the n-dimensional Young diagram associated to S/I and μ is the n-dimensional Young diagram associated to S/J then, by Lemma 8, we see that $S/\operatorname{Ann} N$ corresponds to the n-dimensional Young diagram $\lambda \cup \mu$. Consequently, $\dim S/\operatorname{Ann} N$ is the number of boxes in $\lambda \cup \mu$ which we denote by $|\lambda \cup \mu|$.

Example 5 In Example 1, $\lambda \cup \mu$ is the 4-dimensional Young diagram with five boxes labeled by monomials 1, x_1, x_2, x_3, x_4.

Let N be a module determined by gluing λ to μ along ν_λ, ν_μ as explained above. Let $\nu := \nu_\lambda$. Then, $\dim N = |\lambda| + |\mu| - |\nu|$ and $\dim S/\operatorname{Ann} N = |\lambda \cup \mu|$. So, we have

$$\dim S/\operatorname{Ann} N \le \dim N \iff |\lambda \cup \mu| \le |\lambda| + |\mu| - |\nu| \iff |\nu| \le |\lambda \cap \mu|. \quad (6)$$

Example 6 Continuing Example 1, we see that $|\lambda \cap \mu| = 1$, while $|\nu| = 2$. Thus, we have $|\nu| > |\lambda \cap \mu|$.

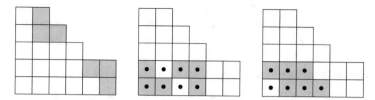

Fig. 1 Start with μ and $\nu_\mu = \nu_1 \cup \nu_2$ as in the left diagram. The grey boxes in the middle diagram are copies of ν_1, ν_2 after they have been shifted vertically down to the x-axis and then left to the origin so that ν_1 and ν_2 are next to one another with no columns in between. The boxes with bullets are the boxes in the smallest Young diagram containing all the grey boxes. The rightmost diagram consists of the ν_i ordered from largest to smallest along the x-axis, and the boxes with the bullets indicate those boxes in η

Continuing Example 4, we have $|\lambda \cap \mu| = 16$ while $|\nu| = 6$, and so $|\nu| \le |\lambda \cap \mu|$.

Example 7 ($|\nu| \le |\lambda \cap \mu|$ in 2-Dimensions) Gerstenhaber's theorem implies that for $\lambda, \mu, \nu \subseteq \mathbb{N}^2$, the inequality $|\nu| \le |\lambda \cap \mu|$ holds. It is also not difficult to prove this directly.

Let $\nu := \nu_\lambda$ and let $\nu_1 \cup \cdots \cup \nu_r$ be the decomposition of ν into skew-diagrams (see Sect. 4.1). For each ν_i, let $H_0(\nu_i)$ be the height of the smallest rectangle that fits the shape ν_i. More generally, let $H_j(\nu_i)$ be the height of the smallest rectangle which fits the skew shape obtained by deleting the leftmost j columns of ν_i. We place a lexicographical order on the ν_i in ν: we say $\nu_i = \nu_j$ if ν_i and ν_j have the same shape. Otherwise, there exists some smallest $m \ge 0$ where $H_m(\nu_i) \ne H_m(\nu_j)$, in which case we say that $\nu_i > \nu_j$ if $H_m(\nu_i) > H_m(\nu_j)$. Arrange the ν_i in ν along the x-axis from largest to smallest in our order so that the largest ν_i touches both the x and y axes, and there are no columns between subsequent ν_j's, and there are no columns that contain boxes from more than one ν_i. Let η denote the smallest Young diagram which contains this configuration of ν_j's.

Now each column of λ contains boxes from at most one ν_i in ν_λ. So, we may shift all the ν_i's down to sit on the x-axis and then shift them left so that one ν_i touches both the x and y axes, and there are no columns between subsequent ν_j's, and no columns that contain boxes from more than one ν_i. Observe that λ contains the smallest Young diagram which fits this arrangement of the ν_i, and this smallest Young diagram contains η. Thus $\eta \subseteq \lambda$. Similarly $\eta \subseteq \mu$. As η contains at least as many boxes as $|\nu|$, we have $|\nu| \le |\lambda \cap \mu|$.

See Fig. 1 for an example of the shifting processes described above.

Question 1 Does the inequality

$$|\nu| \le |\lambda \cap \mu|$$

hold for all possible 3-dimensional λ, μ, ν as above? In other words, does the inequality $\dim S/\operatorname{Ann} N \le \dim N$ always hold when N is a $k[x_1, x_2, x_3]$-module as in (5)?

Despite the simplicity of its 2-dimensional analog, Question 1 seems quite difficult in general. In the next section, we address it in the special case where ν is a union of corners of λ. Note that the standard four dimensional counter-example above has this form.

We end this section with two easy cases where the answer to Question 1 is "yes".

Example 8 (The Case $\nu_\lambda = \nu_1$) Let λ and μ be n-dimensional Young diagrams and suppose that ν_λ contains just one skew-diagram ν_1. Then inequality (6) holds. To see this, let $\mathbf{e}_i$ denote the i-th standard basis vector and let $\nu' \subseteq \mathbb{N}^n$ denote the unique skew-diagram isomorphic to ν_λ with the property that $\nu' - \mathbf{e}_i \not\subseteq \mathbb{N}^n$ for each i. In other words, ν' is obtained from ν_λ by translating as far as possible in all $-\mathbf{e}_i$ directions. Note that $\nu' \subseteq \lambda$. Similarly, $\nu' \subseteq \mu$ and hence $\nu' \subseteq \lambda \cap \mu$, proving that $|\nu_\lambda| = |\nu'| \le |\lambda \cap \mu|$.

Example 9 (The Case Where $\lambda, \mu \subseteq \mathbb{N}^3$ and Each Is Supported in a Plane) In the four variable counter-example discussed in Example 1, we saw $\lambda, \mu \subseteq \mathbb{N}^4$ were each supported in a 2-dimensional plane and $|\lambda \cap \mu| < |\nu|$. Here we see that this does not happen if $\lambda, \mu \subseteq \mathbb{N}^3$. Indeed, if λ and μ are in the same plane, we are reduced to the case of Example 7. So suppose that they are in different planes. Then each ν_i is a single box and we are in the case where we glue corners, which is proven more generally in the next section.

4.3   The Gerstenhaber Problem Where We Glue Corners

As explained in the introduction, (GP_n) is false for $n \ge 4$ due to Example 1. As further noted, this example is obtained by gluing λ to μ along corners. In contrast, we show in Theorem 4 that for $n < 4$, every 2-generated combinatorial module obtained by gluing corners does satisfy (GP_n).

We say that $(\lambda, \mu, \nu_\lambda, \nu_\mu)$ is a *counter-example* if it violates inequality (6). We say it is a *minimal counter-example* if it is a counter-example and $(\lambda', \mu', \nu_{\lambda'}, \nu_{\mu'})$ is not a counter-example whenever $\lambda' \subseteq \lambda$, $\mu' \subseteq \mu$, $\nu_{\lambda'} \subseteq \nu_\lambda$, $\nu_{\mu'} \subseteq \nu_\mu$, and $(\lambda', \mu', \nu_{\lambda'}, \nu_{\mu'}) \ne (\lambda, \mu, \nu_\lambda, \nu_\mu)$.

If additionally, each connected component of ν_λ is a singleton box, then we say $(\lambda, \mu, \nu_\lambda, \nu_\mu)$ is a *counter-example for gluing corners*, respectively a *minimal counter-example for gluing corners*.

Lemma 9 *If $(\lambda, \mu, \nu_\lambda, \nu_\mu)$ is a minimal counter-example, then*

1. $\operatorname{Soc}(\nu_\lambda) \cap (\lambda \cap \mu) = \varnothing = \operatorname{Soc}(\nu_\mu) \cap (\lambda \cap \mu)$,
2. $\operatorname{Soc}(\lambda) = \operatorname{Soc}(\nu_\lambda)$ *and* $\operatorname{Soc}(\mu) = \operatorname{Soc}(\nu_\mu)$.

Proof To prove the first assertion, assume to the contrary that $s_\lambda \in \operatorname{Soc}(\nu_\lambda) \cap (\lambda \cap \mu)$ and let $s_\mu \in \operatorname{Soc}(\nu_\mu)$ be the element to which s_λ is glued. Let $\lambda' = \lambda \setminus \{s_\lambda\}$ and $\mu' = \mu \setminus \{s_\mu\}$. Then

$$\lambda' \cap \mu' = (\lambda \cap \mu) \setminus \{s_\lambda, s_\mu\}.$$

So, $|\lambda' \cap \mu'|$ is either equal to $|\lambda \cap \mu| - 1$ or $|\lambda \cap \mu| - 2$, depending on whether s_μ is in $\lambda \cap \mu$. In either case,

$$|\lambda' \cap \mu'| \leq |\lambda \cap \mu| - 1.$$

Now, by minimality, we know $(\lambda', \mu', v_\lambda \setminus s_\lambda, v_\mu \setminus s_\mu)$ satisfies inequality (6), i.e. $|v_\lambda| - 1 \leq |\lambda' \cap \mu'| \leq |\lambda \cap \mu| - 1$. So, $|v_\lambda| \leq |\lambda \cap \mu|$, contradicting the fact that $(\lambda, \mu, v_\lambda, v_\mu)$ violates inequality (6). We have therefore shown that any minimal example must have the property that $\mathrm{Soc}(v_\lambda) \cap (\lambda \cap \mu) = \varnothing = \mathrm{Soc}(v_\mu) \cap (\lambda \cap \mu)$.

For the second assertion, if $s \in \mathrm{Soc}(\lambda) \setminus v_\lambda$, then let $\lambda' = \lambda \setminus s$. We have $\lambda' \cap \mu \subseteq \lambda \cap \mu$; in fact $|\lambda' \cap \mu| = |\lambda \cap \mu| - 1$ if $s \in \mu$, and $|\lambda' \cap \mu| = |\lambda \cap \mu|$ if $s \notin \mu$. Since $s \notin v_\lambda$, we can glue λ' to μ along v_λ, and by minimality, we know $(\lambda', \mu, v_\lambda, v_\mu)$ is not a counter-example. So, $|v| \leq |\lambda' \cap \mu| \leq |\lambda \cap \mu|$, and hence $(\lambda, \mu, v_\lambda, v_\mu)$ is also not a counter-example. $\square$

We now turn to the case of gluing corners. The following notion will play a central role.

Definition 1 We say λ is *jagged* if $|\mathrm{Soc}(\lambda \setminus s)| < |\mathrm{Soc}(\lambda)|$ for all $s \in \mathrm{Soc}(\lambda)$.

Remark 1 Let $e_i = (0, \ldots, 0, 1, 0, \ldots, 0)$ be the i-th standard basis vector. Notice that λ is jagged if and only if for all $1 \leq i \leq n$ and each $s \in \mathrm{Soc}(\lambda)$, we have $s - e_i \notin \mathrm{Soc}(\lambda \setminus s)$. Equivalently, λ is jagged if and only if for each such i and s, there exists $j \neq i$ such that $s - e_i + e_j \in \lambda$.

Example 10 The standard set of $(x_1, \ldots, x_n)^m$ is jagged for every m and n. Similarly, the standard set of $(x_1, x_2)^6 + x_3(x_1, x_2)^4 + x_3^2(x_1, x_2)^3 + x_3^3(x_1, x_2)$ is jagged; notice that this is obtained by "stacking" copies of $(x_1, x_2)^{m_i}$ on top of one another. However, not every jagged λ is obtained in this manner, e.g. the standard set of $(x_1^2, x_2^2) + x_3(x_1, x_2)^2$ is also jagged.

Corollary 3 *If $(\lambda, \mu, v_\lambda, v_\mu)$ is a minimal counter-example for gluing corners, then*

1. $\mathrm{Soc}(\lambda) \cap (\lambda \cap \mu) = \varnothing = \mathrm{Soc}(\mu) \cap (\lambda \cap \mu)$,
2. λ *and* μ *are jagged.*

Proof The first assertion follows immediately from Lemma 9 as $v_\lambda = \mathrm{Soc}(v_\lambda)$ and $v_\mu = \mathrm{Soc}(v_\mu)$.

For the second assertion, suppose $|\mathrm{Soc}(\lambda \setminus s)| \geq |\mathrm{Soc}(\lambda)|$ for some $s \in \mathrm{Soc}(\lambda)$. Let $\lambda' = \lambda \setminus s$ and choose some $s' \in \mathrm{Soc}(\lambda') \setminus \mathrm{Soc}(\lambda)$. By Lemma 9 (2), we know $\mathrm{Soc}(\lambda) = v_\lambda$, so let $s_\mu \in v_\mu$ be the box to which s is glued. Let $v_{\lambda'} = (v_\lambda \setminus s) \cup \{s'\}$ and note that we can glue λ' to μ along $v_{\lambda'}$ and v_μ; we simply glue s' to s_μ instead of gluing s to s_μ. By minimality, $(\lambda', \mu, v_{\lambda'}, v_\mu)$ is not a counter-example, so $|v_{\lambda'}| \leq |\lambda' \cap \mu|$. Since $s \notin \lambda \cap \mu$, we have $\lambda' \cap \mu = \lambda \cap \mu$, so

$$|v_\lambda| = |v_{\lambda'}| \leq |\lambda' \cap \mu| = |\lambda \cap \mu|,$$

which contradicts the fact that $(\lambda, \mu, v_\lambda, v_\mu)$ is a counter-example. $\square$

We next prove a result characterizing jagged 2-dimensional Young diagrams, and proving the key property of jaggedness that we need in 3 dimensions. We introduce the following terminology.

Definition 2 Let λ be an n-dimensional Young diagram. For each $1 \le s \le n$ and $t \ge 0$, we let

$$\lambda^{s,t} := \{(c_1, \ldots, c_n) \in \lambda \mid c_s = t\}$$

and refer to it as the *t-th slice of λ in the x_s-direction*; it is denoted simply as λ^t when s is understood. If t is maximal such that $\lambda^{s,t} \ne \varnothing$, we refer to $\lambda^{s,t}$ as the *top slice of λ in the x_s-direction*.

Remark 2 Notice that if λ is an n-dimensional Young diagram, then each slice $\lambda^{s,t}$ can be viewed naturally as an $(n-1)$-dimensional Young diagram.

Proposition 3 *Let λ be an n-dimensional Young diagram.*

1. *If $n = 2$, then λ is jagged if and only if it is the standard set of $(x_1, x_2)^k$ for some k.*
2. *If $n = 3$ and λ is jagged, then the top slice λ^t in the x_3-direction, when viewed as a 2-dimensional Young diagram, is the standard set of $(x_1, x_2)^k$ for some k.*

Proof Observe that (2) follows immediately from (1) since jaggedness of λ implies jaggedness of λ^t.

We now turn to (1). It is clear that the standard set of $(x_1, x_2)^k$ is jagged. Conversely, suppose λ is jagged and let $s = x_1^a x_2^b \in \text{Soc}(\lambda)$. By Remark 1, we see: (i) if $a > 0$ then $x_1^{a-1} x_2^{b+1} \in \lambda$, and (ii) if $b > 0$ then $x_1^{a+1} x_2^{b-1} \in \lambda$. Statement (i) implies that λ contains a socle in every column, i.e. for each j with $\lambda^{1,j} \ne \varnothing$, there exists k such that $x_1^j x_2^k \in \text{Soc}(\lambda)$. Similarly, statement (ii) implies that λ contains a socle in every row. Together these statements imply that λ is the standard set of $(x_1, x_2)^k$ for some k. $\square$

We can now answer Question 1 when we glue λ and μ along corners.

Theorem 4 (GP_3) *holds for 3-dimensional Young diagrams glued along corners, i.e. if the connected components of v_λ are singleton boxes then, inequality (6) holds.*

Proof If there is a counter-example for gluing corners, then there is a minimal such counter-example $(\lambda, \mu, v_\lambda, v_\mu)$. By Corollary 3 (2), we know λ and μ are jagged. Let λ^t be the top slice of λ in the x_3-direction, and $\mu^{t'}$ the top slice of μ in the x_3-direction. Without loss of generality, $t \le t'$.

By Proposition 3 (2), we know λ^t is of the form $(x_1, x_2)^k$ when it is viewed as 2-dimensional Young diagram. Since $\text{Soc}(\lambda) \cap \mu = \varnothing$ by Corollary 3 (1), when we view the x_3-slice μ^t as a 2-dimensional Young diagram, we must have $\mu^t \subseteq (x_1, x_2)^{k-1}$. In particular, $\mu^t \subsetneq \lambda^t$.

Next, choose $x_1^a x_2^b$ in the socle of the 2-dimensional Young diagram μ^t, and let $s = x_1^a x_2^b x_3^t \in \mu$. By definition, $s + e_1, s + e_2 \notin \mu$. Since $\mu^t \subsetneq \lambda^t$ and $\text{Soc}(\mu) \cap \lambda =$

$\varnothing$ by Corollary 3 (1), we must have $s \notin \mathrm{Soc}(\mu)$. As a result, $s + e_3 \in \mu$. Let m be maximal such that $s' := s + m e_3 \in \mu$. Then $s' \in \mathrm{Soc}(\mu)$. However, this contradicts jaggedness of μ, since $s' - e_3 \in \mathrm{Soc}(\mu \setminus s')$. $\hfill\square$

5 Addressing the Gerstenhaber Problem in the Monomial Ideal Case

Let I and J be two finite colength ideals in $S = k[x_1, \ldots, x_n]$ with $I \subseteq J$. Let $M = J/I$. Then, we have that $\mathrm{Ann}\, M = (I : J)$, and so

$$\dim S/\mathrm{Ann}\, M \leq \dim M \iff \dim S/(I : J) \leq \dim S/I - \dim S/J. \qquad (7)$$

We do not know if the rightmost inequality in (7) is true in general. In this section, we show it is true for monomial ideals I and J in any number of variables. We thank Alexander Yong for the key observation that shifting overlapping n-dimensional Young diagrams appropriately can only increase the number of boxes in their intersection (see Lemma 10).

We begin with some notation. Let ν be an n-dimensional Young diagram (e.g. associated to some S/I for a monomial ideal $I \subseteq S$, see Sect. 4.1). Given $\mathbf{a} = (a_1, \ldots, a_n)$, let $\nu_{\mathbf{a}}$ be the following shift of ν:

$$\nu_{\mathbf{a}} := \{\mathbf{c} \in \mathbb{N}^n \mid \mathbf{c} - \mathbf{a} \in \nu\}.$$

We can partition $\nu_{\mathbf{a}}$ into slices in the x_s direction. As in Definition 2, if the plane $x_s = t$ intersects $\nu_{\mathbf{a}}$ non-trivially, we define the t-slice of $\nu_{\mathbf{a}}$ to be the set of all $\mathbf{c} \in \nu_{\mathbf{a}}$ such that $c_s = t$. We refer to the $t = a_s$ slice as the *bottom slice* of $\nu_{\mathbf{a}}$ in the x_s direction. Let $\mathbf{e}_s$ be the sth standard basis vector, and note that if $\mathbf{c} \in \nu_{\mathbf{a}}$ is not in the bottom slice in the x_s direction, then $\mathbf{c} - \mathbf{e}_s$ is still an element of $\nu_{\mathbf{a}}$.

Lemma 10 *Let $\nu^1, \ldots, \nu^r$ be n-dimensional Young diagrams, and let $a(1), \ldots, a(r) \in \mathbb{N}^n$. Fix some $1 \leq s \leq n$ and assume that*

$$a(1)_s = a(2)_s = \cdots = a(l)_s > a(l+1)_s \geq \cdots \geq a(r)_s, \qquad (8)$$

for some $1 \leq l \leq r$. Then,

$$\left| \bigcup_{i=1}^{r} \nu^i_{a(i)} \right| \geq \left| \bigcup_{i=1}^{l} \nu^i_{a(i)-\mathbf{e}_s} \cup \bigcup_{i=l+1}^{r} \nu^i_{a(i)} \right|.$$

Proof Let $\nu^{(1)} = \bigcup_{i=1}^{l} \nu^i_{a(i)}$, $\nu^{(2)} = \bigcup_{i=l+1}^{r} \nu^i_{a(i)}$ and $\nu^{(1)} - \mathbf{e}_s = \bigcup_{i=1}^{l} \nu^i_{a(i)-\mathbf{e}_s}$. Then $|\nu^{(1)}| = |\nu^{(1)} - \mathbf{e}_s|$ since $\nu^{(1)} - \mathbf{e}_s$ is just a shift of $\nu^{(1)}$ in the $-\mathbf{e}_s$ direction. So, to prove the lemma, it suffices to show that $|v^{(1)} \cap v^{(2)}| \leq |(v^{(1)} - \mathbf{e}_s) \cap v^{(2)}|$. To do this, we will show that for each $\mathbf{b} \in v^{(1)} \cap v^{(2)}$, we have $\mathbf{b} - \mathbf{e}_s \in (v^{(1)} - \mathbf{e}_s) \cap v^{(2)}$.

If $\mathbf{b} \in v^{(1)} \cap v^{(2)}$ then $\mathbf{b}$ is simultaneously in $v^i_{\mathbf{a}(i)}$, for some $1 \leq i \leq l$, and in $v^j_{\mathbf{a}(j)}$, for some $l + 1 \leq j \leq r$. Then, it is clear by definition that $\mathbf{b} - \mathbf{e}_s$ is in $v^i_{\mathbf{a}(i) - \mathbf{e}_s}$. To see that $\mathbf{b} - \mathbf{e}_s \in v^j_{\mathbf{a}(j)}$, recall that $a(i)_s > a(j)_s$ by the assumption (8). Thus $\mathbf{b}$ is not in the bottom slice of $v^j_{\mathbf{a}(j)}$ in the x_s direction. Hence, $\mathbf{b} - \mathbf{e}_s$ is still in $v^j_{\mathbf{a}(j)}$ as noted above the statement of the present lemma. $\square$

Proposition 4 *Let I and J be monomial ideals in $k[x_1, \ldots, x_n]$ with $I \subseteq J$. Then (GP_n) is true for J/I.*

Proof We first prove the following general combinatorial statement: if $v^1, \ldots, v^r$ are n-dimensional Young diagrams and $\mathbf{a}(1), \ldots, \mathbf{a}(r) \in \mathbb{N}^n$, then

$$\left| \bigcup_{i=1}^r v^i \right| \leq \left| \bigcup_{i=1}^r v^i_{\mathbf{a}(i)} \right|. \tag{9}$$

We proceed by induction on the maximum distance of a vector $\mathbf{a}(i)$ to a coordinate hyperplane. More precisely, we induct on

$$\max\{t \in \mathbb{N} \mid \exists s \in [n] \text{ and } i \in [r] \text{ such that } \mathbf{a}(i)_s = t\}.$$

If $t = 0$, then $v^i_{\mathbf{a}(i)} = v^i$ for all i, and so (9) holds trivially. So, suppose $t > 0$, and choose any s, i such that $\mathbf{a}(i)_s = t$. After possibly re-labelling we may assume $t = \mathbf{a}(1)_s \geq \mathbf{a}(2)_s \geq \cdots \geq \mathbf{a}(r)_s$. If all of these inequalities are equalities, then define $\mathbf{a}'(i) = \mathbf{a}(i) - \mathbf{e}_s$ for each $1 \leq i \leq r$. Observe that (9) holds if and only if it holds upon replacing each $\mathbf{a}(i)$ by $\mathbf{a}'(i)$, as $\bigcup_{i=1}^r v^i_{\mathbf{a}'(i)}$ is just a shift of $\bigcup_{i=1}^r v^i_{\mathbf{a}(i)}$ backwards by one unit in the x_s direction.

If not all inequalities are equality then there is a first occurrence of a strict inequality $\mathbf{a}(l)_s > \mathbf{a}(l + 1)_s$ at some point in the chain. In this case, define $\mathbf{a}'(i) = \mathbf{a}(i) - \mathbf{e}_s$, for $1 \leq i \leq l$, and $\mathbf{a}'(i) = \mathbf{a}(i)$, for $l + 1 \leq i \leq r$. Then, Lemma 10 implies that (9) holds if it holds upon replacing each $\mathbf{a}(i)$ by $\mathbf{a}'(i)$.

In either of the above two cases, the maximum distance t' of an $\mathbf{a}'(i)$ to a coordinate hyperplane is still at most t. If it happens that $t' < t$, then the induction hypothesis yields the desired result. If $t' = t$, we can repeat the above process of shifting the various $v^i_{\mathbf{a}'(i)}$ until the maximum distance to a coordinate hyperplane does drop. It eventually will drop since there are only finitely many coordinate directions in which to shift. Hence (9) holds by induction.

The statement of the Proposition now follows: let $J = (x^{\mathbf{a}(1)}, \ldots, x^{\mathbf{a}(r)})$. Let $v_J := \{\mathbf{c} \in \mathbb{N}^n \mid x^{\mathbf{c}} \text{ is nonzero in } J/I\}$ and observe that $v_J = \bigcup_{i=1}^r v^i_{\mathbf{a}(i)}$, where

$$v^i_{\mathbf{a}(i)} = \{\mathbf{c} \in \mathbb{N}^n \mid x^{\mathbf{c}} \text{ is nonzero in } ((x^{\mathbf{a}(i)}) + I)/I\},$$

and v^i is the shift of $v^i_{\mathbf{a}(i)}$ to the origin, that is, $v^i = \{\mathbf{c} - \mathbf{a}(i) \mid \mathbf{c} \in v^i_{\mathbf{a}(i)}\}$. Let $\tilde{v} :=$ $\bigcup_{i=1}^{r} v^i$. Then, $\dim(J/I) = |v_J|$ and $\dim S/\operatorname{Ann}(J/I) = \dim S/(I : J) = |\tilde{v}|$. The above induction argument implies that $|\tilde{v}| \leq |v_J|$ as desired. □

Acknowledgements We thank Kathryn Lindsey for a useful conversation, Alexander Yong for the idea of the proof of Lemma 10, and the anonymous referee for helpful comments.

This paper is the outcome of an NSERC-USRA project. We thank NSERC for their support through the USRA program, and also through Rajchgot's NSERC grant RGPIN-2017-05732 and Satriano's NSERC grant RGPIN-2015-05631.

References

1. George M. Bergman, *Commuting matrices, and modules over artinian local rings*, arXiv:1309.0053, 2013.
2. José Barría and P. R. Halmos, *Vector bases for two commuting matrices*, Linear and Multilinear Algebra **27** (1990), no. 3, 147–157. MR 1064891
3. Murray Gerstenhaber, *On dominance and varieties of commuting matrices*, Ann. of Math. (2) **73** (1961), 324–348. MR 0132079
4. J. Holbrook and K. C. O'Meara, *Some thoughts on Gerstenhaber's theorem*, Linear Algebra Appl. **466** (2015), 267–295. MR 3278252
5. Thomas J. Laffey and Susan Lazarus, *Two-generated commutative matrix subalgebras*, Linear Algebra Appl. **147** (1991), 249–273. MR 1088666
6. Ezra Miller and Bernd Sturmfels, *Combinatorial commutative algebra*, Graduate Texts in Mathematics, vol. 227, Springer-Verlag, New York, 2005. MR 2110098
7. T. S. Motzkin and Olga Taussky, *Pairs of matrices with property L. II*, Trans. Amer. Math. Soc. **80** (1955), 387–401. MR 0086781
8. Jenna Rajchgot and Matthew Satriano, *New classes of examples satisfying the three matrix analog of Gerstenhaber's theorem*, J. Algebra **516** (2018), 245–270. MR 3863478
9. B. A. Sethuraman, *The algebra generated by three commuting matrices*, Math. Newsl. **21** (2011), no. 2, 62–67. MR 3013206
10. Adrian R. Wadsworth, *The algebra generated by two commuting matrices*, Linear and Multilinear Algebra **27** (1990), no. 3, 159–162. MR 1064892

Structure of Semigroup Rings (Survey)

Hema Srinivasan

1 Notations

Throughout this article, $\mathbb{N}$ will denote the non negative integers which is a commutative semigroup under addition. It is a subsemigroup of the group $\mathbb{Z}$ of integers. Similarly, $\mathbb{N}^n$ denotes a subsemigroup under addition of the group $\mathbb{Z}^n$ under addition.

Given a subsemigroup G of $\mathbb{N}$, we define $k[G] = k[t^a | a \in G]$ to be the semigroup ring associated to G. Thus, $k[G]$ is the subring of the polynomial ring $k[t]$. So, $k[G]$ is a one dimensional integral domain, unless $G = \{0\}$, when it is just k. Let G be minimally generated by a subset S of $\mathbb{N}$. Suppose d is the greatest common divisor of S. Then every element of G is a multiple of d. Hence G is isomorphic to $G/d = \{a/d | a \in G\}$ as a semigroup and $k[G]$ is isomorphic to $k[G/d]$ as a ring. Hence we may assume that S is relatively prime. When G is generated by a set of relatively prime positive integers then it is called a numerical semigroup. Further, when S generates G, $k[G] = k[S]$.

This can be generalized to subsemigroups of $\mathbb{N}^n$ as follows. For a vector $\mathbf{a_1} = (a_{11}, \ldots, a_{n1})^T \subset \mathbb{N}^n$, we write $\mathbf{t^{a_1}} = \prod_{i=1}^{n} t_i^{a_{i1}}$.

If G is a subsemigroup of $\mathbb{N}^n$, then $k[G]$, the semigroup ring associated to G is defined as follows. The semigroup ring $k[G] = k[\mathbf{t^a} | \mathbf{a} \in G]$ which is a subring of the polynomial ring $k[t_1, \ldots, t_n]$. As such, $k[G]$ is an integral domain. We say that $k[G]$ or G is nondegenerate if dimension of $k[G]$ is n. As in the case of $\mathbb{N}$, we may also assume that G is generated by a subset $S \subset \mathbb{N}^n$ which has no common factor other than 1.

H. Srinivasan (✉)
University of Missouri, Columbia, MO, USA
e-mail: srinivasanh@missouri.edu

© The Author(s) and the Association for Women in Mathematics 2020
B. Acu et al. (eds.), *Advances in Mathematical Sciences*, Association for
Women in Mathematics Series 21, https://doi.org/10.1007/978-3-030-42687-3_13

We say that the set S or the semigroup $< S >$ generated by it is Cohen–Macaulay or Complete Intersection or Gorenstein respectively when the semigroup ring $k[S]$ is Cohen–Macaulay or Complete Intersection or Gorenstein.

1.1 Semigroup Rings Presented as Quotients of Polynomial Rings

Let G be a semigroup generated by a subset $A = \{a_1, \ldots, a_s\}$. Then $\phi_A :$ $k[x_1, \ldots, x_s] \to k[t_1 \ldots, t_n]$ is a ring homomorphism given by $\phi_A(x_j) = t^{a_j}, 1 \le j \le s$. The image of ϕ_A is precisely the semigroup ring $k[G]$ and the kernel of ϕ_A is a prime ideal I_A and thus $k[G] \cong k[x, \ldots, x_s]/I_A$. The embedding dimension of $k[G]$ is s if and only if S minimally generate G. We will always consider the case when S is a minimal set of generators for G. When $n = 1$, in the numerical semigroup case, I_A has height $s - 1$ for the dimension of $k[G]$ is 1. In the general case of semigroups in $\mathbb{N}^n$, the dimension of $k[G]$ equals the rank of the $n \times s$ matrix $A = [a_{ij}]$.

Most of this article, we will concentrate on the case $n = 1$ and will mention when the theorems do generalize as they are stated to larger n. It is not hard to show that the ideal I_A is a binomial ideal, that is generated by binomials.

2 Resolutions of Semigroup Rings

In this section, $A = \{a_1, \ldots, a_p\}$ minimally generates the numerical semigroup $G =< A >$. So, $a_1, \ldots, a_p$ are relatively prime. The embedding dimension of the semigroup ring is p. Since $k[A] = kG] = k[x_1, \ldots, x_p]/I_A$ is a module over $k[x_1, \ldots, x_p] = R_A$, we are interested in the R_A- resolution of $k[A]$ as a method of understanding the structure of the semigroup rings. We write R for R_A where there is no danger of confusion.

We can also give a grading to this ring by setting degree of x_i to be a_i. Thus, I_A becomes homogeneous and $k[A]$ is a Cohen Macaulay graded ring of dimension one and the R_A—free resolution of $k[A]$ can be graded.

Recall that the set A or the semigroup $< A >$ generated by it is said to be Cohen–Macaulay or Complete Intersection or Gorenstein respectively when the semigroup ring $k[A]$ is Cohen–Macaulay or Complete Intersection or Gorenstein.

When $p = 1$, the ring is $k[A] = k[x]$.

When $p = 2$, $A = \{a, b\}$, then the ring $k[A] = k[x, y]/x^b - y^a$. The minimal resolution of $k[A]$ is simply

$$0 \to R \to R[A] \to 0$$

where the map is given by multiplication by $y^b - x^a$ and hence is a complete intersection.

Recall that the ideal I_A is a binomial ideal. In fact, $\prod_{i=1}^{p} x_i^{\alpha_i} - \prod_{i=1}^{p} x_i^{\beta_i} \in I_A, \alpha_i, \beta_i \in \mathbb{N}$ if and only if $\sum_i \alpha_i a_i = \sum_i \beta_i$.

When $p = 3$, the problem of finding I_A was completely solved by J. Herzog [6].

Theorem 1 ([6]) *Let* $A = \{a, b, c\}$ *minimally generate the numerical semigroup* $\mathbb{N}^n$. $k[A] = k[x, y, z]/I_A$. *Then there are exactly two cases.*

1. *Two of the three numbers in A, have a gcd common factor d, say,* $(a, b) = d$ *and the third c is in the semigroup generated by* $a/d, b/d$. *In this case,* $k[A]$ *is a complete intersection and hence* I_A *is generated by exactly two binomials.*
2. *Otherwise,* I_A *is generated by exactly three binomials which are necessarily the* 2×2 *minors of a* 2×3 *matrix.*

When $p = 4$, there is not even an upper bound for the minimal number of generators for the ideal I_A by the example of Brezinsky and Hoa.

Theorem 2 ([1]) *Let a be an even number other than 2. Then the minimal number of generators for the set of ideals* $\{I_\mathbf{a} | \mathbf{a} = (a^2 - a, a^2 - 1, a^2 + 2a - 1, a^2 + a)\}$ *is unbounded. In particular, the minimal number of generators for* $I_\mathbf{a}$ *is* $2a$ [7].

When $p = 4$, the best structure theorem we know is for the Gorenstein, non complete intersections given by Bresinsky [1]. There is a strengthening of this in [4] by removing some of the assumptions. This theorem is stated for space monomial curves. The semigroup rings associated to numerical semigroups are the homogenous coordinate rings of the monomial curves. Space monomial curves, thus correspond to semigroup rings with embedding dimension four. When $p = 4$, if the semigroup ring R_A is Gorenstein, then I_A is a height 3 Gorenstein ideal and hence by the structure theorem of Buchsbaum and Eisenbud, it must be given by the pfaffians of a skew symmetric matrix. One direction of the following theorem is that in the case of numerical semigroup rings, this results in I_A being generated by 5 binomials unless it is a complete intersection in which case it is generated by 3 principal binomials.

Theorem 3 ([1, 4]) *Let A be a* 4×4 *matrix of the form*

$$A = \begin{bmatrix} -c_1 & 0 & d_{13} & d_{14} \\ d_{21} & -c_2 & 0 & d_{24} \\ d_{31} & d_{32} & -c_3 & 0 \\ 0 & d_{42} & d_{43} & -c_4 \end{bmatrix}$$

with $c_i \geq 2$ *and* $d_{ij} > 0$ *for all* $1 \leq i, j \leq 4$, *and all the columns summing to zero. Then the first column of the adjoint of A (after removing the signs) defines a monomial curve provided these entries are relatively prime.*

For $p \geq 4$, we know the resolutions for some special cases of semigroup rings. For instance, for any $p \geq 4$, if the sequence $A = \{a_1, \ldots, a_p\}$ is an arithmetic sequence, then we have the explicit resolution for the semigroup rings and derive formulae for all the invariants [3].

The resolution is obtained as an iterated mapping cone. Instead of writing the entire resolution, we will now give the formula for the Betti numbers. For the explicit resolution, with the maps, we refer to [3]

Theorem 4 ([3]) *Let* $A = \{a_1, \ldots, a_{n+1}\}$ *be an Arithmetic sequence of length* $n + 1$. *Let* $k[A]$ *be the semigroup ring associated to* A. *Then the Betti numbers of* $k[A]$ *depend only on* a_1 *modulo* n. *Suppose* $a_1 = na + b, 1 \leq b \leq n$. *The Betti Numbers are given by*
$\beta_0 = 1$ *and*

$$
\beta_j = j \binom{n}{j+1} +
\begin{cases}
(n - b + 2 - j) \binom{n}{j-1} & \text{if } 1 \leq j \leq n - b + 1, \\
(j - n + b - 1) \binom{n}{j} & \text{if } n - b + 1 < j \leq n,
\end{cases}
\tag{1}
$$

Thus, the C-M type of $k[A]$ *is* $b - 1$. *In particular, the C-M type determines all the Betti numbers!*

Further, the regularity is given by

$$
reg k[A] =
\begin{cases}
d \binom{n}{2} + a_1(a + d) + n(a_1 - 1) & \text{if } b = 1, \\
d \left(\binom{n}{2} + b - 1 \right) + a_1(a + d + 1) + n(a_1 - 1) & \text{if } b \geq 2,
\end{cases}
\tag{2}
$$

Remark 1 The above situation is special for arithmetic sequences and will in general not be true. It is easy to check even in embedding dimension 4.

As can be seen in the above theorem, the Betti numbers of the semigroup ring R_A depends only on the Cohen–Macaulay type of $k[A]$ which in turn is determined by a_1 modulo n. In fact, the entire resolution, with the maps also depend only on a_1 modulo n, the height of I_A.

In the next section, we construct the resolution of semigroup rings that are obtained by gluing two semigroups. Some results on Betti numbers [8] and a survey can be found in [8] and [12].

3 Glued Semigroups

A numerical semigroup $< C >$ is obtained by *gluing* two semigroups $< A >$ and $< B >$ if its minimal generating set C can be written as the disjoint union of two subsets, $C = k_1 A \sqcup k_2 B$, where

1. A and B are numerically independent, i.e, minimally generate the semigroups $< A >$ and $< B >$ respectively and
2. k_1, k_2 are relatively prime positive integers such that k_1 is in the semigroup $< B >$ but not in B and $k_2 \in < A > \backslash A$.

When this occurs, we also say that C is *decomposable*, or that C is a *gluing of A and B*.

This notion of decomposition of the minimal generating set of a numerical semigroup already appears in the classical paper by Delorme [2] where it is used to characterize complete intersection numerical semigroups. In,1980, Rosales introduced the concept of gluing for finitely generated subsemigroups of $\mathbb{N}^n$ and showed that, for numerical semigroups, his definition coincides with the decomposition of Delorme [10].

Before we state the theorem constructing these resolutions, we will need more notations.

Let C have a decomposition $C = k_1 A \sqcup k_2 B$.
Let $A = \{a_1, \ldots a_p\}$, $B = \{b_1, \ldots, b_q\}$
Let $k[A] = k[x_1, \ldots, x_p]/I_A$ and
$k[B] = k[y_1, \ldots, y_q]/I_B$.
Let $R_A = k[x_1, \ldots, x_p]$, $R_B = k[y_1, \ldots, y_q]$.
Then $k[C] = k[x_1, \ldots, x_p, y_1, \ldots, y_q]/I_C$
Let us denote by $R = R_C = k[x_1, \ldots, x_p, y_1, \ldots, y_q] = R_A \otimes_k R_B$.

Lemma 1 *Let $C = k_1 A \sqcup k_2 B$ as above. Then the following are easy to check.*

Fact 1. If A and B are numerically independent, then C is numerically independent unless $k_1 \in B$ or $k_2 \in A$.

Fact 2: Since $k_1 \in < B >$ and $k_2 \in < A >$, there exist non negative integers α_i, β_i such that $k_1 = \sum_{j=1}^q \beta_j b_j$ and $k_2 = \sum_{i=1}^p \alpha_i a_i$.

Fact 3. The ideal I_C is minimally generated by the ideals I_A, I_B and exactly one other element

$$\rho = \prod_{i=1}^p x_i^{\alpha_i} - \prod_{j=1}^q y_j^{\beta_j} \in R.$$

Fact 4: ρ is homogeneous of degree $k_1 k_2$ if one gives to each variable in R the corresponding weight in $C = \{k_1 a_1, \ldots, k_1 a_p, k_2 b_1, \ldots, k_2 b_q\}$.

Remark 2 In fact, Rosales [11] defines gluing of semigroup rings in general by the Fact 3. That is, let $< A >$ be a subsemigroup of $\mathbb{N}^n$ minimally generated by a subset A and suppose that $A = B \sqcup C$. Then A is a gluing of B and C precisely if $I_A = I_B + I_C + (\rho)$ where ρ is a binomial of the form $f_X - g_Y$, where $f_X \in R_B$ and $g_Y \in R_C$.

Now, we are ready to state the resolution of the numerical semigroup ring R_C.

Theorem 5 ([5])) *Suppose* $C = k_1 A \sqcup k_2 B$.

1. $F_A \otimes F_B$ *is a minimal graded free resolution of* $R/(I_A R + I_B R)$.
2. *A minimal graded free resolution of the semigroup ring* $k[C]$ *can be obtained as the mapping cone of the map of complexes* $\rho : F_A \otimes F_B \to F_A \otimes F_B$, *where* ρ *is induced by multiplication by* ρ. *(In fact, all the maps in the map of complexes are multiplication by* ρ.) *In particular,* $(I_A R + I_B R :_R \rho) = I_A R + I_B R$.

So, we now can give formulae for the invariants. We collect all the consequences in one corollary. The consequence listed as 3 and 4, namely the type and the Hilbert Series have also been obtained by other methods by H. Nari.

Corollary 1 ([5]) *Let* $C = k_1 A \sqcup k_2 B$ *be a gluing as above, where* $A = \{a_1, \ldots a_p\}$ *and* $B = \{b_1, \ldots, b_q\}$ *minimally generate the corresponding semigroups.*
Then,

1. *The* ith *Betti number* β_i *is given by the formula*

$$\forall i \geq 0, \; \beta_i(C) = \sum_{i'=0}^{i} \beta_{i'}(A)[\beta_{i-i'}(B) + \beta_{i-i'-1}(B)].$$

2.

$$\beta_i(C) = \sum_{i'=0}^{i} \beta_{i'}(B)[\beta_{i-i'}(A) + \beta_{i-i'-1}(A)].$$

3. *The Cohen Macaulay type, is given by* $Type(C) = Type(A)Type(B)$ *This result is also obtained in [9]*
4. *The Hilbert series* H_C *of* R_C *is given by the formula*

$$H_C(t) = (1 - t^{k_1 k_2}) H_A(t^{k_1}) H_B(t^{k_2})$$

5. *The regularity is given by*

$$reg(C) = k_1(reg(A)) + k_2 reg(B) + (p-1)(k_1-1) + (q-1)(k_2-1) + k_1 k_2 - 1.$$

This formula is obtained by other methods in [9]
6. $k[C]$ *is Gorenstein, respectively a complete intersection, if and only if* $k[A]$ *and* $k[B]$ *are both Gorenstein, respectively complete intersections.*
7. *If neither* $k[A]$ *nor* $k[B]$ *is Gorenstein, then the Cohen–Macaulay type of* $k[C]$ *is not prime.*
8. *The graded Betti numbers are given by the formula*

$$\beta_{i,j}(C) = \sum_{i'=0}^{i} \left(\sum_{r,s/k_1 r + k_2 s = j} \beta_{i'r}(A)[\beta_{i-i',s}(B) + \beta_{i-i'-1,s-k_1}(B)] \right).$$

9. *The minimal graded resolution of R_C does admit an DG-algebra structure and hence is an associative, graded commutative differential graded algebra provided the minimal resolution of R_A and R_B do. In fact, the multiplication of R_C can be written explicitly in terms of the multiplication in R_A and R_B.*

There are examples in the published papers to illustrate the theorems.

Now, we will quickly summarize some of what we can do in higher dimension in the next remark. Following Rosales, we say a matrix $A = B \sqcup C$ is a gluing of B and C if there exists a $\rho \in R_A$, such that $I_A = I_B + I_C + (\rho)$. In the general case of semigroups in $\mathbb{N}^n$, we no longer have a nice situation like the Lemma 1. In fact, it is an open question to determine a criterion for gluing similar to Delorme's criterion for $n \geq 2$.

Remark 3 For semigroups contained in $\mathbb{N}^n$, we prove a generalization of Theorem 5 for $A = B \sqcup C$ provided at least one of the matrices B and C has rank one. This result will be in a forthcoming paper of Gimenez and Srinivasan. In fact, it can be shown that if B and C are both Cohen–Macaulay, we cannot glue them to get a Cohen–Macaulay semigroup ring R_A if $n \geq 2$. Thus, there are severe restrictions on what can be glued in higher dimension.

References

1. H. Bresinsky, Symmetric semigroups of integers generated by 4 elements, *Manuscrpita Math.* **17** (1975), 205–219.
2. C. Delorme, Sous-monoïdes d'intersection complète de N, *Ann. Sci. École Norm. Sup.* (4) **9** (1976), 145–154.
3. P. Gimenez, I. Sengupta and H. Srinivasan, Minimal graded free resolutions for monomial curves defined by arithmetic sequences, *Journal of Algebra* **388** (2013), 294–310.
4. P. Gimenez and H. Srinivasan, A note on Gorenstein monomial curves, *Bull. Braz. Math. Soc. New Series* **45** (2014), 671–678.
5. P. Gimenez and H. Srinivasan, Structure of Semigroup rings obtained by Gluing,*Journal of Pure and Applied Algebra* **223**(2019),1411–1426
6. J. Herzog, Generators and relations of abelian semigroups and semigroup rings, *Manuscripta Math.* **3** (1970), 175–193.
7. J. Herzog, D.I. Stamate, On the defining equations of the tangent cone of a numerical semigroup ring. *J. Algebra* Vol 418, **15** (2014), 8–28.
8. A. V. Jayanthan and H. Srinivasan, Periodic Occurance of Complete Intersection Monomial Curves, *Proc. Amer. Math. Soc.* **141** (2013), 4199–4208.
9. H. Nari, Symmetries on almost arithmetic numerical semigroups, Semigroup Forum **86** (2013), 140–154.
10. J. C. Rosales, On presentations of subsemigroups of $\mathbb{N}^n$, Semigroup Forum **55** (1997), 152–159.
11. J. C. Rosales and P. A. García-Sánchez, *Numerical Semigroups*, Developments in Mathematics **20**, Springer, 2009.
12. D.I. Stamate, Betti numbers for numerical semigroup rings, Multigraded Algebra and Applications *Springer Proceedings in Mathematics & Statistics, 133–157, (V. Ene, E. Miller, Eds.)*

Part V
Analysis, Probability, and PDEs

Using Monte Carlo Particle Methods to Estimate and Quantify Uncertainty in Periodic Parameters (Research)

Andrea Arnold

1 Introduction

Estimating and quantifying uncertainty in system parameters remains a big challenge in applied and computational mathematics. A subset of these problems includes estimating parameters that vary periodically with time but have unknown or uncertain time evolution models. Examples of periodic, time-varying parameters in dynamical systems arising from life sciences applications include the seasonal transmission in modeling the spread of infectious diseases [1, 9, 17] and the external voltage in modeling the spiking dynamics of neurons [34].

While most traditional algorithms aim at estimating constant parameters, the challenge in estimating time-varying parameters lies in accurately accounting for their time evolution without observations or known evolution models. In the case of periodic parameters, the resulting time series estimates should also maintain periodicity. Along with their time series, the period of these parameters may also be unknown and therefore may need to be estimated. This is particularly true in real data applications where a reasonable approximation of the period may not be clear from the available information.

The aim of this paper is to address the periodic parameter estimation problem, with particular focus on exploring the uncertainty associated with estimating periodic, time-varying parameters. In particular, this work uses sequential Monte Carlo particle methods (or nonlinear filtering methods) [13, 16, 26, 27, 29] to estimate the time series of periodic parameters. Note that while the term "sequential Monte Carlo" sometimes refers exclusively to particle filters, in this work the term more generally refers to sequential-in-time, Monte Carlo-based particle methods,

A. Arnold (✉)
Department of Mathematical Sciences, Worcester Polytechnic Institute, Worcester, MA, USA
e-mail: anarnold@wpi.edu

© The Author(s) and the Association for Women in Mathematics 2020
B. Acu et al. (eds.), *Advances in Mathematical Sciences*, Association for
Women in Mathematics Series 21, https://doi.org/10.1007/978-3-030-42687-3_14

including both particle filters and ensemble Kalman-type filters. In the Bayesian family of parameter estimation algorithms, Monte Carlo particle methods naturally account for uncertainty in the resulting parameter estimates by treating the unknowns as random variables with probability distributions describing their most likely values.

Both parameter tracking [20, 30, 34] and piecewise function approximations [8] of periodic parameters are considered, highlighting aspects of parameter uncertainty in each approach when considering factors such as the frequency of available data and the number of piecewise segments used in the approximation. Estimation of the period of the periodic parameters and related uncertainty is also analyzed in the piecewise formulation. As is demonstrated in the numerical results, while the parameter tracking method is efficient in tracking the overall behavior of slowly-varying parameters, it is unable to guarantee that periodicity is maintained in resulting parameter estimates. Pros and cons of each approach are discussed as applied to a numerical example estimating the external voltage parameter in the FitzHugh–Nagumo system for modeling neuron spiking dynamics.

The paper is organized as follows. Section 2 gives a review of the parameter estimation inverse problem and the Bayesian solution using sequential Monte Carlo particle methods, specifically outlining the augmented ensemble Kalman filter. Section 3 describes the parameter tracking and piecewise function approaches to estimating periodic parameters and discusses aspects of uncertainty relating to each approach. Section 4 gives numerical results on estimating the external voltage parameter in the FitzHugh–Nagumo model, and Sect. 5 provides discussion and future work.

2   Parameter Estimation and Monte Carlo Particle Methods

The parameter estimation inverse problem can be summarized as estimating unknown or uncertain system parameters given some discrete, noisy observations of (possibly a subset or some function of) the states of the system. More specifically, assume that an ordinary differential equation (ODE) model of the form

$$\frac{dx}{dt} = f(t, x, \theta), \qquad x(0) = x_0 \tag{1}$$

describes the dynamics of a system, which involves states $x = x(t) \in \mathbb{R}^d$ and unknown (or poorly known) parameters $\theta \in \mathbb{R}^q$. While the model function $f : \mathbb{R} \times \mathbb{R}^d \times \mathbb{R}^q \to \mathbb{R}^d$ is assumed to be known, the initial value $x_0 \in \mathbb{R}^d$ may also be unknown—in this case, x_0 may also be estimated along with the parameters θ. Further, assume the discrete, noisy observations $y_k \in \mathbb{R}^m$, $k = 1, 2, \ldots, T$, have the form

$$y_k = g(x(t_k), \theta) + w_k, \qquad 0 < t_1 < t_2 < \ldots < t_T \tag{2}$$

where $g : \mathbb{R}^d \times \mathbb{R}^q \to \mathbb{R}^m$, $m \leq d$, is a known observation function and w_k represents the observation error. The inverse problem is therefore to estimate the parameters θ and states $x(t)$ at some discrete times from the observations y_k.

From the Bayesian perspective, the unknown parameters θ, states x, and observations y are treated as random variables with probability distributions $\pi(\cdot)$, and the solution to the inverse problem is the joint posterior density

$$\pi(x, \theta \mid y) \propto \pi(y \mid x, \theta)\pi(x, \theta) \tag{3}$$

which follows from Bayes' theorem. The likelihood $\pi(y \mid x, \theta)$ indicates how likely it is that the data y are observed if the states x and parameters θ were known, and the prior density $\pi(x, \theta)$ encodes any information known about the states and parameters before accounting for the data.

There are various approaches to solving Bayesian inverse problems, including both sequential and nonsequential methods. Nonsequential methods, such as Markov chain Monte Carlo (MCMC)-type schemes [3, 18, 19], sample the posterior density by taking into account the full time series of data at once. Sequential Monte Carlo particle methods [16, 26, 27], on the other hand, make use of stochastic evolution-observation models to sequentially update the posterior using a two-step, predictor-corrector-type scheme, accounting for each data point as it arrives in time. A variety of Monte Carlo particle methods are available in the literature, including particle filters [6, 24, 29, 32] and ensemble Kalman-type filters [7, 11–13]. For a recent review, see [14].

Given the set $D_k = \{y_1, y_2, \ldots, y_k\}$ of observations up to time t_k, sequential Monte Carlo particle methods update the posterior distribution from time t_k to time t_{k+1} as follows:

$$\pi(x_k, \theta \mid D_k) \longrightarrow \pi(x_{k+1}, \theta \mid D_k) \longrightarrow \pi(x_{k+1}, \theta \mid D_{k+1}) \tag{4}$$

The first step (i.e., the prediction step) in the scheme predicts the values of the states at time t_{k+1} without knowledge of the data, while the second step (i.e., the analysis step) updates the predictions by taking into account the data at time t_{k+1}. Note that if there is no data observed at t_{k+1}, then $D_{k+1} = D_k$ and the prediction density $\pi(x_{k+1}, \theta \mid D_k)$ is equivalent to the posterior $\pi(x_{k+1}, \theta \mid D_{k+1})$. Starting with a prior density $\pi(x_0, \theta_0 \mid D_0)$, $D_0 = \emptyset$, this updating scheme is repeated until the final posterior density is obtained when $k = T$.

2.1 Augmented Ensemble Kalman Filter

The ensemble Kalman filter (EnKF) [11, 12] is a sequential particle approach that, unlike other particle methods that require importance sampling, moves (or pushes) particles forward in time based on the prediction and correction steps of the filter. Assume that the current density $\pi(x_k, \theta \mid D_k)$ is represented by a discrete ensemble

$$\mathcal{S}_{k|k} = \left\{ \left(x_{k|k}^n, \theta_{k|k}^n \right) \right\}_{n=1}^{N} \tag{5}$$

comprising N joint samples of the states $x_{k|k}^n$ and parameters $\theta_{k|k}^n$ at time k. In the prediction step of the filter, the state ensemble is updated using the equation

$$x_{k+1|k}^n = F(x_{k|k}^n, \theta_{k|k}^n) + v_{k+1}^n, \qquad v_{k+1}^n \sim \mathcal{N}(0, C_{k+1}) \tag{6}$$

for each $n = 1, \ldots, N$, where F is the numerical solution to the ODE system (1) from time k to $k + 1$. Note that the parameter samples $\theta_{k|k}^n$ are not updated in the prediction step.

To prepare for the analysis step, in which both the states and parameter values will be updated, the predicted state ensemble is combined with the current parameter ensemble into the augmented vectors

$$z_{k+1|k}^n = \begin{bmatrix} x_{k+1|k}^n \\ \theta_{k|k}^n \end{bmatrix} \in \mathbb{R}^{d+q}, \qquad n = 1, \ldots, N \tag{7}$$

and ensemble statistics formulas are used to compute the augmented ensemble mean $\bar{z}_{k+1|k}$ and covariance $\Gamma_{k+1|k}$. The covariance matrix $\Gamma_{k+1|k}$ contains cross-correlation information between the states and parameters that is used to update the parameter values in the next step.

In the analysis step, an observation ensemble

$$y_{k+1}^n = y_{k+1} + w_{k+1}^n, \qquad w_{k+1}^n \sim \mathcal{N}(0, D_{k+1}), \qquad n = 1, \ldots, N \tag{8}$$

is generated around the observation y_{k+1} to prevent the resulting posterior ensemble from having too low a variance [11]. The observation ensemble is then compared to the observation model predictions

$$\widehat{y}_{k+1}^n = g(x_{k+1|k}^n, \theta_{k|k}^n), \qquad n = 1, \ldots, N \tag{9}$$

with g as in (2) in the updating equation

$$z_{k+1|k+1}^n = z_{k+1|k}^n + K_{k+1}\left(y_{k+1}^n - \widehat{y}_{k+1}^n \right), \qquad n = 1, \ldots, N. \tag{10}$$

To accommodate nonlinear observations [31], the Kalman gain K_{k+1} in (10) is computed by

$$K_{k+1} = S_{k+1}^{z\widehat{y}} \left(S_{k+1}^{\widehat{y}\widehat{y}} + D_{k+1} \right)^{-1} \tag{11}$$

where $S_{k+1}^{z\widehat{y}}$ gives the cross-correlation between the augmented predictions $z_{k+1|k}^n$ in (7) and observation model predictions $\widehat{y}_{k+1}^n$ in (9), $S_{k+1}^{\widehat{y}\widehat{y}}$ is the forecast error covariance, and D_{k+1} is the observation noise covariance as in (8). The above algorithm, known as the augmented EnKF for combined state and parameter estimation [7, 13], is repeated until the joint posterior density is obtained at $k = T$.

3 Estimating Periodic Parameters and the Role of Uncertainty

In the traditional Monte Carlo particle methods described in Sect. 2, the parameters θ are assumed to be constant (or static) parameters, i.e., $d\theta/dt = 0$, and are artificially evolved over time as the posterior is updated. Depending on the implementation of the method used, the parameter values may be updated during both the prediction and analysis steps, or only in the analysis step via their correlation with the state predictions. In particular, the augmented EnKF outlined in Sect. 2.1 updates the parameter estimates only in the analysis step at each data arrival through the use of cross-correlation information encoded in the Kalman gain (11). The periodic parameters of interest in this work, however, are known to vary with time but do not have known time evolution models. The main challenges in this problem therefore lie in accurately accounting for the time evolution of these parameters while also maintaining the periodic structure.

One approach is to consider parameter tracking algorithms [20, 30, 34], which can trace the dynamics of slowly-changing parameters over time by allowing for a drift in the parameter values during the prediction step of sequential Monte Carlo. More specifically, the predicted change in the parameter $\theta(t)$ is modeled as a random walk

$$\theta_{k+1|k}^n = \theta_{k|k}^n + \xi_{k+1}^n, \qquad \xi_{k+1}^n \sim \mathcal{N}(0, \mathsf{E}_{k+1}), \qquad n = 1, \ldots, N \qquad (12)$$

where E_{k+1} defines the covariance of the drift term ξ_{k+1}^n. Note that inclusion of the drift term in (12) is crucial in allowing the algorithm to track the underlying dynamics of the time-varying parameter. While parameter tracking algorithms are straightforward to implement, the drift covariance E_{k+1}, which is typically modeled as $\mathsf{E}_{k+1} = \sigma_\xi^2 \mathsf{I}$ for some constant σ_ξ, must be chosen carefully in order to avoid filter divergence [2, 10, 21, 23, 37] and result in a useful parameter estimate. The drift covariance also plays a direct role in the uncertainty of the resulting parameter estimate, thereby affecting the corresponding model output predictions [5]. Moreover, in the case of estimating periodic parameters, parameter tracking algorithms do not guarantee that periodicity is maintained throughout the estimation process.

An alternative approach that maintains periodicity in the parameter estimation is to model the periodic parameter $\theta(t)$ as a piecewise function

$$\theta(t) = \begin{cases} \theta_1(t), & t \in \left[0, \dfrac{p}{\ell}\right) \\[2mm] \theta_2(t), & t \in \left[\dfrac{p}{\ell}, \dfrac{2p}{\ell}\right) \\[2mm] \vdots & \vdots \\[2mm] \theta_\ell(t), & t \in \left[\dfrac{(\ell-1)p}{\ell}, p\right) \end{cases} \qquad (13)$$

where each $\theta_i(t)$, $i = 1, \ldots, \ell$, is a function relying on some unknown constant coefficients, repeated each period p, that can be estimated using traditional Monte Carlo particle methods. A similar piecewise formulation using nonlinear filtering was presented in [8] assuming that the period p was known and fixed during the estimation process. However, in general the period of the parameter may not be known a priori and may need to be estimated along with the other unknown system parameters. Therefore, in this work, the period p is assumed to be unknown and is estimated along with the unknown piecewise function coefficients.

The formulation in (13) can accommodate estimation using piecewise constant functions or splines of various order. In this study and the numerical experiments that follow in Sect. 4, we employ a continuous linear interpolating spline (of degree 1) where

$$\theta_i(t) = a_i + b_i(t - t_{i-1}), \quad t \in [t_{i-1}, t_i) = \left[\frac{(i-1)p}{\ell}, \frac{ip}{\ell}\right) \qquad (14)$$

for $i = 1, \ldots, \ell$, with constant coefficients a_i and b_i denoting the y-intercept and slope of the line $\theta_i(t)$, respectively. Note that the spline knots t_j, $j = 0, \ldots, \ell$, in (14) depend on both the period p and number of spline segments ℓ. Continuity dictates that $\theta_i(t_i) = \theta_{i+1}(t_i)$ for $i = 1, \ldots, \ell - 1$, and it follows from definition of the linear spline in (14) that $\theta_i(t_{i-1}) = a_i$ for $i = 1, \ldots, \ell$. Since the slope coefficients b_i can be computed directly from the y-intercepts a_i, $i = 1, \ldots, \ell + 1$, via the formula

$$b_i = \frac{a_{i+1} - a_i}{t_i - t_{i-1}} = \frac{a_{i+1} - a_i}{p/\ell} = \frac{\ell}{p}(a_{i+1} - a_i) \qquad (15)$$

it suffices to estimate only the values for a_i, $i = 1, \ldots, \ell + 1$, along with the period p. Therefore the parameter estimation problem consists of estimating $L = \ell + 1$ spline coefficients and the period p, for a total of $L + 1$ unknown static parameters relating to the periodic parameter of interest.

Various factors must be considered in analyzing the uncertainty relating to the piecewise formulation (13)–(14). In this study, we consider how the frequency of the data in time and the number of linear spline segments ℓ used in the estimation affects the resulting periodic parameter estimates and corresponding uncertainty.

4 Numerical Example: External Voltage in FitzHugh–Nagumo

As a numerical example, we consider synthetic data generated from the FitzHugh–Nagumo system [15] which acts as a simplified version of the Hodgkin–Huxley system [22] for modeling the spiking dynamics of single neurons. The FitzHugh–Nagumo equations are given by

$$\frac{dx_1}{dt} = c\left(x_2 + x_1 - \frac{x_1^3}{3} + v(t)\right) \tag{16}$$

$$\frac{dx_2}{dt} = -\frac{1}{c}\left(x_1 - a + bx_2\right) \tag{17}$$

where the state variable $x_1(t)$ represents the measurable membrane potential of the neuron, while $x_2(t)$ denotes an unobservable combined effect of various ionic currents. The parameters a, b, and c are commonly fixed to some known values a priori, but the external voltage $v(t)$ is an unknown, time-varying parameter.

Figure 1 shows the synthetic data and underlying system states generated from (16)–(17) using initial values $x_1(0) = 1$ and $x_2(0) = 0.5$ and fixed parameters $a = 0.7, b = 0.8$, and $c = 3$, along with the time-varying external voltage parameter modeled as a periodic, sinusoidal function $v(t) = 0.5\sin(\omega t + \pi/2) - 1$ with frequency $\omega = 0.1$. Therefore, in this example, $v(t)$ plays the role of the periodic parameter $\theta(t)$ described in Sect. 3. This choice of $v(t)$ varies more slowly than the system dynamics, making it amenable to particle methods with parameter tracking. A similar example was considered in [5], where the focus was to study the effects of uncertainty in parameter tracking estimates and their corresponding model output predictions. The data was generated by observing $x_1(t)$ at 1257 equidistant time instances over the interval $[0, 251.2]$, covering four periods of $v(t)$, and corrupting the observations with zero-mean Gaussian noise. The standard deviation of the noise was taken to be 20% of the standard deviation of $x_1(t)$ over the full time interval.

For the first numerical experiment, we consider estimating the periodic parameter $v(t)$ using the piecewise formulation (13)–(14) with $\ell = 10$ spline segments ($L = 11$ knots) and estimating the $L = 11$ unknown linear spline coefficients $a_1, \ldots, a_{11}$, along with the unknown period p. While various particle methods could be applied, we employ an augmented EnKF in the style of [7] with $N = 150$ ensemble members to estimate the system states $x_1(t)$ and $x_2(t)$ along with the parameter vector $\theta = (a_1, \ldots, a_{11}, p) \in \mathbb{R}^{12}$ as described; see [7] for implementation details of the filter beyond those given in Sect. 2.1.

Assuming that the initial values of the system are not fully known, the prior ensemble of states is drawn uniformly from 0.5 to 1.5 times the value of the first observed value of $x_1(t)$ and set to 0 for $x_2(t)$ (unobserved). The prior ensemble of parameter values is drawn uniformly from $\mathcal{U}(-2, 1)$ for each of the spline coefficients $a_1, \ldots, a_{11}$ and from $\mathcal{U}(55, 75)$ for the period p. Throughout the

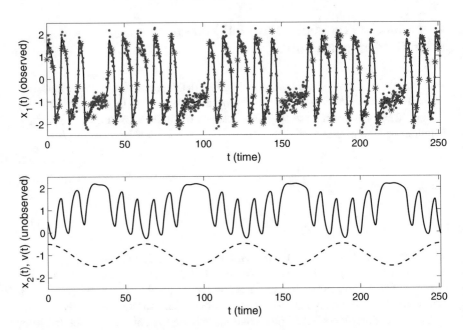

Fig. 1 Noisy observations of the membrane potential $x_1(t)$ (top, blue and purple markers) from the FitzHugh–Nagumo system (16)–(17), along with the unobserved lumped ionic current $x_2(t)$ (bottom, solid black) and external voltage parameter $v(t)$ (bottom, dashed black). In the top panel, the blue (dots) and purple (asterisks) markers together represent noisy observations taken every 0.2 time units, while the purple markers alone show noisy observations every 2 time units

estimation process, a positivity constraint is placed on the period such that $p_{k|k}^n > 0$ for all n and k, and time integration is performed using the Adams-Moulton linear multistep methods of orders 1 and 2 [25, 28].

Figure 2 shows the resulting linear spline estimates of $v(t)$ computed using the estimated parameter means and ± 2 standard deviation values for the spline coefficients $a_1, \ldots, a_{11}$ and period p, repeated over four periods. The corresponding estimates of p are listed in Table 1. For comparison with the piecewise approach, Fig. 2 also shows the resulting mean and ± 2 standard deviation curves estimating $v(t)$ using the augmented EnKF with parameter tracking, where the drift term has prescribed standard deviation $\sigma_\xi = 0.01$. Note that the uncertainty in the resulting estimates of $v(t)$ is much smaller in the piecewise formulation; however, some parts of the true $v(t)$ curve are not captured within the uncertainty bounds, specifically near the beginning of the estimated period (≈ 63.7212) at each repetition. While the uncertainty in the parameter drift estimate is able to almost fully capture the underlying true $v(t)$, the mean estimate does not fully maintain the periodicity intrinsic to $v(t)$.

The next numerical experiment explores how the uncertainty in the parameter estimates using both the piecewise linear spline formulation and parameter tracking is affected by the frequency of available data over time. To that end, the synthetic

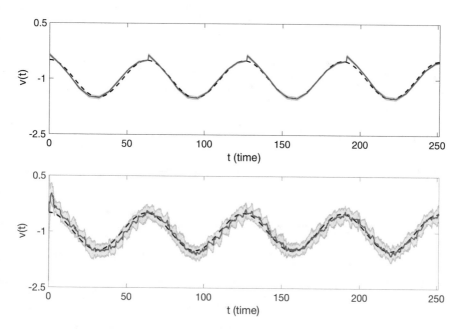

Fig. 2 Parameter estimates of the external voltage parameter $v(t)$ in the FitzHugh–Nagumo system (16)–(17) computed using piecewise linear splines with estimated spline coefficients when $\ell = 10$ and period p repeated over four periods (top panel) and parameter tracking (bottom panel). In the top panel, the linear spline using the augmented EnKF mean estimates of the spline coefficients and period is shown in solid red, while the linear splines computing using the ± 2 standard deviation parameter estimates are shown in dark grey, filled with light grey. In the bottom panel, the mean parameter tracking estimating using the augmented EnKF is shown in solid red, while the ± 2 standard deviation curves are shown in dark grey, filled with light grey. In both panels, the true $v(t)$ used in generating the synthetic data is shown in dashed black. Estimates were obtained using the full synthetic data shown in Fig. 1

Table 1 Augmented EnKF mean and ± 2 standard deviation parameter estimates of the period p of the piecewise linear spline estimate of the external voltage parameter $v(t)$ for different numbers of spline segments ℓ

# of spline segments	Estimated p (mean $\pm$ 2 std)	Relative error (mean)
$\ell = 2$	62.7699 $\pm$ 0.0523	0.0009
$\ell = 5$	63.1169 $\pm$ 0.0620	0.0045
$\ell = 10$	63.7212 $\pm$ 0.0881	0.0142
$\ell = 15$	65.9231 $\pm$ 0.0540	0.0492
$\ell = 20$	73.8291 $\pm$ 0.0189	0.1750

The relative error between the mean estimate and true value of p in each case is computed using the formula in (18). Values in the table are reported to four decimal places

data is subsampled, taking every 10 data points for a total of 126 noisy observations of $x_1(t)$ at equidistant time instances over the interval $[0, 251.2]$. Figure 1 displays the subsampled data in purple markers (asterisks) on the top panel. Figure 3 shows

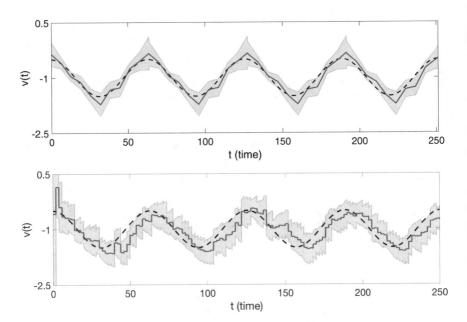

Fig. 3 Parameter estimates of the external voltage parameter $v(t)$ in the FitzHugh–Nagumo system (16)–(17) computed using piecewise linear splines with estimated spline coefficients when $\ell = 10$ and period p repeated over four periods (top panel) and parameter tracking (bottom panel). In the top panel, the linear spline using the augmented EnKF mean estimates of the spline coefficients and period is shown in solid red, while the linear splines computing using the ± 2 standard deviation parameter estimates are shown in dark grey, filled with light grey. In the bottom panel, the mean parameter tracking estimating using the augmented EnKF is shown in solid red, while the ± 2 standard deviation curves are shown in dark grey, filled with light grey. In both panels, the true $v(t)$ used in generating the synthetic data is shown in dashed black. Estimates were obtained using the subsampled synthetic data shown in Fig. 1

the resulting parameter estimates, using both piecewise linear splines and parameter tracking, initialized as in the previous numerical experiment. Note that less frequent observations result in significantly more uncertainty in both the piecewise linear spline and parameter tracking estimates of $v(t)$. While the linear spline estimate is able to fairly well approximate and fully capture the true $v(t)$ within the uncertainty bounds, the parameter tracking algorithm has more difficulty tracking $v(t)$ in this case—the mean estimate does not maintain periodicity and is also noticeably out of phase with the true $v(t)$.

The last numerical experiment considered in this paper studies the effect of the number of linear spline segments ℓ (corresponding to $L = \ell + 1$ spline knots) on the piecewise estimation of $v(t)$ and corresponding estimate for the period p. To this end, the piecewise linear spline estimation is performed using five different choices of ℓ (namely, $\ell = 2, 5, 10, 15, 20$) and the full synthetic data in Fig. 1. Figure 4 shows the resulting linear spline estimates for each ℓ over one estimated period. Table 1 gives the corresponding estimates of the period p in each case, along

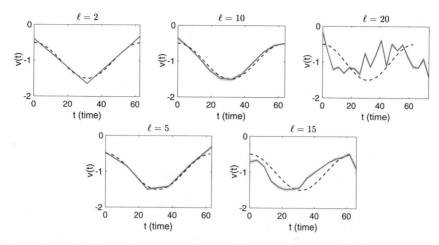

Fig. 4 Parameter estimates of the external voltage parameter $v(t)$ in the FitzHugh–Nagumo system (16)–(17) computed using piecewise linear splines with estimated spline coefficients when $\ell = 2, 5, 10, 15$ and 20, respectively, and period p, shown over one period. In each panel, the linear spline using the augmented EnKF mean estimates of the spline coefficients and period is shown in solid red, while the linear splines computing using the ± 2 standard deviation parameter estimates are shown in dark grey, filled with light grey. The true $v(t)$ used in generating the synthetic data is shown in dashed black. Estimates were obtained using the full synthetic data shown in Fig. 1. Corresponding period estimates are given in Table 1

with the relative error comparing the EnKF mean estimate of the period with the true period used in generating the synthetic data. The relative error in each case is computed via the formula

$$\text{relative error} = \left| \frac{p_{\text{true}} - p_{\text{est}}}{p_{\text{true}}} \right| \tag{18}$$

where p_{est} is the augmented EnKF mean estimate of the period and $p_{\text{true}} \approx 62.8319$ is the true period (up to four decimal places).

The amount of uncertainty in each spline estimate is low, similar to the results seen in Fig. 2 using the full time series of data. It is interesting to note that as the number of spline segments ℓ increases, the EnKF estimate of the period p tends to increase. While the fit of the spline improves from $\ell = 2$ to $\ell = 10$, adding more spline segments eventually starts to degrade the fit along with overestimating the period, as is the case when $\ell = 15$ and $\ell = 20$.

5   Discussion

This paper addresses the problem of estimating and quantifying uncertainty in periodic, time-varying parameters using sequential Monte Carlo parameter estimation techniques. Estimation approaches using both particle methods with parameter tracking and piecewise linear spline (with spline coefficients and periods estimated using particle methods) are considered, and the role of uncertainty is highlighted in each. In particular, uncertainty relating to the frequency of available time series data and the number of spline segments used in the linear spline estimates is tested via numerical experiment on an electrophysiology example estimating the external voltage parameter in the FitzHugh–Nagumo system for modeling the spiking dynamics of single neurons.

As demonstrated in the numerical results in Sect. 4, there are pros and cons to using each of the presented approaches for estimating periodic parameters. One clear computational advantage of the parameter tracking algorithm is its straightforward implementation in the sequential Monte Carlo framework and flexibility in approximating the shape of the parameter of interest. Nothing is assumed about periodicity a priori, and there is only one parameter to track over time. However, since nothing is assumed about periodicity in the parameter tracking, the periodicity of the parameter is therefore neglected in the estimation process and periodicity is not maintained. The choice of the drift variance also has a significant impact on the resulting parameter tracking estimate, in terms of both accuracy and uncertainty. Moreover, the numerical experiments show that the parameter tracking algorithm has more difficulty tracking the periodic parameter as less frequent time series data is available.

The piecewise linear spline formulation maintains periodicity by prescribing a periodic form to the parameter a priori, then estimating the coefficients and period that best fit the available data via a particle approach. The numerical experiments show that the frequency of available time series data has a direct impact on the uncertainty relating to the linear spline estimates, with more frequent data resulting in tighter uncertainty bounds. The number of spline segments ℓ also has a significant effect on both the fit of the resulting spline and the corresponding period estimation. An interesting problem would be to consider estimating ℓ along with the period p; however, this is not straightforward, as ℓ and p depend on one another in the piecewise formulation in (13). Instead, one could interpret the problem of choosing ℓ as a model selection problem and could apply available methods for model selection; see, e.g., [4, 33, 35, 36]. This remains as future work.

Note that while the linear splines shown in Figs. 2 and 3 are formulated to be continuous within a given period, the piecewise formulation in (13)–(14) does not necessarily guarantee continuity of the spline between periods. In order to maintain continuity between periods, an additional constraint that $a_1 = a_\ell$ is required (but not considered in this work). Regarding the period estimation, note that special care must be taken in the implementation to avoid, e.g., negative or inappropriate values being assigned for the period of the periodic parameter. A simple approach used in the numerical results in this paper is to apply a positivity constraint within the Monte Carlo particle algorithm to retain $p_{k|k}^n > 0$ for all n and k. Further, since periodic parameters are subset of all possible parameters in (1), additional constant parameters, such as the initial states of the dynamical system, may be estimated simultaneously in both the parameter tracking and piecewise formulations.

While it is possible to use a variety of parameter estimation techniques, the use of Monte Carlo particle methods in this work provides a natural framework for analyzing the time series data typically available in applications where time-varying parameters are relevant. Moreover, Monte Carlo methods provide a natural measure of uncertainty in the parameter estimation, which can be used for model prediction and uncertainty quantification. Future work includes the design and analysis of parameter tracking-type Monte Carlo particle algorithms that incorporate structural characteristics like periodicity into the sequential estimation without relying on a piecewise functional form for the time-varying parameters.

Acknowledgement This work is supported by the National Science Foundation under grant number NSF/DMS-1819203.

References

1. Altizer, S., Dobson, A., Hosseini, P., Hudson, P., Pascual, M., Rohani, P.: Seasonality and the dynamics of infectious diseases. Ecology Letters **9**, 467–484 (2006)
2. Anderson, J.L.: An ensemble adjustment Kalman filter for data assimilation. Mon Weather Rev **129**, 2884–2903 (2001)
3. Andrieu, C., Thoms, J.: A tutorial on adaptive MCMC. Statistics and Computing **18**(4), 343–373 (2008)
4. Arlot, S., Celisse, A.: A survey of cross-validation procedures for model selection. Statistics Surveys **4**, 40–79 (2010)
5. Arnold, A.: Exploring the effects of uncertainty in parameter tracking estimates for the time-varying external voltage parameter in the FitzHugh-Nagumo model. In: P. Nithiarasu, M. Ohta, M. Oshima (eds.) 6th International Conference on Computational and Mathematical Biomedical Engineering – CMBE2019, pp. 512–515 (2019)
6. Arnold, A., Calvetti, D., Somersalo, E.: Linear multistep methods, particle filtering and sequential Monte Carlo. Inverse Problems **29**(8), 085007 (2013)
7. Arnold, A., Calvetti, D., Somersalo, E.: Parameter estimation for stiff deterministic dynamical systems via ensemble Kalman filter. Inverse Problems **30**(10), 105008 (2014)
8. Arnold, A., Lloyd, A.L.: An approach to periodic, time-varying parameter estimation using nonlinear filtering. Inverse Problems **34**(10), 105005 (2018)
9. Aron, J., Schwartz, I.: Seasonality and period-doubling bifurcations in an epidemic model. J Theor Biol **110**, 665–679 (1984)
10. Berry, T., Sauer, T.: Adaptive ensemble Kalman filtering of non-linear systems. Tellus A **65**, 20331 (2013)
11. Burgers, G., van Leeuwen, P., Evensen, G.: Analysis scheme in the ensemble Kalman filter. Mon Weather Rev **126**(6), 1719–1724 (1998)
12. Evensen, G.: Sequential data assimilation with a nonlinear quasi-geostrophic model using Monte Carlo methods to forecast error statistics. J Geophys Res **99**(C5), 10143–10162 (1994)
13. Evensen, G.: The ensemble Kalman filter for combined state and parameter estimation. IEEE Control Syst Mag **29**(3), 83–104 (2009)
14. Fearnhead, P., Kunsch, H.R.: Particle filters and data assimilation. Annual Review of Statistics and Its Application **5**, 421–449 (2018)
15. FitzHugh, R.: Impulses and physiological states in theoretical models of nerve membrane. Biophys J **1**, 445–466 (1961)
16. Gordon, N.J., Salmond, D.J., Smith, A.F.M.: Novel approach to nonlinear/non-Gaussian Bayesian state estimation. IEEE Proceedings-F **140**(2), 107–113 (1993)

17. Grassly, N., Fraser, C.: Seasonal infectious disease epidemiology. Proc R Soc B **273**, 2541–2550 (2006)
18. Haario, H., Laine, M., Mira, A., Saksman, E.: DRAM: Efficient adaptive MCMC. Statistics and Computing **16**, 339–354 (2006)
19. Haario, H., Saksman, E., Tamminen, J.: An adaptive Metropolis algorithm. Bernoulli **7**, 223–242 (2001)
20. Hamilton, F., Berry, T., Peixoto, N., Sauer, T.: Real-time tracking of neuronal network structure using data assimilation. Physical Review E **88**, 052715 (2013)
21. Harlim, J., Majda, A.J.: Catastrophic filter divergence in filtering nonlinear dissipative systems. Commun Math Sci **8**(27–43) (2010)
22. Hodgkin, A., Huxley, A.: A quantitative description of membrane current and its application to conduction and excitation in nerve. J Physiol **117**, 500–544 (1952)
23. Houtekamer, P.L., Mitchell, H.L.: Data assimilation using an ensemble Kalman filter technique. Mon Weather Rev **126**, 796–811 (1998)
24. Ionides, E., Breto, C., King, A.: Inference for nonlinear dynamical systems. PNAS **103**(49), 18438–18443 (2006)
25. Iserles, A.: A First Course in the Numerical Analysis of Differential Equations, 2 edn. Cambridge Texts in Applied Mathematics. Cambridge University Press, New York (2009)
26. Kantas, N., Doucet, A., Singh, S.S., Maciejowski, J., Chopin, N.: On particle methods for parameter estimation in state-space models. Statistical Science **30**(3), 328–351 (2015)
27. Kitagawa, G.: A self-organizing state-space model. Journal of the American Statistical Association **93**(443), 1203–1215 (1998)
28. LeVeque, R.J.: Finite Difference Methods for Ordinary and Partial Differential Equations. SIAM, Philadelphia (2007)
29. Liu, J., West, M.: Combined parameter and state estimation in simulation-based filtering. In: A. Doucet, N. de Freitas, N. Gordon (eds.) Sequential Monte Carlo Methods in Practice, pp. 197–223. Springer, New York (2001)
30. Matzuka, B.: Nonlinear filtering methodologies for parameter estimation and uncertainty quantification in noisy, complex biological systems. Ph.D. thesis, North Carolina State University (2014)
31. Moradkhani, H., Sorooshian, S., Gupta, H., Houser, P.: Dual state-parameter estimation of hydrological models using ensemble Kalman filter. Adv Water Resour **28**(2), 135–147 (2005)
32. Pitt, M., Shephard, N.: Filtering via simulation: auxiliary particle filters. J Amer Statist Assoc **94**, 590–599 (1999)
33. Toni, T., Welch, D., Strelkowa, N., Ipsen, A., Stumpf, M.P.H.: Approximate Bayesian computation scheme for parameter inference and model selection in dynamical systems. J R Soc Interface **6**, 187–202 (2009)
34. Voss, H., Timmer, J., Kurths, J.: Nonlinear dynamical system identification from uncertain and indirect measurements. International Journal of Bifurcation and Chaos **14**(6), 1905–1933 (2004)
35. Vyshemirsky, V., Girolami, M.A.: Bayesian ranking of biochemical system models. Bioinformatics **24**(6), 833–839 (2007)
36. Wasserman, L.: Bayesian model selection and model averaging. Journal of Mathematical Psychology **44**, 92–107 (2000)
37. Whitaker, J.S., Hamill, T.M.: Ensemble data assimilation without perturbed observations. Mon Weather Rev **130**, 1913–1924 (2002)

A Note on Singularity Formation for a Nonlocal Transport Equation (Research)

Vu Hoang and Maria Radosz

1 Introduction

One of the most fundamental equations in modeling the motion of fluids and gases is the transport equation

$$\omega_t + u \cdot \nabla \omega = 0, \tag{1}$$

here written using the vorticity ω. The velocity u may depend on ω, in which case (1) is called an active scalar equation.

When the relationship $u[\omega]$ is specified, (1) gives rise to many important models in fluid dynamics. $u[\omega]$ is often called a Biot-Savart law. Here are some examples of particular transport equations, which are also active scalar equations. Take for example,

$$u = \nabla^{\perp}(-\Delta)^{-1}\omega, \tag{2}$$

where $\nabla^{\perp} = (-\partial_y, \partial_x)$ is the perpendicular gradient. Equations (1) and (2) are the vorticity form of 2D Euler equations. As another example, one can take

$$u = \nabla^{\perp}(-\Delta)^{-\frac{1}{2}}\omega.$$

Then (1) becomes the surface quasi-geostrophic (SQG) equation. The SQG equation has important applications in geophysics and atmospheric sciences [13]. Moreover it serves as a toy model for the 3D-Euler equations (see [4] for more details).

V. Hoang (✉) · M. Radosz
Department of Mathematics, University of Texas at San Antonio, San Antonio, TX, USA
e-mail: duynguyenvu.hoang@utsa.edu

© The Author(s) and the Association for Women in Mathematics 2020
B. Acu et al. (eds.), *Advances in Mathematical Sciences*, Association for
Women in Mathematics Series 21, https://doi.org/10.1007/978-3-030-42687-3_15

A question of great importance is whether solutions for these equations form singularities in finite time.

A game-changing observation in dealing with some two- and three dimensional models for fluid motion is that imposing certain symmetries on the solution of (1) on simple domains like a half-disc or a quadrant creates a special kind of flow called a *hyperbolic flow*. We will not go into details here but refer to [8–11] for more information. In this hyperbolic flow scenario, it seems that the behavior of the fluid on and near the domain boundary plays the most important part in creating either blowup or strong gradient growth. One-dimensional models capturing this behavior are therefore an essential tool for investigating possible blowup mechanisms without the additional complications of more-dimensional equations. We refer to [1] for discussion of the aspects relating to the hyperbolic flow scenario. For 1d models of fluid equations, see also [1–3, 5].

In this paper, we will study a 1D model of (1) on $\mathbb{R}$ with the following Biot-Savart law:

$$u = (-\Delta)^{-\frac{\alpha}{2}}\omega = -c_\alpha \int_{\mathbb{R}} |y - x|^{-(1-\alpha)}\omega(t, y)\, dy.$$

This model is called α-patch model and has also been treated in [6] with an additional viscosity term causing dissipation. Local existence of the solution and the existence of a blowup for the viscous α-patch model were given in [6]. The existence of blowup was obtained by using energy methods. In contrast, this paper deals with more geometric aspects of singularity formation for the inviscid model— such as the final profile of the solution at the singular time.

Regularity-wise, the α-patch model is less regular than 1D Euler $u_x = H\omega$ and more regular than the Córdoba-Córdoba-Fontelos (CCF) model $u = H\omega$ (see [5]). The latter is a 1D analogue of the SQG equation.

2 One Dimensional α-Patch Model and Main Results

We study the transport equation

$$\omega_t + u[\omega]\omega_x = 0 \tag{4}$$

in one space dimension for the unknown function $\omega(t, x) : [0, T] \times \mathbb{R} \to \mathbb{R}$ with sufficiently smooth initial data $\omega(\cdot, 0) = \omega_0$. The velocity field is given by the nonlocal Biot-Savart law

$$u(x, t) = (-\Delta)^{-\alpha/2}\omega(x, t) = -c_\alpha \int_{\mathbb{R}} |y - x|^{-(1-\alpha)}\omega(t, y)\, dy, \qquad \alpha \in (0, 1).$$

The α-patch model becomes the 1d model for the 2d Euler equation in the limit $\alpha \to 1$ with velocity field given by

$$u(x, t) = (-\Delta)^{-1/2} \omega(x, t).$$

For convenience, we will assume the constant c_α associated with the fractional Laplacian is 1, and we write $\gamma = 1 - \alpha$.

We consider classical solutions where $\omega(\cdot, x)$ is odd in x and such that

$$\|\omega(t, \cdot)\|_q + \|\omega_x(t, \cdot)\|_{q-1} < \infty$$

for all $t \in [0, T]$. Here, the norm $\| \cdot \|_s$ is defined by

$$\|\omega\|_s := \sup_{x \geq 0} |\omega(x)|(1 + x)^{-s}.$$

Our main concern will be the question if more information can be deduced about the nature of the singularity formation of (4). In particular, we are interested in the formation of an *odd cusp* (see Fig. 1), i.e. the possibility that a smooth solution becomes singular at the time $t = T_s > 0$ in a way such that

$$\omega \sim \text{sign}(x)|x|^p \quad (t \to T_s)$$

with some power $p \in (0, 1)$. The sense in which this holds will be made clear below. Another result on cusp formation can be found in [7].

We shall take odd C^2-smooth initial data:

$$\omega(0, -x) = -\omega(0, x) = \omega_0(x) \quad (x \in \mathbb{R})$$

This implies that $\omega(t, \cdot)$ is odd for all t (as long as a smooth solution exists) and also that $u[\omega](t, \cdot)$ is odd as well. Moreover, ω_0 is such that

$$\|\omega_0\|_q + \|\partial_x \omega_0\|_{q-1} + \|\partial_x^2 \omega_0\|_{q-2} < \infty \tag{5}$$

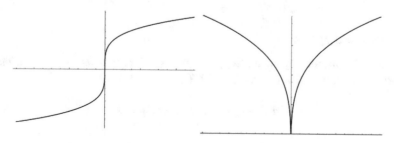

Fig. 1 Left: odd cusp, Right: (even) cusp

for some $0 < q < \gamma$. Note that due to the symmetry, the velocity field is well-defined for ω satisfying (5). This is because

$$u[\omega](t, x) = - \int_0^\infty K(x, y)\omega(t, y)\, dy \qquad (6)$$

where

$$K(x, y) = \left(\frac{1}{|y - x|^\gamma} - \frac{1}{|x + y|^\gamma} \right) \geq 0$$

and hence $|K(x, y)| \leq C(x)|y|^{-1-\gamma}$ for $y \geq 2x$. This implies that (6) converges for all $x > 0$. Note in particular that $u[\omega](x) \leq 0$ for $x \geq 0$, if $\omega(x) \geq 0$ for $x \geq 0$.

Theorem 1 (Local Existence and Uniqueness) *Given C^2-initial data ω_0 satisfying (5), there exists a $T > 0$ and a unique solution ω of (4) defined on $[0, T] \times \mathbb{R}$ so that*

- $\omega(0, \cdot) = \omega_0(\cdot)$
- $\omega(t, \cdot)$ *is odd for all $t \in [0, T)$.*
- $\omega \in C^1([0, T] \times \mathbb{R})$

Define now

$$\phi(t, x) = a^p(t) f\left(\frac{x}{a(t)} \right) \qquad (7)$$

with $f(z) = (z + 1)^p - 1$. The function ϕ serves a barrier for solutions of (4), as shown by our main result below. The function $a(t)$ controls the evolution of the barrier's shape in time and will also be specified in Theorem 2. Suppose that $a(t) \to 0$, as $t \to T^*$. Then as $t \to T^*$

$$\phi(t, x) \to x^p$$

pointwise for $x > 0$ and also uniformly on $x \geq c$ for any $c > 0$. Note moreover that

$$\phi(t, x) \leq (x + a_0)^p \quad (t \geq 0, x \geq 0) \qquad (8)$$

provided $0 < a(t) \leq a_0$.

Theorem 2 (Singularity Formation) *Let $p = \frac{1}{2}\gamma$. There exists a $c_0 > 0$ such that the following implication is true: If $a : [0, T(a_0)] \to \mathbb{R}$ solves*

$$\dot{a} = -c_0 a^{1-p}, \quad a(0) = a_0 > 0$$

with $a_0 < 1$ and $T(a_0) > 0$ being the unique time such that

$$a(T(a_0)) = 0, \quad a(t) > 0 \text{ for } t < T(a_0).$$

and if ω is a smooth solution of (4) such that

$$\|\omega(0, \cdot)\|_p + \|\omega_x(0, \cdot)\|_{p-1} < \infty,$$
$$\omega(0, x) > (1 + \epsilon)\phi(0, x) \quad (x > 0)$$

for $\epsilon > 0$ satisfying

$$\epsilon > (1 - a_0^p)^{-1} - 1. \tag{9}$$

Then $\bar{T} \le T(a_0)$ and

$$\omega(t, x) > \phi(t, x) \quad (x > 0)$$

for $0 \le t < \bar{T}$, where $\bar{T} > 0$ denotes the maximal lifetime of the smooth solution. Provided ω does not break down earlier, ω forms at least a cusp at time $t = T(a_0)$ (or a potentially stronger singularity, see Fig. 2).

Note that a function $a(t)$ with the properties in the Theorem exists for any given $a(0) > 0$. Our theorem does not exclude the possibility of a singularity forming before $T(a_0)$. However, we offer the following conjecture.

Conjecture 1 If $\omega_{xx}(0, x) < 0$ for $x > 0$, then singularity formation can only happen at $x = 0$, i.e. if $\sup_{0 \le t < \hat{T}} |\omega_x(t, 0)| < \infty$ for some $\hat{T} > 0$, then the smooth solution can be continued past $t = \hat{T}$. In this case we have that the profile of $\omega(t, \cdot)$ converges to an odd cusp at the singularity.

Fig. 2 Possible singularity formations. Grey: Odd cusp, Black: Shock

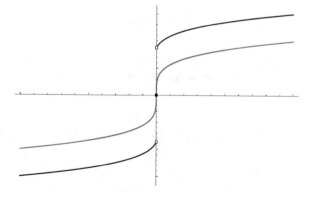

3   Proofs of Main Results

Proposition 1 *For all $0 < q < \gamma$, we have the estimates*

$$\|u[\omega]\|_{1-\gamma+q} \leq C\|\omega\|_q$$

$$\|\partial_x u[\omega]\|_{-\gamma+q} \leq C\|\omega_x\|_{q-1} \tag{10}$$

$$\|\partial_{xx} u[\omega]\|_{-\gamma+q-1} \leq C\|\omega_{xx}\|_{q-2}.$$

with a universal constant $C > 0$.

Proof We estimate

$$|u[\omega]| \leq \|\omega\|_q \int_0^\infty K(x, y)(1 + y)^q dy$$

$$= \|\omega\|_q \int_0^\infty x^{-\gamma} K\left(1, \frac{y}{x}\right) x^q \left(x^{-1} + \frac{y}{x}\right)^q dy$$

$$= \|\omega\|_q x^{1-\gamma+q} \int_0^\infty K(1, z)(x^{-1} + z)^q dz$$

after making the substitution $y = xz$. Now note that for $x \geq 1$, $(x^{-1}+z)^q \leq (1+z^q)$, hence with

$$C := \int_0^\infty K(1, z)(1 + z)^q dz$$

the estimate $\sup_{x \geq 1} |u[\omega]| x^{-(1-\gamma+q)} \leq C\|\omega\|_q$ holds. Note that $C < \infty$ on account of $0 < q < \gamma$. Straightforward computations show that $|u[\omega](x)| \leq C\|\omega\|_q$ for $0 \leq x \leq 1$, hence the first line of (10) holds. To continue with the second line of (10), we first note

$$u[\omega]_x = -\int_\mathbb{R} \frac{\omega_x(y)}{|x - y|^\gamma} dy = -\int_0^\infty \omega_x(y)\left(|x - y|^{-\gamma} + |x + y|^{-\gamma}\right) dy$$

where ω_x is an even function and the integral is absolutely convergent. Similar estimations now show the second line and third line of (10).

3.1   Proof of Theorem 1 (Local Existence and Uniqueness)

The proof consists of two parts. We first construct global solutions of of an approximate problem. The following a-priori bounds are crucial:

Proposition 2 *Suppose* $0 < p < \gamma$, $\omega_0 \in C^2(\mathbb{R})$ *is odd and*

$$\|\omega_0\|_p + \|\partial_x \omega_0\|_{p-1} + \|\partial_{xx}\omega_0\|_{p-2} < \infty.$$

Let $\omega : [0, \infty) \times \mathbb{R} \to \mathbb{R}$ *solve the equation* $\omega_t + v\omega_x = 0$ *where the velocity field satisfies*

$$\|v(t, \cdot)\|_{1-\gamma+p} \le K\|\omega(t, \cdot)\|_p,$$

$$\|\partial_x v(t, \cdot)\|_{-\gamma+p} \le K\|\omega_x(t, \cdot)\|_{p-1}$$

$$\|\partial_{xx} v(t, \cdot)\|_{-\gamma+p-1} \le K\|\omega_{xx}(t, \cdot)\|_{p-2}$$

for all $t \ge 0$ *and some* $K > 0$. *Then there exists a time* $T^* > 0$ *and a* $C > 0$ *depending only on* ω_0 *and* K *such that*

$$\sup_{0 \le t \le T^*} \{\|\omega(t, \cdot)\|_p + \|\partial_x \omega(t, \cdot)\|_{p-1} + \|\partial_{xx}\omega(t, \cdot)\|_{p-2}\} \le C < \infty \qquad (11)$$

holds.

Proof Along any particle trajectory $X(t)$ we compute

$$\frac{d}{dt}\left((1 + X(t))^{-p}\omega(t, X(t))\right)$$

$$= -p(1 + X(t))^{-p-1}v(t, X(t))\omega(t, X(t))$$

$$\quad + (1 + X(t))^{-p}\frac{d}{dt}\omega(t, X(t))$$

$$= -p(1 + X(t))^{-p-1}v(t, X(t))\omega(t, X(t))$$

$$\le (1 + X(t))^{-p-1}K\|\omega(t, \cdot)\|_p(1 + X(t))^{1-\gamma+p}|\omega(t, X(t))|$$

$$= K\|\omega(t, \cdot)\|_p(1 + X(t))^{-\gamma+p}(1 + X(t))^{-p}|\omega(t, X(t))|$$

$$\le K\|\omega(t, \cdot)\|_p^2 \sup_{x \ge 0}(1 + x)^{-\gamma+p}$$

$$\le K\|\omega(t, \cdot)\|_p^2$$

where we have used $\gamma > p$. A similar computation shows that

$$-\frac{d}{dt}\left((1 + X(t))^{-p}\omega(t, X(t))\right) \ge -K\|\omega(t, \cdot)\|_p^2$$

and hence there exists a $T^* > 0$ depending only on $\|\omega_0\|_p$, K such that $\|\omega(t, \cdot)\|_p$ is bounded by a constant on $[0, T^*]$. To prove a similar bound for $\partial_x \omega(t, \cdot)$, we observe that

$$\frac{d}{dt}\left((1+X(t))^{-p+1}\omega_x(t, X(t))\right)$$

$$= (-p+1)(1+X(t))^{-p}v(t, X(t))\omega_x(t, X(t))$$

$$- (1+X(t))^{-p+1}(\partial_x v)(t, X(t))\omega_x(t, X(t))$$

$$\leq K\|\omega_x(t, \cdot)\|_{p-1}(1+X(t))^{-1}(1+X(t))^{1-\gamma+p}\|\omega(t, \cdot)\|_p$$

$$+ K(1+X(t))^{-p+1}(1+X(t))^{-\gamma+p}\|\omega_x(t, \cdot)\|_{p-1}|\omega_x(t, X(t))|$$

$$\leq K\|\omega(t, \cdot)\|_p\|\omega_x(t, \cdot)\|_{p-1}$$

$$+ K(1+X(t))^{-p+1}(1+X(t))^{-\gamma+p}\|\omega_x(t, \cdot)\|_{p-1}(1+X(t))^{p-1}\|\omega_x(t, X(t))\|_{p-1}$$

$$\leq CK\left(\|\omega(t, \cdot)\|_p\|\omega_x(t, \cdot)\|_{p-1} + \|\omega_x(t, \cdot)\|_{p-1}^2\right)$$

with some universal $C > 0$. A similar lower bound for $-\frac{d}{dt}(1+X(t))^{-p+1}\omega_x$ exists. We hence get an a-priori bound for $\|\partial_x\omega(t, \cdot)\|_{p-1}$. A similar argument for $\partial_{xx}\omega$ completes (11).

We now define a family of regularized problems. Set

$$k_\epsilon(z) = \eta_\epsilon^{-\gamma}(|z|)$$

where $\eta_\epsilon(z) = \epsilon\eta(z/\epsilon)$ with η being a smooth, nonincreasing function with the properties

$$\eta(z) = \frac{3}{4} \quad (z \in [0, \frac{3}{4}])$$

$$\eta(z) = z \quad (z \geq 1).$$

Now define for odd ω

$$v_\epsilon[\omega] = -\int_{\mathbb{R}} k_\epsilon(x-y)\omega(y, t)\, dt$$

which can also be written as

$$v_\epsilon[\omega](t, x) = -\int_0^\infty (k_\epsilon(x-y) - k_\epsilon(x+y))\,\omega(t, y)\, dy. \tag{12}$$

We note the following estimates:

$$\|v_\epsilon[\omega]\|_{1-\gamma+p} \leq K\|\omega\|_p$$

$$\|\partial_x v_\epsilon[\omega]\|_{-\gamma+p} \leq K\|\omega_x\|_{p-1}$$

$$\|\partial_{xx} v_\epsilon[\omega]\|_{-\gamma+p-1} \leq K\|\omega_{xx}\|_{p-2}.$$

with some universal $K > 0$. This is shown similarly as in Proposition 1, noting that the regularized kernel k_ϵ is bounded by the original kernel $|z|^{-\gamma}$. On the other hand, we have estimates of the form

$$\|\partial_x v_\epsilon[\omega]\|_{-\gamma+p} \leq C(\epsilon)\|\omega\|_p$$
$$\|\partial_{xx} v_\epsilon[\omega]\|_{-\gamma+p-1} \leq C(\epsilon)\|\omega\|_p. \tag{13}$$

with $C(\epsilon) \to \infty$ as $\epsilon \to 0$.

Proposition 3 *For all $\epsilon > 0$, the regularized problems*

$$\omega_t + v_\epsilon[\omega]\omega_x = 0, \quad \omega(0, x) = \omega_0 \tag{14}$$

have solutions $\omega \in C^2([0, \infty) \times \mathbb{R})$.

Proof The first step is to show the local-in-time existence of solutions using the *particle trajectory method* (see [12]). The flow map $\Phi = \Phi(t, z)$ satisfies the following equation:

$$\frac{d\Phi}{dt}(t, z) = v_\epsilon[\omega_\Phi](\Phi(t, z), t), \quad \Phi(z, 0) = z.$$

or equivalently

$$\Phi(t, z) = z + \int_0^t v_\epsilon[\omega_\Phi(\cdot, s)](\Phi(s, z), s) \, ds. \tag{15}$$

Here, for a given a flow map Φ, we define

$$\omega_\Phi(t, y) := \omega_0(\Phi^{-1}(t, y)). \tag{16}$$

This means, (15) is an equation for Φ with velocity field given by (12) and ω_Φ given by (16). Moreover, a solution of (15) translates into a solution of (14) via relation (16).

We define the operator $\mathcal{G}$ formally by

$$\mathcal{G}_\epsilon[\Phi](x, t) := x + \int_0^t v_\epsilon[\omega_\Phi](\Phi(s, x), s) \, ds$$

with v_ϵ defined by the expression (12) where ω_Φ is given by (16). Then solving (14) is equivalent to solving the fixed point equation

$$\mathcal{G}_\epsilon[\Phi] = \Phi.$$

Next we need to introduce a suitable metric space on which $\mathcal{G}$ is well defined and a contraction. To ease notation, we now fix $\epsilon > 0$ and henceforth drop the subscript ϵ.

Definition 1 Let $\mathcal{B}$ be the set of all $\Phi \in C([0, T], C^2([0, \infty)))$ with the following properties:

$$\Phi(t, 0) = 0$$

$$\Phi(t, \mathbb{R}^+) \subseteq \mathbb{R}^+. \tag{17}$$

Moreover, Φ is of the form

$$\Phi = \mathrm{Id} + \widehat{\Phi}, \quad \|\widehat{\Phi}\| \leq \zeta$$

where Id means the mapping $\mathrm{Id}(t, z) = z$ and

$$\|\widehat{\Phi}\| := \sup_{t \in [0,T]} \left(\|\widehat{\Phi}(t, \cdot)\|_{1-\gamma+p} + \|\widehat{\Phi}_z(t, \cdot)\|_{-\gamma+p} + \|\widehat{\Phi}_{zz}(t, \cdot)\|_{-1-\gamma+p} \right).$$

$\mathcal{B}$ is a complete metric space with metric

$$d(\mathrm{Id} + \widehat{\Phi}, \mathrm{Id} + \widehat{\Psi}) = \|\widehat{\Phi} - \widehat{\Psi}\|.$$

Note that for sufficiently small $\zeta > 0$ and for any $\Phi \in \mathcal{B}$ and $t \in [0, T]$

$$\Phi(t, \cdot) : \mathbb{R}^+ \to \mathbb{R}^+$$

is a diffeomorphism of $(0, \infty)$ onto $(0, \infty)$. To show that, we first note that $\Phi(t, (0, \infty)) \subset (0, \infty)$ by (17). The derivative $\partial_x \Phi(t, x)$ is given by $1 - \partial_x \widehat{\Phi}(t, x)$ and is also uniformly bounded away from zero for small $\zeta > 0$. We also see now that $\mathcal{G}[\Phi]$ is well-defined.

The rest of the proof is standard. First one shows that for sufficiently small ζ, T that $\mathcal{G}$ maps $\mathcal{B}$ into $\mathcal{B}$ and is a contraction. Note that the ϵ-dependent estimates (13) are crucial for the self-mapping and contraction properties. By the contraction mapping theorem there exists a unique solution Φ of (15) on some small time-interval $[0, T]$.

The local-in-time solution of the regularized problem (for any fixed $\epsilon > 0$) is easily extended to $t \in [0, \infty)$ by standard arguments, noting that the a-priori bound

$$\|v_x\|_{-\gamma+p} \leq C(\epsilon)\|\omega\|_p$$

holds independent of the length of the time interval $[0, T]$.

In view of the bounds in Proposition 2, we can now complete the proof of Theorem 1 by using the Arzelá–Ascoli theorem on the sequence of solutions of the regularized problem as $\epsilon \to 0$. This gives a solution to (4), which is $C^1([0, T] \times \mathbb{R})$.

3.2 Proof of Theorem 2 (Singularity Formation)

3.2.1 Preliminaries

We need a few preliminary propositions first.

Proposition 4 *Let* $a : [0, T) \mapsto \mathbb{R}$ *be a smooth function with* $a_0 := a(0) > 0$ *and define*

$$\phi(t, x) = a^p(t) f\left(\frac{x}{a(t)}\right)$$

with $f(z) = (z + 1)^p - 1$. *Then*

$$-u[\phi(t, \cdot)] = a(t)^{1-\gamma+p} U\left(\frac{x}{a(t)}\right) \tag{18}$$

where

$$U(z) = \int_0^\infty \left(\frac{1}{|y - z|^\gamma} - \frac{1}{|y + z|^\gamma}\right) f(y)\, dy.$$

Moreover,

- $U'(0) > 0$
- $U(x) > 0, \quad (x > 0)$
- $U(x) \sim C x^{1-\gamma+p}$ *as* $x \to \infty$ *with some* $C > 0$.

Proof We have

$$-u[\phi(t, \cdot)](x) = a^p \int_0^\infty \left(\frac{1}{|y - x|^\gamma} - \frac{1}{|x + y|^\gamma}\right) f\left(\frac{y}{a}\right) dy$$

$$= a^{1-\gamma+p} \int_0^\infty \left(\frac{1}{|w - \frac{x}{a}|^\gamma} - \frac{1}{|w + \frac{x}{a}|^\gamma}\right) f(w)\, dw$$

$$= a^{1-\gamma+p} U\left(\frac{x}{a}\right).$$

after substituting $y = aw$ in the integral. Hence the representation (18) holds. From the form of U, we directly see that $U(x) > 0$, since the integrand is > 0. To see that $U(x) \sim C x^{1-\gamma+p}$ we compute

$$U(x) = x^{1-\gamma+p} \left\{ \int_0^\infty \left(\frac{1}{|z - 1|^\gamma} - \frac{1}{|z + 1|^\gamma}\right) \left(\left(z + \frac{1}{x}\right)^p - \frac{1}{x^p}\right) dz \right\}$$

and note that the integral in curly brackets converges to a positive constant as $x \to \infty$, as can be seen using the dominated convergence theorem. To see $U'(0) > 0$, we write

$$U'(0) = 2 \int_0^\infty |y|^{-\gamma} f'(y) \, dy$$

and note $f'(y) > 0$.

Lemma 1 *Let ω satisfy all the assumptions of Theorem 2. Let $\bar{T}$ be the maximal life-time of the smooth solution ω. Then for all $t \in [0, \min\{\bar{T}, T(a_0)\})$ and all $x \geq 1$,*

$$\omega(t, x) > \phi(t, x).$$

Moreover, there exists a $\delta > 0$ so that $\omega(t, x) > \phi(t, x)$ for $0 \leq t \leq \delta, 0 < x < \infty$.

Proof To prove the statement referring to $x \geq 1$, we first note that for any $t < \bar{T}, x \geq 1$, there exists a particle trajectory $t \mapsto X(t)$

$$\dot{X}(t) = u[\omega(t, \cdot)](X(t), t), \quad X(0) = X_0$$

such that $X(t) = x$. The assumptions of Theorem 2 imply in particular that $\omega(t, x)$ is always non-negative, so that $u[\omega]$ is non-positive for $x > 0$. Hence the particle trajectory originates from a point $X_0 \geq x \geq 1$ and we have, by (8),

$$\omega(t, x) = \omega(0, X_0) \geq (1 + \epsilon)\phi(0, X_0) > (X_0 + a_0)^p \geq (x + a_0)^p \geq \phi(t, x)$$

since our choice of ϵ (see (9)) guarantees the inequality

$$\frac{\epsilon}{1 + \epsilon} > \frac{a_0^p}{(1 + a_0)^p}$$

implying $(1 + \epsilon)\phi(0, X_0) > (X_0 + a_0)^p$ for all $X_0 \geq 1$.

To argue that the second statement of the Lemma holds, we show first the existence of an $0 < \delta_1$ and an $0 < b < 1$ such that $\omega > \phi$ on $0 \leq t \leq \delta, 0 \leq x \leq b$. The assumption $\omega(0, x) > (1 + \epsilon)\phi(0, x)$ implies $\partial_x \omega(0, x) > \partial_x \phi(0, x)$ for $x \in [0, b]$ for some small $b > 0$. Smoothness of ω in time implies that $\partial_x \omega(t, x) > \partial_x \phi(t, x)$ for $(t, x) \in [0, \delta_1] \times [0, b]$ and some $\delta_1 > 0$. Because of $\omega(t, 0) = \phi(t, 0) = 0$, we then conclude by integrating with respect to x that $\omega(t, x) > \phi(t, x)$ on $[0, \delta_1] \times [0, b]$. To complete the proof of the second part of the proposition, we choose a $\delta_2 > 0$ such that $\omega > \phi$ on $[0, \delta_2] \times [b, 1]$ and set $\delta := \min\{\delta_1, \delta_2\}$.

Lemma 2 *Let all the assumptions of Theorem 2 hold. Define a time T^* by*

$$T^* = \sup\{0 \leq t < \min\{\bar{T}, T(a_0)\} : \omega(\tau, x) > \phi(\tau, x) \text{ for all } (\tau, x) \in [0, t] \times (0, \infty)\}.$$

Suppose also for this proposition that

$$\phi_t + u[\phi]\phi_x < 0 \quad (x > 0). \tag{19}$$

Then if $T^* < \bar{T}$

$$\partial_x \omega(T^*, 0) > \partial_x \phi(T^*, 0). \tag{20}$$

As a consequence, there exists a $b > 0$ such that $\omega(T^*, x) > \phi(T^*, x)$ for all $0 < x < b$.

Proof The supremum defining T^* is > 0 because of Lemma 1. The Eq. (4) and $u[\omega](t, 0) = 0$ imply for short times

$$\frac{d}{dt} \ln \partial_x \omega(t, 0) = -(\partial_x u[\omega])(t, 0). \tag{21}$$

because $\partial_x \omega(t, 0) > 0$ for small $t > 0$. Observe that $\omega(t, \cdot) > \phi(t, \cdot)$ for all $t < T^*$. Using this, we get for $t < T^*$

$$-(\partial_x u[\omega])(t, 0) = -\lim_{x \to 0^+} \frac{u[\omega](t, x)}{x} \geq \lim_{x \to 0^+} \frac{1}{x} \int_0^\infty K(x, y)\phi(t, y) \, dy$$

$$\geq -\lim_{x \to 0^+} \frac{u[\phi](t, x)}{x} = -(\partial_x u[\phi])(t, 0). \tag{22}$$

Moreover the assumption (25) implies, on account of $\phi_t(t, 0) = 0$ and (19),

$$\frac{1}{x} \int_0^x \partial_x \phi_t(t, y) \, dy = \frac{1}{x} \phi_t(t, x) < -\frac{u[\phi](t, x)\phi_x(t, x)}{x}$$

from which by taking the limit $x \to 0$ and using $u[\phi](t, 0) = 0$ we get

$$\partial_t \phi_x(t, 0) \leq -(\partial_x u[\phi])(t, 0)\phi_x(t, 0). \tag{23}$$

Combining (21), (22) and (23), we get for small $t > 0$

$$\frac{d}{dt} \ln \omega_x(t, 0) \geq \frac{d}{dt} \ln \phi_x(t, 0) > 0. \tag{24}$$

The inequality (24) remains valid as long as $\omega_x(t, 0) > 0$. By direct calculation, one finds that $\frac{d}{dt} \ln \phi_x(t, 0) > 0$ for all $t < T(a_0)$ and hence the inequality holds up to T^*. By taking into account that $\omega_x(t, 0) > \phi_x(t, 0)$ for small positive $t > 0$ and integrating (24) up to T^* we arrive at (20).

Proposition 5 *Let a_0, ϕ be as in Proposition 4. Suppose ω is a smooth, odd solution of (4) and that all the assumptions of Theorem 2 hold. Suppose moreover for now that*

$$\phi_t + u[\phi]\phi_x < 0 \quad (x > 0). \tag{25}$$

Then $\omega(t, x) > \phi(t, x)$ for all $x > 0$ and for times $t < \min\{\bar{T}, T(a_0)\}$.

Proof Define T^* as in Lemma 2 and assume that the conclusion of the Proposition is false, i.e. $T^* < \min\{\bar{T}, T(a_0)\}$. Then there exists a sequence (τ_n, x_n) with $\tau_n \to T^*$, $\omega(\tau_n, x_n) \le \phi(\tau_n, x_n)$ and by Lemmas 1 and 2, $0 < b \le x_n \le 1$. By passing to a subsequence, we have $x_n \to x^* \in (b, 1]$. As a consequence, we have $\omega(T^*, x^*) = \phi(T^*, x^*)$ for some $x^* > 0$.

Let $X(t)$ denote any particle trajectory defined by

$$\dot{X}(t) = u(X(t), t), \quad X(0) = X_0$$

where $X_0 > 0$ and such that $X(T^*) = x^*$. Observe that

$$\frac{d}{dt}\omega(t, X(t))\Big|_{t=T^*} \le \frac{d}{dt}\phi(t, X(t))\Big|_{t=T^*}$$

since otherwise by backtracking the trajectory we see that at all times $T^* - \eta$ with small $\eta > 0$ and positions $X(T^*-\eta)$, $\omega(T^*-\eta, X(T^*-\eta)) < \phi(T^*-\eta, X(T^*-\eta))$ holds, in contradiction to the definition of T^*. Then,

$$0 = \frac{d}{dt}\omega(t, X(t))\Big|_{t=T^*} \le \frac{d}{dt}\phi(t, X(t))\Big|_{t=T^*} = (\phi_t + u[\omega]\phi_x)|_{t=T^*}$$

$$\le (\phi_t + u[\phi]\phi_x)(T^*, x^*) < 0$$

where we have used $\omega(T^*, x) \ge \phi(T^*, x)$ for all $x \ge 0$ to conclude

$$u[\omega] \le u[\phi].$$

Hence in summary we get at (T^*, x^*) the relationship

$$0 = (\phi_t + u[\phi]\phi_x)(T^*, x^*) < 0,$$

a contradiction.

Proposition 6 Let $p = \frac{1}{2}\gamma$. There exists a positive constant $c > 0$ such that

$$0 < c \le \frac{U(z)f'(z)}{-pf(z) + zf'(z)} \quad (z > 0)$$

Proof We calculate $-pf(z) + zf'(z) = -p((z+1)^p - 1) + zp(z+1)^{p-1}$ and note that $-pf(z) + zf'(z) > 0$ for all $z > 0$. Now observe that by Proposition 4, $U(z) \sim c_1 z$ for small $z > 0$ with some $c_1 > 0$ and furthermore that $-pf(z) + zf'(z) \sim p(1-p)z$ for small z. Hence there exists a $z_1 > 0$ such that

$$c_2 \le \frac{U(z)f'(z)}{-pf(z) + zf'(z)} \quad (0 < z \le z_1)$$

for some $c_2 > 0$. As $z \to \infty$, $U(z) \sim Cz^{1-\gamma+p}$, $f'(z) \sim z^{p-1}$ and $-pf(z) + zf'(z) \to p$. Using again $p = \frac{1}{2}\gamma$, we conclude the existence of an $z_2 > z_1$ so that

$$c_3 \leq \frac{U(z)f'(z)}{-pf(z) + zf'(z)} \quad (z_2 \leq z).$$

for some $c_3 > 0$. The statement of the Proposition now follows since $\frac{U(z)f'(z)}{-pf(z)+zf'(z)}$ is continuous in z and never zero in $[z_1, z_2]$.

3.2.2 Proof of Theorem 2

We now turn to Theorem 2. We need to check the following: $\phi(t, x)$ satisfies

$$\phi_t + u[\phi]\phi_x < 0 \quad (x > 0). \tag{26}$$

To prove this, we first compute the left hand side using (7) and Proposition 4:

$$\phi_t + u[\phi]\phi_x = \left[\dot{a}a^{p-1}(pf(z) - zf'(z)) - a^{2p-\gamma}U(z)f'(z) \right]_{z=\frac{x}{a}}$$

and so $\phi_t + u[\phi]\phi_x < 0$ is equivalent to

$$(-\dot{a}) < a^{1-\gamma+p} \frac{U(z)f'(z)}{-pf(z) + zf'(z)} \tag{27}$$

for all $z > 0$. By Proposition 6, (27) is implied by

$$(-\dot{a}) < ca^{1-\gamma+p}$$

so $\dot{a} = -\frac{1}{2}ca^{1-p}$ is sufficient for (26) to hold, since $p = \frac{1}{2}\gamma$. By applying Proposition 5, we see that $\omega > \phi$ as long as $t < \min\{\bar{T}, T(a_0)\}$. Note that now necessarily $\bar{T} \leq T(a_0)$, since in the case $T(a_0) < \bar{T}$ the inequality $\omega(T(a_0), x) \geq \phi(T(a_0), x) = x^p$ and $\omega(t, 0) = \phi(t, 0)$ would imply that ω has an infinite slope at $t = T(a_0)$, $x = 0$. This completes the proof of Theorem 2.

Acknowledgements The authors would like to thank the anonymous reviewer for a careful reading of this manuscript, many helpful comments, and in particular for pointing out a gap in the first version. We would also like to thank D. Li for very helpful comments on a preprint version of this paper, indicating a related gap. Vu Hoang wishes to thank the National Science Foundation for support under grants NSF DMS-1614797 and NSF DMS-1810687.

References

1. K. Choi, T.Y. Hou, A. Kiselev, G. Luo, V. Šverák and Y. Yao, On the Finite-Time Blowup of a One-Dimensional Model for the Three-Dimensional Axisymmetric Euler Equations. Communications on Pure and Applied Mathematics, vol. 70, no. 11, Nov. 2017, pp. 2218–43. Scopus, doi: https://doi.org/10.1002/cpa.21697.
2. K. Choi, A. Kiselev and Y. Yao, Finite time blow up for a 1d model of 2d Boussinesq system. *Comm. Math. Phys.*, 334(3):1667–1679, 2015.
3. P. Constantin, P.D. Lax and A. Majda, A simple one-dimensional model for the three-dimensional vorticity equation, *Comm. Pure Appl. Math.*, 38:715–724, 1985.
4. P. Constantin, A. Majda and E. Tabak Formation of strong fronts in the 2-d quasigeostrophic thermal active scalar, *Nonlinearity*, 7(6):1495–1533, 1994.
5. A. Córdoba, D. Córdoba and M.A. Fontelos, Formation of singularities for a transport equation with nonlocal velocity, *Ann. of Math.(2)*, 162(3):1377–1389, 2005.
6. Hongjie Dong and Dong Li. On a one-dimensional α-patch model with nonlocal drift and fractional dissipation. *Trans. Amer. Math. Soc.*, 366(4):2041–2061, 2014.
7. V. Hoang, M. Radosz, Cusp formation for a nonlocal evolution equation, Archive for Rational Mechanics and Analysis, June 2017, Volume 224, Issue 3, pp 1021–1036, https://doi.org/10.1007/s00205-017-1094-3
8. V. Hoang, B. Orcan-Ekmeckci, M. Radosz and H. Yang, Blowup with vorticity control for a 2D model of the Boussinesq equations. Journal of Differential Equations, Vol. 264 (12), p.7328–7356 (2018).
9. T. Y. Hou and, G. Luo, On the finite-time blowup of a 1D model for the 3D incompressible Euler equations, *arXiv:1311.2613*.
10. T.Y. Hou and G. Luo, Potentially singular solutions of the 3D axisymmetric Euler equations. *PNAS*, vol. 111 no. 36, 12968–12973, *DOI https://doi.org/10.1073/pnas.1405238111*.
11. A. Kiselev and V. Šverák. Small scale creation for solutions of the incompressible two dimensional Euler equation. *Ann. of Math.(2)*, 180(3):1205–1220, 2014.
12. A. Madja, A. Bertozzi, Vorticity and Incompressible Flow, Cambridge University Press, 2002.
13. J. Pedlosky, Geophysical fluid dynamics, Springer Verlag, New York (1979)

Prescribing Initial Values for the Sticky Particle System (Survey)

Ryan Hynd

1 Introduction

The sticky particle system (SPS) is a system of partial differential equations which describes the motion of a collection of particles in $\mathbb{R}^d$ which move freely and interact only through perfectly inelastic collisions. Denoting ρ as the density of particles and v as an associated velocity field, the SPS consists of the *conservation of mass*

$$\partial_t \rho + \nabla \cdot (\rho v) = 0 \tag{1}$$

along with the *conservation of momentum*

$$\partial_t (\rho v) + \nabla \cdot (\rho v \otimes v) = 0. \tag{2}$$

These equations hold in $\mathbb{R}^d \times (0, \infty)$ and were first derived by the astronomer Yakov Zel'dovich in his work on the expansion of matter without pressure [17].

In this note, we will be concerned with determining whether or not solution pairs ρ and v exist for given initial conditions. In particular, we would like to prescribe an initial mass distribution ρ_0 and an initial velocity field v_0

$$\rho|_{t=0} = \rho_0 \quad \text{and} \quad v|_{t=0} = v_0 \tag{3}$$

and use the SPS to describe the evolution of the mass distribution ρ and associated velocity field v at later times. To this end, we will define a generalized solution and phrase our initial value problem using this notion.

R. Hynd (✉)
University of Pennsylvania, Philadelphia, PA, USA
e-mail: rhynd@math.upenn.edu

© The Author(s) and the Association for Women in Mathematics 2020
B. Acu et al. (eds.), *Advances in Mathematical Sciences*, Association for
Women in Mathematics Series 21, https://doi.org/10.1007/978-3-030-42687-3_16

Since the total mass of any physical system we consider will be conserved, we will assume throughout that it is always equal to 1. As a result, it will be natural for us to employ the space $\mathcal{P}(\mathbb{R}^d)$ of Borel probability measures on $\mathbb{R}^d$. We recall that this space has a natural topology: $(\mu^k)_{k \in \mathbb{N}} \subset \mathcal{P}(\mathbb{R}^d)$ converges *narrowly* to $\mu \in \mathcal{P}(\mathbb{R}^d)$ provided

$$\lim_{k \to \infty} \int_{\mathbb{R}^d} g \, d\mu^k = \int_{\mathbb{R}^d} g \, d\mu$$

for each continuous and bounded $g : \mathbb{R}^d \to \mathbb{R}$. The notion of solution that we will use throughout this paper is as follows.

Definition 1 Suppose $\rho_0 \in \mathcal{P}(\mathbb{R}^d)$ and $v_0 : \mathbb{R}^d \to \mathbb{R}^d$ is continuous and bounded. A narrowly continuous $\rho : [0, \infty) \to \mathcal{P}(\mathbb{R}^d); t \mapsto \rho_t$ and Borel measurable $v : \mathbb{R}^d \times [0, \infty) \to \mathbb{R}^d$ is a *weak solution pair* of the SPS with initial conditions (3) provided the following hold.

- For each $T > 0$,

$$\int_0^T \int_{\mathbb{R}^d} |v|^2 d\rho_t dt < \infty.$$

- For each $\psi \in C_c^\infty(\mathbb{R}^d \times [0, \infty))$,

$$\int_0^\infty \int_{\mathbb{R}^d} (\partial_t \psi + \nabla \psi \cdot v) d\rho_t dt + \int_{\mathbb{R}^d} \psi(\cdot, 0) d\rho_0 = 0. \tag{4}$$

- For each $\varphi \in C_c^\infty(\mathbb{R}^d \times [0, \infty); \mathbb{R}^d)$,

$$\int_0^\infty \int_{\mathbb{R}^d} (\partial_t \varphi \cdot v + \nabla \varphi \, v \cdot v) d\rho_t dt + \int_{\mathbb{R}^d} \varphi(\cdot, 0) \cdot v_0 d\rho_0 = 0. \tag{5}$$

It is not hard to check that this definition extends the usual notion of a smooth solution pair. Indeed, if $\rho : \mathbb{R}^d \times [0, \infty) \to [0, \infty)$ and $v : \mathbb{R}^d \times [0, \infty) \to \mathbb{R}^d$ are continuously differentiable and satisfy the SPS, we can multiply (1) by $\psi \in C_c^\infty(\mathbb{R}^d \times [0, \infty))$ and (2) by $\varphi \in C_c^\infty(\mathbb{R}^d \times [0, \infty); \mathbb{R}^d)$ and integrate by parts in order to derive (4) and (5), respectively. It is also useful to have a more flexible notion of solution as classical solutions may not exist for a given pair of smooth initial conditions.

The problem that motivated this work is as follows.

Problem 1 Suppose $\rho_0 \in \mathcal{P}(\mathbb{R}^d)$ and $v_0 : \mathbb{R}^d \to \mathbb{R}^d$ is continuous and bounded. Determine whether or not there is a weak solution pair ρ and v of the SPS with these initial conditions, respectively.

This initial value problem was resolved in one spatial dimension ($d = 1$) in the pioneering works by E, Rykov and Sinai [8] and by Brenier and Grenier [3]. Much less progress has been made when $d > 1$. In this article, we will survey a few known results and explain some challenges with the SPS.

2 The Method of Characteristics

Suppose that $\rho_0 \in \mathcal{P}(\mathbb{R}^d)$ and $v_0 : \mathbb{R}^d \to \mathbb{R}^d$ are given. The simplest setting in which we can solve the SPS is when

$$\mathrm{id}_{\mathbb{R}^d} + tv_0 \text{ is invertible for each } t > 0.$$

This occurs, for example, if v_0 is continuously differentiable and monotone. Also note that this assumption implies that the rays $[0, \infty) \ni t \mapsto x + tv_0(x)$ and $[0, \infty) \ni t \mapsto y + tv_0(y)$ do not intersect for $x \neq y$. Let us show how this assumption leads to a solution pair.

For each $t \geq 0$, define $\rho_t \in \mathcal{P}(\mathbb{R}^d)$ via the formula

$$\int_{\mathbb{R}^d} g(y)d\rho_t(y) := \int_{\mathbb{R}^d} g(x + tv_0(x))d\rho_0(x) \tag{6}$$

for all continuous and bounded $g : \mathbb{R}^d \to \mathbb{R}$. The measure ρ_t is simply the mass distribution ρ_0 transported along the family of nonintersecting rays $[0, \infty) \ni t \mapsto x + tv_0(x)$ for $x \in \mathbb{R}^d$. Let us also specify $v : \mathbb{R}^d \times [0, \infty) \to \mathbb{R}^d$ implicitly by the equation

$$v(x + tv_0(x), t) = v_0(x).$$

This tells us that the velocity is constant along the straight line trajectories. It is straightforward to check that ρ and v is a weak solution pair with initial conditions ρ_0 and v_0.

Suppose in addition that ρ_0 has a smooth density, which we will identify with ρ_0 itself, and that v_0 is continuously differentiable. Using the change of variables theorem in (6), we find that the SPS admits the classical solution pair

$$\begin{cases} \rho(\cdot, t) := \left[\dfrac{\rho_0}{\det \nabla(\mathrm{id}_{\mathbb{R}^d} + tv_0)} \right] \circ (\mathrm{id}_{\mathbb{R}^d} + tv_0)^{-1} \\[4mm] v(\cdot, t) := v_0 \circ (\mathrm{id}_{\mathbb{R}^d} + tv_0)^{-1}. \end{cases}$$

Unfortunately, once $\mathrm{id}_{\mathbb{R}^d} + tv_0$ fails to be injective, these formulae are no longer valid.

3 Sticky Particle Trajectories

Suppose there are N particles in $\mathbb{R}^d$ with masses $m_1, \ldots, m_N$ that sum to 1. In addition, suppose that these point masses move freely in space until they collide; when particles collide, they undergo perfectly inelastic collisions. For example, if the particles with masses $m_1, \ldots, m_k$ have respective velocities $v_1, \ldots, v_k \in \mathbb{R}^d$ just before they collide, they will join to form a single particle of mass $m_1 + \cdots + m_k$ which has velocity

$$\frac{m_1 v_1 + \cdots + m_k v_k}{m_1 + \cdots + m_k}$$

right after the collision.

We will denote *the sticky particle trajectories* $\gamma_1, \ldots, \gamma_N : [0, \infty) \to \mathbb{R}^d$ as the piecewise linear paths that track the position of the respective point masses discussed above. Specifically, $\gamma_i(t)$ is the location of the particle with mass m_i at time t. Note that this particle could be by itself or part of a larger mass if it collided with another particle before time t. It is not hard to show that these paths are well defined and satisfy the sticky particle condition

$$\gamma_i(s) = \gamma_j(s) \implies \gamma_i(t) = \gamma_j(t) \tag{7}$$

for $t \geq s$ and $i, j = 1, \ldots, N$ (Proposition 2.1 and Corollary 2.4 of [10]).

One of the most important properties of sticky particle trajectories is as follows.

Proposition 1 (Proposition 2.5 of [10]) *Assume* $g : \mathbb{R}^d \to \mathbb{R}^d$. *Then*

$$\sum_{i=1}^{N} m_i g(\gamma_i(t)) \cdot \dot{\gamma}_i(t+) = \sum_{i=1}^{N} m_i g(\gamma_i(t)) \cdot \dot{\gamma}_i(s+)$$

for $0 \leq s \leq t$.

This averaging property embodies the conservation of momentum that particles experience in between and during collisions. To see this, we define

$$\rho_t := \sum_{i=1}^{N} m_i \delta_{\gamma_i(t)} \in \mathcal{P}(\mathbb{R}^d) \tag{8}$$

for each $t \geq 0$. Note that since $\gamma_1, \ldots, \gamma_N$ are continuous paths, $\rho : [0, \infty) \to \mathcal{P}(\mathbb{R}^d); t \mapsto \rho_t$ is narrowly continuous. Using (7), we may also set

$$v(x, t) = \begin{cases} \dot{\gamma}_i(t+), & x = \gamma_i(t) \\ 0, & \text{otherwise.} \end{cases} \tag{9}$$

It is a simple exercise to check that $v : \mathbb{R}^d \times [0, \infty) \to \mathbb{R}^d$ is Borel measurable. Moreover, the following assertion holds.

Proposition 2 (Corollary 2.6 and Section 2.3 of [10]) *Suppose*

$$\rho_0 := \sum_{i=1}^{N} m_i \delta_{\gamma_i(0)} \tag{10}$$

and $v_0 : \mathbb{R}^d \to \mathbb{R}^d$ *is continuous with*

$$v_0(\gamma_i(0)) = \dot{\gamma}_i(0+)$$

for $i = 1, \ldots, N$. *Then* ρ *and* v *defined in* (8) *and* (9), *respectively, is a weak solution pair of the SPS with initial conditions* (3). *Moreover,*

$$\int_{\mathbb{R}^d} \frac{1}{2} |v(x, t)|^2 d\rho_t(x) \leq \int_{\mathbb{R}^d} \frac{1}{2} |v(x, s)|^2 d\rho_s(x)$$

for each $0 \leq s \leq t$.

When $d = 1$, we have the additional property. We call it the *quantitative sticky particle property* as it quantifies (7).

Proposition 3 (Corollary 2.8 of [10]) *Suppose* $d = 1$. *For* $0 < s \leq t < \infty$ *and* $i, j = 1, \ldots, N$,

$$\frac{1}{t} |\gamma_i(t) - \gamma_j(t)| \leq \frac{1}{s} |\gamma_i(s) - \gamma_j(s)|.$$

By this quantitative sticky particle property,

$$\frac{d}{dt} \frac{1}{t^2} |\gamma_i(t) - \gamma_j(t)|^2$$

$$= \frac{2}{t^2} \left[(v(\gamma_i(t), t) - v(\gamma_j(t), t))(\gamma_i(t) - \gamma_j(t)) - \frac{1}{t} |\gamma_i(t) - \gamma_j(t)|^2 \right] \leq 0$$

for almost every $t > 0$ and each $i, j = 1, \ldots N$. As a result,

$$(v(x, t) - v(y, t))(x - y) \leq \frac{1}{t}(x - y)^2 \tag{11}$$

for almost every $t > 0$ and each $x, y \in \text{supp}(\rho_t)$. We emphasize this entropy inequality only holds for $d = 1$.

4   Large Particle Limit

Suppose $\rho_0 \in \mathcal{P}(\mathbb{R}^d)$ is a given initial mass distribution. We may select a sequence $(\rho_0^k)_{k \in \mathbb{N}} \subset \mathcal{P}(\mathbb{R}^d)$ for which

$$\begin{cases} \text{each } \rho_0^k \text{ is of the form (10), and} \\ \rho_0^k \to \rho_0 \text{ narrowly as } k \to \infty \end{cases}$$

(Remark 5.1.2 in [1]). Let us additionally suppose $v_0 : \mathbb{R}^d \to \mathbb{R}^d$ is continuous and bounded. Using sticky particle trajectories, we can produce a weak solution pair ρ^k and v^k of the SPS with initial conditions ρ_0^k and v_0 for each $k \in \mathbb{N}$.

It is now natural to ask if there are subsequences $(\rho^k)_{k \in \mathbb{N}}$ and $(v^k)_{k \in \mathbb{N}}$ which converge in some sense to a weak solution pair ρ and v of the SPS with initial conditions ρ_0 and v_0. For this to work, we would need to send $k \to \infty$ along a subsequence in

$$\int_0^\infty \int_{\mathbb{R}^d} (\partial_t \psi + \nabla \psi \cdot v^k) d\rho_t^k dt + \int_{\mathbb{R}^d} \psi(\cdot, 0) d\rho_0^k = 0$$

for each $\psi \in C_c^\infty(\mathbb{R}^d \times [0, \infty))$ and in

$$\int_0^\infty \int_{\mathbb{R}^d} (\partial_t \varphi \cdot v^k + \nabla \varphi \, v^k \cdot v^k) d\rho_t^k dt + \int_{\mathbb{R}^d} \varphi(\cdot, 0) \cdot v_0 d\rho_0^k = 0$$

for each $\varphi \in C_c^\infty(\mathbb{R}^d \times [0, \infty); \mathbb{R}^d)$. The only estimate we have at our disposal is

$$\int_{\mathbb{R}^d} \frac{1}{2} |v^k(x, t)|^2 d\rho_t^k(x) \le \int_{\mathbb{R}^d} \frac{1}{2} |v_0(x)|^2 d\rho_0^k(x)$$

for $t \ge 0$ and $k \in \mathbb{N}$.

It turns out that this strategy only works in dimension 1, where we have the additional entropy estimate (11)

$$(v^k(x, t) - v^k(y, t))(x - y) \le \frac{1}{t}(x - y)^2$$

for each $x, y \in \mathrm{supp}(\rho_t^k)$ and $t > 0$. We may interpret this estimate informally as the one sided derivative bound

$$\partial_x v^k(x, t) \le \frac{1}{t}$$

for ρ_t^k almost every $x \in \mathbb{R}$. This estimate is just enough to ensure that $(\rho^k)_{k \in \mathbb{N}}$ and $(v^k)_{k \in \mathbb{N}}$ have subsequences which converge in ways which allow us to conclude that their limits ρ and v indeed comprise a weak solution pair of the SPS with the desired initial data [10, 15].

The following existence theorem was first deduced by E, Rykov and Sinai [8] and by Brenier and Grenier [3]. We also note that there have been many other significant contributions to the initial value problem for the SPS in one spatial dimension including [2, 5, 6, 9–14].

Theorem 1 *Suppose $d = 1$, $\rho_0 \in \mathcal{P}(\mathbb{R})$ and $v_0 : \mathbb{R} \to \mathbb{R}$ continuous and bounded. There is a weak solution pair ρ and v of the SPS with initial conditions $\rho|_{t=0} = \rho_0$ and $v|_{t=0} = v_0$. Moreover, for almost every $t > 0$ and each $x, y \in \mathrm{supp}(\rho_t)$,*

$$(v(x, t) - v(y, t))(x - y) \le \frac{1}{t}(x - y)^2;$$

and for almost every $0 \le s \le t < \infty$,

$$\int_{\mathbb{R}} \frac{1}{2} v(x, t)^2 d\rho_t(x) \le \int_{\mathbb{R}} \frac{1}{2} v(x, s)^2 d\rho_s(x).$$

We remark that the uniqueness of a weak solution pair which satisfies the entropy inequality was first proved by Huang and Wang [9]. We also note that Nguyen and Tudorascu proved that there is a unique entropy solution pair provided the pth moment of ρ_0 is finite and $v_0 \in L^p(\rho_0)$ for $p \ge 2$ [15].

5   Instability

It seems the main issue with solving the initial value problem in several spatial dimensions is that the solutions to the SPS are unstable. For example, suppose the rays $[0, \infty) \ni t \mapsto x_1 + tv_1$ and $[0, \infty) \ni t \mapsto x_2 + tv_2$ intersect at time $s > 0$. If these rays initially describe the paths of two colliding particles with masses m_1 and m_2, respectively, the corresponding sticky particle trajectories are

$$\gamma_i(t) = \begin{cases} x_i + tv_i, & t \in [0, s] \\ z + (t - s)(m_1 v_1 + m_2 v_2), & t \in [s, \infty) \end{cases} \tag{12}$$

for $i = 1, 2$. Here $z = x_1 + sv_1 = x_2 + sv_2$ is the point where the particles collide.

When $d \ge 2$, we can replace x_2 with $\tilde{x}_2 \ne x_2$ and obtain two rays $[0, \infty) \ni t \mapsto x_1 + tv_1$ and $[0, \infty) \ni t \mapsto \tilde{x}_2 + tv_2$ which do not intersect. Furthermore, we can do so in a way that $\tilde{x}_2$ is as close to x_2 as desired. Therefore, a small change in initial conditions results in solutions which do not appear to be close to each other. This example also shows that the limit of a sequence of solution pairs of the SPS may not be a solution. Indeed if we send $\tilde{x}_2 \to x_2$, we obtain two intersecting rays $[0, \infty) \ni t \mapsto x_1 + tv_1$ and $[0, \infty) \ni t \mapsto x_2 + tv_2$ and not the sticky particle trajectories γ_1 and γ_2 in (12). We believe this simple observation is at the core of the what is preventing us from naively designing an approximation scheme as discussed in the previous section.

6 Lagrangian Variables

Not long after the initial value problem for the SPS was solved in one spatial dimension, Sever developed an interesting approach for the initial value problem in any spatial dimension [16] (see also [7]). This approach is based on the auxiliary initial value problem

$$
\begin{cases}
\dfrac{d}{dt} X(t) = \mathbb{E}_{\rho_0}[v_0 | X(t)], & \text{a.e. } t \geq 0 \\
X_0 = \mathrm{id}_{\mathbb{R}^d}.
\end{cases}
\tag{13}
$$

Here the unknown is an absolutely continuous path $X : [0, \infty) \to L^2(\rho_0; \mathbb{R}^d)$. We also recall that for each $t \geq 0$, the conditional expectation $\mathbb{E}_{\rho_0}[v_0 | X(t)]$ is an $L^2(\rho_0; \mathbb{R}^d)$ map $g(X(t))$, where $g : \mathbb{R}^d \to \mathbb{R}^d$ is Borel and

$$
\int_{\mathbb{R}^d} g(X(t)) \cdot h(X(t)) d\rho_0 = \int_{\mathbb{R}^d} v_0 \cdot h(X(t)) d\rho_0
$$

for every bounded, continuous $h : \mathbb{R}^d \to \mathbb{R}$.

The key to linking this flow equation to the SPS is as follows. First define $\rho : [0, \infty) \to \mathcal{P}(\mathbb{R}^d); t \mapsto \rho_t$ via the formula

$$
\int_{\mathbb{R}^d} h d\rho_t := \int_{\mathbb{R}^d} h \circ X(t) d\rho_0.
$$

Next choose a Borel $v : \mathbb{R}^d \times [0, \infty) \to \mathbb{R}^d$ such that

$$
v(X(t), t) = \mathbb{E}_{\rho_0}[v_0 | X(t)] \quad \text{a.e. } t \geq 0.
$$

Then ρ and v is a weak solution pair of the SPS with initial conditions (3).

Sever also argued that (13) has a solution which satisfies a natural sticky particle property (Theorem 4.2 of [16]). However, Bressan and Nguyen discovered that this result may fail to hold [4]. Specifically, they showed that there are initial conditions for which (13) does not have a solution as described by Sever's theorem. As a result, there is some controversy and much room for clarification with this method.

7 Kinetic Theory

We conclude this note by recalling an initial value problem in kinetic theory related to the SPS. The problem is to find $f : [0, \infty) \to \mathcal{P}(\mathbb{R}^d \times \mathbb{R}^d); t \mapsto f_t$ which satisfies

$$\begin{cases} \partial_t f + \xi \cdot \nabla_x f = 0, \text{ in } \mathbb{R}^d \times \mathbb{R}^d \times (0, \infty) \\ f|_{t=0} = (\mathrm{id}_{\mathbb{R}^d} \times v_0)_{\#}\rho_0 \end{cases}$$

in the weak sense. That is,

$$\int_0^\infty \int_{\mathbb{R}^d \times \mathbb{R}^d} (\partial_t \psi + \xi \cdot \nabla_x \psi) \, df_t(x, \xi) dt + \int_{\mathbb{R}^d} \psi(x, v_0(x), 0) d\rho_0(x) = 0$$

for each $\psi \in C_c^1(\mathbb{R}^d \times \mathbb{R}^d \times [0, \infty))$. The physical interpretation is that f_t is the distribution of particles in position and velocity space $\mathbb{R}^d \times \mathbb{R}^d$ at time $t \geq 0$.

It turns out that this initial value problem can easily be solved for any ρ_0 and v_0. If there is a solution f with

$$\int_{\mathbb{R}^d \times \mathbb{R}^d} g(x, \xi) df_t(x, \xi) = \int_{\mathbb{R}^d} g(x, v(x, t)) d\rho_t(x)$$

for almost every $t \geq 0$, then ρ and v is a weak solution pair of the SPS which satisfies the initial conditions (3). If no such solution exists, we wonder if it is possible to select a solution f which can somehow be associated to the SPS in a most natural way.

Acknowledgements Partially supported by NSF grant DMS-1554130.

References

1. L. Ambrosio, N. Gigli, and G. Savaré. *Gradient flows in metric spaces and in the space of probability measures.* Lectures in Mathematics ETH Zürich. Birkhäuser Verlag, Basel, second edition, 2008.
2. F. Bouchut. On zero pressure gas dynamics. In *Advances in kinetic theory and computing*, volume 22 of *Ser. Adv. Math. Appl. Sci.*, pages 171–190. World Sci. Publ., River Edge, NJ, 1994.
3. Y. Brenier and E. Grenier. Sticky particles and scalar conservation laws. *SIAM J. Numer. Anal.*, 35(6):2317–2328, 1998.
4. A. Bressan and T. Nguyen. Non-existence and non-uniqueness for multidimensional sticky particle systems. *Kinet. Relat. Models*, 7(2):205–218, 2014.
5. F. Cavalletti, M. Sedjro, and M. Westdickenberg. A simple proof of global existence for the 1D pressureless gas dynamics equations. *SIAM J. Math. Anal.*, 47(1):66–79, 2015.

6. A. Dermoune. Probabilistic interpretation of sticky particle model. *Ann. Probab.*, 27(3):1357–1367, 1999.

7. A. Dermoune. d-dimensional pressureless gas equations. *Teor. Veroyatn. Primen.*, 49(3):610–614, 2004.

8. W. E, Yu. G. Rykov, and Ya. G. Sinai. Generalized variational principles, global weak solutions and behavior with random initial data for systems of conservation laws arising in adhesion particle dynamics. *Comm. Math. Phys.*, 177(2):349–380, 1996.

9. F. Huang and Z. Wang. Well posedness for pressureless flow. *Comm. Math. Phys.*, 222(1):117–146, 2001.

10. R. Hynd. A pathwise variation estimate for the sticky particle system. *arXiv preprint*, 2018.

11. P.-E. Jabin and T. Rey. Hydrodynamic limit of granular gases to pressureless Euler in dimension 1. *Quart. Appl. Math.*, 75(1):155–179, 2017.

12. O. Moutsinga. Convex hulls, sticky particle dynamics and pressure-less gas system. *Ann. Math. Blaise Pascal*, 15(1):57–80, 2008.

13. L. Natile and G. Savaré. A Wasserstein approach to the one-dimensional sticky particle system. *SIAM J. Math. Anal.*, 41(4):1340–1365, 2009.

14. T. Nguyen and A. Tudorascu. Pressureless Euler/Euler-Poisson systems via adhesion dynamics and scalar conservation laws. *SIAM J. Math. Anal.*, 40(2):754–775, 2008.

15. T. Nguyen and A. Tudorascu. One-dimensional pressureless gas systems with/without viscosity. *Comm. Partial Differential Equations*, 40(9):1619–1665, 2015.

16. M. Sever. An existence theorem in the large for zero-pressure gas dynamics. *Differential Integral Equations*, 14(9):1077–1092, 2001.

17. Ya. B. Zel'dovich. Gravitational instability: An Approximate theory for large density perturbations. *Astron. Astrophys.*, 5:84–89, 1970.

Part VI
Topology

Towards Directed Collapsibility (Research)

Robin Belton, Robyn Brooks, Stefania Ebli, Lisbeth Fajstrup,
Brittany Terese Fasy, Catherine Ray, Nicole Sanderson,
and Elizabeth Vidaurre

1 Introduction

Spaces that are equipped with a direction have only recently been given more
attention from a topological point of view. The spaces of directed paths are the
defining feature for distinguishing different directed spaces. One reason for studying
directed spaces is their application to the modeling of concurrent programs, where

R. Belton (✉) · B. T. Fasy
Montana State University, Bozeman, MT, USA
e-mail: robin.belton@montana.edu; brittany.fasy@montana.edu

R. Brooks
Tulane University, New Orleans, LA, USA
e-mail: rbrooks3@tulane.edu

S. Ebli
EPFL, Lausanne, Switzerland
e-mail: stefania.ebli@epfl.ch

L. Fajstrup
Aalborg University, Aalborg, Denmark
e-mail: fajstrup@math.aau.dk

C. Ray
Northwestern University, Evanston, IL, USA
e-mail: cray@math.northwestern.edu

N. Sanderson
Lawrence Berkeley National Lab, Berkeley, CA, USA
e-mail: nsanderson@lbl.gov

E. Vidaurre
Molloy College, Rockville Centre, NY, USA
e-mail: evidaurre@molloy.edu

© The Author(s) and the Association for Women in Mathematics 2020
B. Acu et al. (eds.), *Advances in Mathematical Sciences*, Association for
Women in Mathematics Series 21, https://doi.org/10.1007/978-3-030-42687-3_17

standard algebraic topology does not provide the tools needed [4]. Concurrent programming is used when multiple processes need to access shared resources. Directed spaces are models for concurrent program, where paths respecting the time directions represent executions of programs. In such models, executions are equivalent if their execution paths are homotopic through a family of directed paths. This observation has already led to new insights and algorithms. For instance, verification of concurrent programs is simplified by verifying one execution from each connected component of the space of directed paths; see [4, 5].

While equivalence of executions is clearly stated in concurrent programming, equivalence of the directed topological spaces themselves is not well understood. Directed versions of homotopy groups and homology groups are not agreed upon. Directed homeomorphism is too strong; whereas, directed homotopy equivalence is often too weak, to preserve the properties of the concurrent programs. In classical (undirected) topology, the concept of simplifying a space by a sequence of collapses goes back to J.H.C. Whitehead [11], and has been studied in [1, 6], among others. However, a definition for a directed collapse of a Euclidean cubical complex that preserves spaces of directed paths is notably missing from the literature.

In this article, we consider spaces of directed paths in Euclidean cubical complexes. Our objects of study are spaces of directed paths relative to a fixed pair of endpoints. We show how local information of the past links of vertices in a Euclidean cubical complex can provide global information on the spaces of directed paths. As an example, our results are applied to study the spaces of directed paths in the well-known dining philosophers problem. Furthermore, we define directed collapse so that a directed collapse of a Euclidean cubical complex preserves the relevant spaces of directed paths in the original complex. Our theoretical work has applications to simplifying verification of concurrent programs without loops, and better understanding partial executions in those concurrent programs.

We begin, in Sect. 2, with two motivating examples of how the execution of concurrent programs can be modeled by Euclidean cubical complexes and directed path spaces. In Sect. 3, we introduce the notions of spaces of directed paths and Euclidean cubical complexes. Given the directed structure of these Euclidean cubical complexes, we do not consider the link of a vertex but the *past* link of it. In Sect. 4, we give results on the topology of the spaces of directed paths from an initial vertex to other vertices in terms of past links. Theorem 1 gives sufficient conditions on the past links of every vertex of a complex so that spaces of directed paths are contractible. Theorem 2 gives conditions that are sufficient for the spaces of directed paths to be connected. In Theorem 3, we give sufficient conditions on the past link of a vertex so that the space of directed paths from the initial vertex to that vertex is disconnected. In Sect. 5, we describe a method of collapsing one complex into a simpler complex, while preserving the directed path spaces.

2 Concurrent Programs and Directed Path Spaces

We illustrate how to organize possible executions of concurrent programs using Euclidean cubical complexes and directed spaces. An execution is a scheduling of the events that occur in a program in order to compute a specific task. In Example 1, we describe the dining philosophers problem. In Example 2, we illustrate how to model executions of concurrent programs in the context of the dining philosophers problem in the case of two philosophers.

Example 1 (Dining Philosophers) The dining philosophers problem originally formulated by E. Dijkstra [2] and reformulated by T. Hoare [7] illustrates issues that arise in concurrent programs. Consider n philosophers sitting at a round table ready to eat a meal. Between each pair of neighboring philosophers is a chopstick for a total of n chopsticks. Each philosopher must eat with the two chopsticks lying directly to the left and right of her. Once the philosopher is finished eating, she must put down both chopsticks. Since there are only n chopsticks, the philosophers must share the chopsticks in order for all of them to eat. The dining philosopher problem is to design a concurrent program where all n philosophers are able to eat once for some finite amount of time.

A design of a program is a choice of actions for each philosopher. One example of a design of a program is where each of the n philosophers does the following:

1. Wait until the right chopstick is available, then pick it up.
2. Wait until the left chopstick is available, then pick it up.
3. Eat for some finite amount of time.
4. Put down the left chopstick.
5. Put down the right chopstick.

While correct executions of this program are possible (e.g., where the philosophers take turns eating alone), this design has states in which every philosopher has picked up the chopstick to her right and is waiting for the other chopstick. Such a situation exemplifies a *deadlock* in concurrent programming, an execution that gets "stuck" and never finishes.

The design described above also has states that cannot occur. For example, consider the dining philosophers problem when $n = 2$. The state in which both philosophers are finished eating and one is still holding onto chopstick a while the other is holding chopstick b would imply that a philosopher was able to eat with only one chopstick—an example of an *unreachable* state in concurrent programming.

The dining philosophers problem illustrates the difficulties in designing concurrent programs. Difficulties arise since each philosopher must use chopsticks that must be shared with the neighboring philosophers. Analogously, in concurrent programming, multiple processes must access shared resources that have a finite capacity.

The next example illustrates how to model executions of the dining philosophers problem with a Euclidean cubical complex. When the problem consists of two

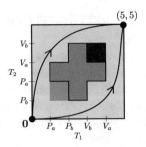

Fig. 1 The Swiss Flag. The pink region is the forbidden region. Any bi-monotone path outside of F is a possible execution. The set of all executions of two processes, T_1 and T_2, is called the *state-space* Two regions in the state space are of particular interest. The black region is the set of all unreachable states, and the blue region is the set of all states that are doomed to never complete. A state is doomed if any path starting at that state leads to a deadlock. The black curves in the figure are two possible paths in this directed space

philosophers, the Euclidean cubical complex used to model the dining philosophers problem is often referred to as the *Swiss Flag*.

Example 2 (Swiss Flag) In the language of concurrent programming, the two philosophers represent two processes denoted by T_1 and T_2. The two chopsticks represent shared resources denoted by a and b. One process is executing the program $P_a P_b V_b V_a$ and the other process is executing the program $P_b P_a V_a V_b$. Here, P means that a process has a *lock* on that resource while V means that a process *releases* a resource. To model this concurrent program with a Euclidean cubical complex, we construct a 5×5 grid where the x-axis is labeled by $P_a P_b V_b V_a$, each a unit apart, and the y-axis is labeled by $P_b P_a V_a V_b$, each also a unit apart (see Fig. 1). The region $[1, 4] \times [2, 3]$ represents when both T_1 and T_2 have a lock on a. In the dining philosophers problem, a single chopstick can only be held by one philosopher at a given time. The mutual exclusion of the chopsticks translates to the shared resources, a and b, each having *capacity* one, where the capacity of a resource is the number of processes that can have access to the resource simultaneously. We call the region $[1, 4] \times [2, 3]$ *forbidden* since T_1 and T_2 cannot have a lock on a at the same time. The region $[2, 3] \times [1, 4]$ represents when both T_1 and T_2 have a lock on b. This region is also forbidden. The set complement of the interior of $[1, 4] \times [2, 3] \cup [2, 3] \times [1, 4]$ in $[0, 5] \times [0, 5]$ is called the Swiss flag and is the Euclidean cubical complex modeling this program design for the dining philosophers problem.

In general, the Euclidean cubical complex modeling a concurrent program is the complement of the interior of the forbidden region. An execution is a directed path from the initial point to the terminal point. Executions are equivalent if they give the same output given the same input, which can be interpreted geometrically as the corresponding paths are dihomotopic in the path space. The Swiss flag has two distinct directed paths up to homotopy equivalence: one corresponding to T_1 using the shared resources first, and the other corresponding to T_2 using the shared resources first. See Fig. 1.

3 Past Links as Obstructions

In this section, we introduce the notions of spaces of directed paths and Euclidean cubical complexes. The (relative) past link of a vertex of a Euclidean cubical complex is defined as a simplicial complex. Studying the contractibility and connectedness of past links gives us insight on the contractibility and connectedness of certain spaces of directed paths.

Definition 1 (d-space) A *d-space* is a pair $(X, \overrightarrow{P}(X))$, where X is a topological space and $\overrightarrow{P}(X) \subseteq P(X) := X^{[0,1]}$ is a family of paths on X (called *dipaths*) that is closed under non-decreasing reparametrizations and concatenations, and contains all constant paths.

For every x, y in X, let $\overrightarrow{P}_x^y(X)$ be the family of *dipaths from x to y*:

$$\overrightarrow{P}_x^y(X) := \{\alpha \in \overrightarrow{P}(X) : \alpha(0) = x \text{ and } \alpha(1) = y\}.$$

In particular, consider the following directed space: the *directed real line* $\overrightarrow{\mathbb{R}}$ is the directed space constructed from the real line whose family of dipaths $\overrightarrow{P}(\mathbb{R})$ consists of all non-decreasing paths. The *Euclidean space* $\overrightarrow{\mathbb{R}^n}$ is the n-fold product $\overrightarrow{\mathbb{R}} \times \cdots \times \overrightarrow{\mathbb{R}}$ with family of dipaths the n-fold product $\overrightarrow{P}(\mathbb{R}^n) = \overrightarrow{P}(\mathbb{R}) \times \cdots \times \overrightarrow{P}(\mathbb{R})$.

Furthermore, we can solely focus on the family of dipaths in a d-space and endow it with the compact open topology.

Definition 2 (Space of Directed Paths) In a d-space $(X, \overrightarrow{P}(X))$, the *space of directed paths* from x to y is the family $\overrightarrow{P}_x^y(X)$ with the compact open topology.

By topologizing the space of directed paths, we may now use topological reasoning and comparison. Since $\overrightarrow{P}_x^y(X)$ does not have directionality, contractibility and other topological features are defined as in the classical case. Moreover, observe that the set $\overrightarrow{P}_x^y(X)$ might have cardinality of the continuum, but is considered trivial if it is homotopy equivalent to a point.

The d-spaces that we consider in this article are constructed from Euclidean cubical complexes. Let $\mathbf{p} = (p_1, \ldots, p_n), \mathbf{q} = (q_1, \ldots, q_n) \in \mathbb{R}^n$. We write $\mathbf{p} \preceq \mathbf{q}$ if and only if $p_i \leq q_i$ for all $i = 1, \ldots, n$. Furthermore, we denote by $\mathbf{q} - \mathbf{p} := (q_1 - p_1, \ldots, q_n - p_n)$ the component-wise difference between $\mathbf{q}$ and $\mathbf{p}$, $|\mathbf{p}| := \sum_{i=1}^n p_i$ is the element-wise sum, or one-norm, of $\mathbf{p}$. Similarly to the one-dimensional case, the interval $[\mathbf{p}, \mathbf{q}]$ is defined as $\{\mathbf{x} \in \mathbb{R}^n : \mathbf{p} \preceq \mathbf{x} \preceq \mathbf{q}\}$.

Definition 3 (Euclidean Cubical Complex) Let $\mathbf{p}, \mathbf{q} \in \mathbb{R}^n$. If $\mathbf{q}, \mathbf{p} \in \mathbb{Z}^n$ and $\mathbf{q} - \mathbf{p} \in \{0, 1\}^n$, then the interval $[\mathbf{p}, \mathbf{q}]$ is an *elementary cube in* $\mathbb{R}^n$ of dimension $|\mathbf{q}-\mathbf{p}|$. A *Euclidean cubical complex* $K \subseteq \mathbb{R}^n$ is the union of elementary cubes.

Remark 1 A Euclidean cubical complex K is a subset of $\mathbb{R}^n$ and it has an associated abstract cubical complex. By a slight abuse of notation, we do not distinguish these.

Every cubical complex K inherits the directed structure from the Euclidean space $\overrightarrow{\mathbb{R}^n}$, described after Definition 1. An elementary cube of dimension d is called a d-cube. The m-skeleton of K, denoted by K_m, is the union of all elementary cubes contained in K that have dimension less than or equal to m. The elements of the zero-skeleton are called the vertices of K. A vertex $\mathbf{w} \in K_0$ is said to be *minimal* (resp., *maximal*) if $\mathbf{w} \preceq \mathbf{v}$ (resp., $\mathbf{w} \succeq \mathbf{v}$) for every vertex $\mathbf{v} \in K_0$.

Following [12], we define the (relative) past link of a vertex of a Euclidean cubical complex as a simplicial complex. Let Δ^{n-1} denote the complete simplicial complex with vertices $\{1, \ldots, n\}$. Simplices of Δ^{n-1} is be identified with elements $\mathbf{j} \in \{0, 1\}^n$. That is, every subset $S \subseteq \{1, \ldots, n\}$ is mapped to the n-tuple with entry 1 in the k-th position if k belongs to S and 0 otherwise. The topological space associated to the simplicial complex Δ^{n-1} is the one given by its geometric realization.

Definition 4 (Past Link) In a Euclidean cubical complex K in $\mathbb{R}^n$, the *past link*, $lk_{K,\mathbf{w}}^-(\mathbf{v})$, of a vertex $\mathbf{v}$, with respect to another vertex $\mathbf{w}$ is the simplicial subcomplex of Δ^{n-1} defined as follows: $\mathbf{j} \in lk_{K,\mathbf{w}}^-(\mathbf{v})$ if and only if $[\mathbf{v} - \mathbf{j}, \mathbf{v}] \subseteq K \cap [\mathbf{w}, \mathbf{v}]$.

Remark 2 While K is a *cubical* complex, the past link of a vertex in K is always a *simplicial* complex.

Remark 3 Often the vertex $\mathbf{w}$ and the complex K are understood. In this case, we denote the past link of $\mathbf{v}$ by $lk^-(\mathbf{v})$.

Remark 4 Other definitions of the (past) link are found in the literature. Unlike Definition 4, (past) links are usually subcomplexes of K. However, the (past) links found in other literature are homeomorphic to the (past) link of Definition 4.

In the following example, we show that a vertex $\mathbf{v}$ can have past links with different homotopy type depending on what the initial vertex $\mathbf{w}$ is. We consider as a Euclidean cubical complex the open top box (Fig. 2) and the past links of the vertex $\mathbf{v} = (1, 1, 1)$, with respect to the vertices $\mathbf{w} = \mathbf{0}$ and $\mathbf{w}' = (0, 0, 1)$.

Example 3 (Open Top Box) Let $L \subset \mathbb{R}^3$ be the Euclidean cubical complex consisting of all of the edges and vertices in the elementary cube $[\mathbf{0}, \mathbf{v}]$ and five of the six two-cubes, omitting the elementary two-cube $[(0, 0, 1), \mathbf{v}]$, i.e., the top of the box. Because the elementary one-cube $[\mathbf{v} - (0, 0, 1), \mathbf{v}] \subseteq L \cap [\mathbf{0}, \mathbf{v}] = L$, $lk_{L,\mathbf{0}}^-(\mathbf{v})$ contains the vertex in Δ^2 corresponding to $\mathbf{j} = (0, 0, 1)$. Similarly, because the elementary two-cube $[\mathbf{v} - (0, 1, 1), \mathbf{v}] \subseteq L$, the past link $lk_{L,\mathbf{0}}^-(\mathbf{v})$ contains the edge in Δ^2 corresponding to $\mathbf{j} = (0, 1, 1)$. However, because the elementary two-cube $[\mathbf{v} - (1, 1, 0), \mathbf{v}]$ is not contained in L, $lk_{L,\mathbf{0}}^-(\mathbf{v})$ does not include the edge corresponding to $\mathbf{j} = (1, 1, 0)$. Instead taking the initial vertex to be $\mathbf{w} = (0, 0, 1)$, we get that $lk_{L,\mathbf{w}}^-(\mathbf{v})$ consists of the two vertices corresponding to $\mathbf{j} = (0, 1, 0)$ and $\mathbf{j}' = (1, 0, 0)$. See Fig. 2.

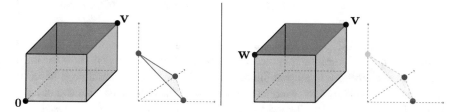

Fig. 2 The Open Top Box. Left: the open top box and the geometric realization of the past link of the red vertex $\mathbf{v} = (1, 1, 1)$, with respect to the black vertex $\mathbf{0}$. The geometric realization of $lk_{L,\mathbf{0}}^{-}(\mathbf{v})$ contains two edges of a triangle, since the two red faces are included in $[\mathbf{0}, \mathbf{v}]$ and three vertices, since the three red edges are included in $[\mathbf{0}, \mathbf{v}]$. Right: the open top box and the geometric realization of the past link of the red vertex $\mathbf{v} = (1, 1, 1)$, with respect to the black vertex $\mathbf{w} = (0, 0, 1)$. The geometric realization of $lk_{L,\mathbf{w}}^{-}(\mathbf{v})$ consists only of two vertices of a triangle, since the two red edges are included in $[\mathbf{w}, \mathbf{v}]$

4   The Relationship Between Past Links and Path Spaces

In this section, we illustrate how to use past links to study spaces of directed paths with an initial vertex of $\mathbf{0}$. In particular, the contractibility and connectedness of all past links guarantees the contractibility and connectedness of spaces of directed paths. We also provide a partial converse to the result concerning connectedness.

Theorem 1 (Contractibility) *Let $K \subset \mathbb{R}^n$ be a Euclidean cubical complex with minimal vertex $\mathbf{0}$. Suppose for all $\mathbf{k} \in K_0$, $\mathbf{k} \neq \mathbf{0}$, the past link $lk_{\mathbf{0}}^{-}(\mathbf{k})$ is contractible. Then, all spaces of directed paths $\overrightarrow{P}_{\mathbf{0}}^{\,\mathbf{k}}(K)$ are contractible.*

Proof By [12, Prop. 5.3], if $\overrightarrow{P}_{\mathbf{0}}^{\,\mathbf{k}-\mathbf{j}}(K)$ is contractible for all $\mathbf{j} \in \{0, 1\}^n$, $\mathbf{j} \neq \mathbf{0}$, and $\mathbf{j} \in lk^{-}(\mathbf{k})$, then $\overrightarrow{P}_{\mathbf{0}}^{\,\mathbf{k}}(K)$ is homotopy equivalent to $lk^{-}(\mathbf{k})$. Hence, it suffices to see that all the spaces $\overrightarrow{P}_{\mathbf{0}}^{\,\mathbf{k}-\mathbf{j}}(K)$ are contractible. This follows by structural induction on the partial order on vertices in K.

The start is at $\overrightarrow{P}_{\mathbf{0}}^{\,\mathbf{0}+\mathbf{e}_i}(K)$, where $\mathbf{e}_i$ is the i-th unit vector, and $\mathbf{0} + \mathbf{e}_i \in K_0$. If the edge $[\mathbf{0}, \mathbf{0} + \mathbf{e}_i]$ is in K, then $\overrightarrow{P}_{\mathbf{0}}^{\,\mathbf{0}+\mathbf{e}_i}(K)$ is contractible. Otherwise, $lk_{\mathbf{0}}^{-}(\mathbf{0} + \mathbf{e}_i)$ is empty, which contradicts the hypothesis that all of the past links are contractible. By structural induction, using also that $\overrightarrow{P}_{\mathbf{0}}^{\,\mathbf{0}}$ is contractible, the theorem now holds.   □

Now, we give an analogous sufficient condition for when spaces of directed paths are connected. We provide two different proofs of Theorem 2. The first proof shows how we can use [9, Prop. 2.20] to get our desired result. The second proof uses notions from category theory and is based on the fact that the colimit of connected spaces over a connected category is connected.

Theorem 2 (Connectedness) *With K as above, suppose all past links $lk_{\mathbf{0}}^{-}(\mathbf{k})$ of all vertices $\mathbf{k} \neq \mathbf{0}$ are connected. Then, for all $\mathbf{k} \in K_0$, all spaces of directed paths $\overrightarrow{P}_{\mathbf{0}}^{\,\mathbf{k}}(K)$ are connected.*

In this first proof we show that [9, Prop. 2.20] is an equivalent condition to all past links being connected.

Proof In [9, Prop. 2.20], a local condition is given that ensures that all spaces of directed paths to a certain final point are connected. Here, we explain how the local condition is equivalent to all past links being connected. Their condition is in terms of the local future; however, we reinterpret this in terms of local past instead of local future. Since we consider all spaces of directed paths *from* a point (as opposed to *to* a point), then reinterpreting the result in terms of local past is the right setting we should look at. The local condition is the following: for each vertex, $\mathbf{v}$, and all pairs of edges $[\mathbf{v} - \mathbf{e}_r, \mathbf{v}]$, $[\mathbf{v} - \mathbf{e}_s, \mathbf{v}]$ in K, there is a sequence of two-cells $\{[\mathbf{v} - \mathbf{e}_{k_i} - \mathbf{e}_{l_i}, \mathbf{v}]\}_{i=1}^m$, each of which is in K such that $l_i = k_{i+1}$ for $i = 1, \ldots, m - 1$, $k_1 = r$ and $l_m = s$. Now, we show that this local condition is equivalent to ours. In the past link considered as a simplicial complex, such a sequence of two-cells corresponds to a sequence of edges from the vertex r to the vertex s. For $x, y \in lk^-(\mathbf{v})$, they are both connected to a vertex via a line. And those vertices are connected. Hence, the past link is connected.

Vice versa: Suppose $lk^-(\mathbf{v})$ is connected. Let p, q be vertices in $lk^-(\mathbf{v})$ and let $\gamma : I \to lk^-(\mathbf{v}) \in \Delta^{n-1}$ be a path from p to q. The sequence of simplices traversed by γ, $S_1, S_2, \ldots, S_k$, satisfies $S_i \cap S_{i+1} \neq \emptyset$. Moreover, the intersection is a simplex. Let $p_i \in S_i \cap S_{i+1}$. A sequence of pairwise connected edges connecting p to q is constructed by such sequences from p_i to p_{i+1} in S_{i+1} thus providing a sequence of two-cells similar to the requirement in [9]. Hence, by [9], if all past links of all vertices are connected, then all $\overrightarrow{P}_{\mathbf{0}}^{\mathbf{k}}$ are connected $\qquad\square$

This second proof of Theorem 2 has a more categorical flavor.

Proof We give a more categorical argument which is closer to the proof of Theorem 1. In [10, Prop. 2.3 and Equation 2.2], the space of directed paths $\overrightarrow{P}_{\mathbf{0}}^{\mathbf{k}}$ is given as a colimit over $\overrightarrow{P}_{\mathbf{0}}^{\mathbf{k}-\mathbf{j}}$. The indexing category is $\mathcal{J}_K$ with objects $\{\mathbf{j} \in \{0, 1\}^n : [\mathbf{k} - \mathbf{j}] \subseteq K\}$ and morphisms $\mathbf{j} \to \mathbf{j}'$ for $\mathbf{j} \geq \mathbf{j}'$ given by inclusion of the simplex $\Delta^{\mathbf{j}} \subset \Delta^{\mathbf{j}'}$. The geometric realization of the index category is the past link which with our requirements is connected. The colimit of connected spaces over a connected category is connected. Hence, by induction as above, beginning with edges from $\mathbf{0}$, the directed paths $\overrightarrow{P}_{\mathbf{0}}^{\mathbf{k}-\mathbf{j}}$ are all connected and the conclusion follows. $\qquad\square$

Remark 5 Our conjecture is that similar results for k-connected past links should follow from the k-connected Nerve Lemma.

Remark 6 The statements of both Theorems 1 and 2 concern past links and path spaces defined with respect to a fixed initial vertex. To see why past links depend on their initial vertex, consider the open top box of Example 3. All past links in L with respect to the initial vertex $\mathbf{0}$ are contractible, but $\overrightarrow{P}_{\mathbf{w}'}^{\mathbf{v}}(L)$, where $\mathbf{w}' = (0, 0, 1)$ and $\mathbf{v} = (1, 1, 1)$, is not contractible. It is in fact two points. Note, this does not contradict Theorem 1, which only asserts that $\overrightarrow{P}_{\mathbf{0}}^{\mathbf{v}}(L)$ is contractible; see Fig. 2.

We now show how Theorems 1 and 2 can be used to study the spaces of the directed paths in slight modifications of the dining philosophers problem.

Example 4 (Three Concurrent Processes Executing the Same Program) We consider a modification of Example 1 where we have three processes and two resources each with capacity two. All processes are executing the program $P_a P_b V_b V_a$. The Euclidean cubical complex modeling this situation has three dimensions, each representing the program of a process. Since each resource has capacity two, it is not possible to have a three way lock on any of the resources. The three processes have a lock on a in the region $[P_a, V_a]^{\times 3}$, which is the cube $[(1, 1, 1), (4, 4, 4)]$. Similarly, the three processes have a lock on b in the region $[P_b, V_b]^{\times 3}$ which is the cube $[(2, 2, 2), (3, 3, 3)]$. The forbidden region is the union of these two sets which is $[(1, 1, 1), (4, 4, 4)]$. We can model this concurrent program as a three-dimensional Euclidean cubical complex and the forbidden region is the inner $3 \times 3 \times 3$ cube.

In order to analyze the connectedness and contractibility of the spaces of directed paths with initial vertex $\mathbf{0}$, we study the past links of the vertices of K. First, we show that not all past links are contractible. Let $\mathbf{v} = (4, 4, 4)$. Then, $lk_{K,\mathbf{0}}^-(\mathbf{v})$ consists of all $\mathbf{j} \in \{0, 1\}^3$ except $(1, 1, 1)$. The past link does not contain $(1, 1, 1)$ because the cube $[(3, 3, 3), (4, 4, 4)]$ is not contained in K, but $[\mathbf{v} - \mathbf{j}, \mathbf{v}] \subset K$ for all other $\mathbf{j}$. Therefore, $lk_{K,\mathbf{0}}^-(\mathbf{v})$ is the boundary of the two simplex (see Fig. 3). Because the boundary of the two simplex is not contractible, the hypothesis of Theorem 1 is not satisfied. Hence, we cannot use Theorem 1 to study the contractibility of the spaces of directed paths.

Next, we show that all past links are connected. If we directly compute the past link $lk_{K,\mathbf{0}}^-(\mathbf{k})$ for all $\mathbf{k} \in K_0$, we find that the past link consists of either a zero simplex, one simplex, the boundary of the two simplex, or a two simplex. All these past links are connected. Theorem 2 implies that for all $\mathbf{k} \in K_0$, the space of directed paths, $\overrightarrow{P}_{\mathbf{0}}^{\mathbf{k}}(K)$ is connected.

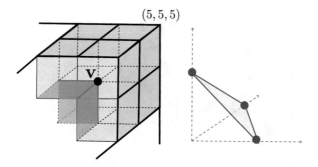

Fig. 3 Three processes, same program. Illustrating $lk_{K,\mathbf{0}}^-(\mathbf{v})$ where K is the cube $[\mathbf{0}, (5, 5, 5)]$ minus the inner cube, $[(1, 1, 1), (4, 4, 4)]$, and $\mathbf{v} = (4, 4, 4)$. The geometric realization of the simplicial complex $lk_{K,\mathbf{0}}^-(\mathbf{v})$ is the boundary of the two simplex since the three pink faces and edges are included in $[\mathbf{0}, \mathbf{v}]$

We can generalize this example to n processes and two resources with capacity $n - 1$ where all processes are executing the program $P_a P_b V_b V_a$. For all n, Theorem 2 shows that all spaces of directed paths are connected.

The converse of Theorem 2 is not true. To see this, and give the conditions under which the converse does hold, we need to introduce the following definition:

Definition 5 (Reachable) The point $x \in K$ is *reachable* from $\mathbf{w} \in K_0$ if there is a path from $\mathbf{w}$ to x. A subcomplex of K is induced by the set of points that are reachable from a vertex $\mathbf{w}$.

Example 5 (Boundary of the $3 \times 3 \times 3$ Cube with Top Right Cube) Let K be the Euclidean cubical complex that is the boundary of the $3 \times 3 \times 3$ cube along with the cube $[(2, 2, 2), (3, 3, 3)]$. Observe that all spaces of directed paths with initial vertex $\mathbf{0}$ are connected. However, K has a disconnected past link at $\mathbf{v} = (3, 2, 2)$. If we consider the subcomplex $\hat{K}$ that is reachable from $\mathbf{0}$, then $\hat{K}$ is the boundary of the $3 \times 3 \times 3$ cube. The past links of all vertices in $\hat{K}$ are connected. This motivates the conditions given in Theorem 3 of removing the unreachable points of a Euclidean cubical complex. The connected components of a disconnected past link in the remaining complex can then be represented by directed paths from the initial point and not only locally (Fig. 4).

Theorem 3 (Realizing Obstructions) *Let K be a Euclidean cubical complex with initial vertex $\mathbf{0}$. Let $\hat{K} \subset K$ be the subcomplex reachable from $\mathbf{0}$. If for $\mathbf{v} \in \hat{K}_0$, the past link in $\hat{K}$ is disconnected, then the path space $\overrightarrow{P}_{\mathbf{0}}^{\mathbf{v}}(K)$ is disconnected.*

Proof Let $\mathbf{v}$ be a vertex such that $lk_{\hat{K},\mathbf{0}}^{-}(\mathbf{v})$ is disconnected and let $\mathbf{j}_1, \mathbf{j}_2$ be vertices in $lk_{\hat{K}}^{-}(v)$ in different components. The edges $[\mathbf{v} - \mathbf{j}_i, \mathbf{v}]$ are then in $\hat{K}$ and, in particular, $\mathbf{v} - \mathbf{j}_i \in \hat{K}_0$. Hence, there are paths $\mu_i : \overrightarrow{I} \to \hat{K}$ such that $\mu_i(0) = \mathbf{0}$ and $\mu_i(1) = \mathbf{v} - \mathbf{j}_i$.

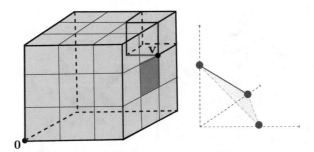

Fig. 4 Motivating reachability condition. Let K be the boundary of the $3 \times 3 \times 3$ cube union with $[(2, 2, 2), (3, 3, 3)]$. Then, the geometric realization of the simplicial complex $lk_{\hat{K},\mathbf{0}}^{-}(\mathbf{v})$ is an edge and a point since the three pink edges and one face are included in $[(0, 0, 0), \mathbf{v}]$

By [3], there are $\hat{\mu}_i$ which are dihomotopic to μ_i and such that $\hat{\mu}_i$ is combinatorial, i.e., a sequence of edges in $\hat{K}$. Let γ_i be the concatenation of $\hat{\mu}_i$ with the edge $[\mathbf{v} - \mathbf{j}_i, \mathbf{v}]$.

Suppose for contradiction that γ_1 and γ_2 are connected by a path in $\overrightarrow{P}_{\mathbf{0}}^{\mathbf{v}}(K)$. Let $H : \overrightarrow{I} \times I \to K$ be such a path with $H(t, 0) = \gamma_1(t)$ and $H(t, 1) = \gamma_2(t)$. Since $H(t, s)$ is reachable from $\mathbf{0}$, H maps to $\hat{K}$.

By [3], there is a combinatorial approximation $\hat{H} : \overrightarrow{I} \times I \to \hat{K}_2$ to the 2-skeleton of $\hat{K} \subset K$. Let B be the open ball centered around $\mathbf{v}$ with radius $1/2$. Since $\hat{H}$ is continuous, the inverse image of B under $\hat{H}$ is a neighborhood of $\{1\} \times I \subset \overrightarrow{I} \times I$. For $0 < \epsilon < 1/2$, this neighborhood contains a strip $(1 - \epsilon, 1] \times I$ (by compactness of I). Then $\hat{H}(1 - \epsilon/2 \times I)$ gives a path connecting the two edges $[\mathbf{v}-\mathbf{j}_i, \mathbf{v}]$. This path traverses a sequence of 2-cubes (the carriers). These correspond to a sequence of edges in the past link that connect $\mathbf{j}_1$ and $\mathbf{j}_2$, which contradicts the assumption that they are in different components. Therefore, γ_1 and γ_2 correspond to two points in $\overrightarrow{P}_{\mathbf{0}}^{\mathbf{v}}(K)$ that are not connected by a path. □

In general, the reachability condition in Theorem 3 eliminates the spurious disconnected past links that could appear in the unreachable parts of a Euclidean cubical complex.

Example 6 To see how Theorem 3 can be applied, consider Example 2, the Swiss flag. The Swiss flag has two vertices with disconnected past links with respect to $\mathbf{0}$ namely $(4, 3)$ and $(3, 4)$. These disconnected past links imply that Theorem 2 is inconclusive. If the unreachable section of the Swiss flag is removed, we obtain a new Euclidean cubical complex in which the vertex $\mathbf{v} = (4, 4)$ has a disconnected past link, consisting of two points. By Theorem 3, the path space $\overrightarrow{P}_{\mathbf{0}}^{\mathbf{v}}(K)$ is also disconnected. In fact, $\overrightarrow{P}_{\mathbf{0}}^{\mathbf{v}}(K)$ has two points, representing the dihomotopy classes of paths which pass above the forbidden region, and those paths which pass below.

The disconnected path space, $\overrightarrow{P}_{\mathbf{0}}^{\mathbf{v}}(K)$, found in the previous example helps illustrate the following: given two vertices $\mathbf{w}$ and $\mathbf{v}$ in a Euclidean cubical complex K, if the path space $\overrightarrow{P}_{\mathbf{w}}^{\mathbf{v}}(K)$ is disconnected, then there exists a vertex in $[\mathbf{w}, \mathbf{v}]$ that has a disconnected past link with respect to $\mathbf{w}$ (the vertices $(4, 3)$ and $(3, 4)$ in the Swiss flag). If $\mathbf{w} = \mathbf{0}$, then we get the contrapositive of Theorem 2. If K is reachable from $\mathbf{0}$, Theorem 3 allows us to draw conclusions about the space of directed paths.

5 Directed Collapsibility

To simplify the underlying topological space of a d-space while preserving topological properties of the associated space of directed paths, we introduce the process of directed collapse. The criteria we require to perform directed collapse on Euclidean cubical complexes involves the topology of the past links of the vertices of the complex. We defined the past links as simplicial complexes that are not themselves directed, so our topological criteria are in the usual sense.

Definition 6 (Directed Collapse) Let K be a Euclidean cubical complex with initial vertex $\mathbf{0}$. Consider $\sigma, \tau \in K$ such that $\tau \subsetneq \sigma$, σ is maximal, and no other maximal cube contains τ. Let $K' = K \setminus \{\gamma \in K \mid \tau \subseteq \gamma \subseteq \sigma\}$. K' is a *directed (cubical) collapse* of K if, for all $\mathbf{v} \in K'_0$, $lk_K^-(\mathbf{v})$ is homotopy equivalent to $lk_{K'}^-(\mathbf{v})$. The pair τ, σ is then called a *collapsing pair*.

K' is a *directed 0-collapse* of K if for all $\mathbf{v} \in K'_0$, $lk_K^-(\mathbf{v})$ is connected if and only if $lk_{K'}^-(\mathbf{v})$ is connected.

Remark 7 As in the simplicial case, when we remove σ from the abstract cubical complex, the effect on the geometric realization is to remove the interior of the cube corresponding to σ.

Remark 8 Note for finding collapsing pairs, (τ, σ), using Definition 6, with the geometric realization of σ given by the elementary cube, $[\mathbf{w} - \mathbf{j}, \mathbf{w}]$, it is sufficient to only check $\mathbf{v} \in K'_0$ such that $\mathbf{v} = \mathbf{w} - \mathbf{j}'$ where $\mathbf{j} - \mathbf{j}' > 0$. Otherwise the past links, $lk_K^-(\mathbf{v})$ and $lk_{K'}^-(\mathbf{v})$, are equal.

Definition 7 (Past Link Obstruction) Let $\mathbf{w} \in K_0$. A *past link obstruction (type-∞)* in K with respect to $\mathbf{w}$ is a vertex $\mathbf{v} \in K_0$ such that $lk_{K,\mathbf{w}}^-(\mathbf{v})$ is not contractible. A *past link obstruction (type-0)* in K with respect to $\mathbf{w}$ is a vertex $\mathbf{v} \in K_0$ such that $lk_{K,\mathbf{w}}^-(\mathbf{v})$ is not connected.

Directed collapses preserve some topological properties of the space of directed paths. In particular:

Corollary 1 *If there are no type-∞ past link obstructions, then all spaces of directed paths from the initial point are contractible. If there are no type-0 past link obstructions, all spaces of directed paths from the initial point are connected.*

Proof Contractibility is a direct consequence of Theorem 1. Likewise, connectedness follows from Theorem 2.   □

Corollary 2 (Invariants of Directed Collapse) *If we have a sequence of directed collapses from K to K', then there are no obstructions in K iff there are no obstructions in K'.*

Remark 9 (Past Link Obstructions are Inherently Local) The past link of a vertex is constructed using local (rather than global) information from the cubical complex. Therefore, a past link obstruction is also a local property, which is not dependent on the global construction of the cubical complex.

Below, we provide a few motivating examples for our definition of directed collapse. In general, we want our directed collapses to preserve all spaces of directed paths between the initial vertex and any other vertex in our cubical complex. Notice, τ from Definition 6 is a *free face* of K. Performing a directed collapse with an arbitrary free face of a directed space K with minimal element $\mathbf{0} \in K_0$ and maximal element $\mathbf{1} \in K_0$ can modify the individual spaces of directed paths $\overrightarrow{P}{}^{\mathbf{v}}_{\mathbf{0}}(K)$ and $\overrightarrow{P}{}^{\mathbf{1}}_{\mathbf{v}}(K)$ for $\mathbf{v} \in K_0$.

When $\overrightarrow{P}{}^{\mathbf{1}}_{\mathbf{v}}(K) = \emptyset$, we call $\mathbf{v}$ a *deadlock*. When $\overrightarrow{P}{}^{\mathbf{v}}_{\mathbf{0}}(K) = \emptyset$, we call $\mathbf{v}$ *unreachable*. Deadlocks and unreachable vertices are in a sense each others opposites. Notice if we take the same directed space K yet reverse the direction of all dipaths, then deadlocks become unreachable vertices and vice versa. However, as Examples 7 and 8 illustrate, the creation of an unreachable vertex in the process of a directed collapse might result in a past link obstruction at a neighboring vertex while the creation of a deadlock does not.

Example 7 (3×3 Grid, Deadlocks and Unreachability) Let K be the Euclidean cubical complex in $\mathbb{R}^2$ that is the 3×3 grid. Consider the Euclidean cubical complexes K' and K'' obtained by removing (τ, σ) with $\tau = [(1, 3), (2, 3)]$, $\sigma = [(1, 2), (2, 3)]$ and (τ', σ') with $\tau' = [(1, 0), (2, 0)]$, $\sigma' = [(1, 0), (2, 1)]$, respectively. While K' is a directed collapse of K, K'' is not a directed collapse of K because K'' introduces a past link obstruction at $(2, 1)$. So, (τ, σ) is a collapsing pair while (τ', σ') is not. Collapsing K to K' creates a deadlock at $(1, 3)$ but this does not change the space of directed paths from the designated start vertex $\mathbf{0}$ to any of the vertices between $\mathbf{0}$ and the designated end vertex $(3, 3)$ (see K' in Fig. 5). However, collapsing K to K'' creates an unreachable vertex $(2, 0)$ from the start vertex $\mathbf{0}$ (see K'' in Fig. 5) which does change the space of directed paths from $\mathbf{0}$ to $(2, 0)$ to be empty. Hence not all spaces of directed paths starting at $\mathbf{0}$ are preserved. This motivates our definition of directed collapse.

Our next example shows how directed collapses can be performed with collapsing pairs (τ, σ) when τ is of codimension one and greater.

Example 8 (3×3 grid, Edge and Vertex Collapses) Consider again the Euclidean cubical complex K from Example 7. If we allow a collapsing pair (τ, σ) with τ of dimension greater than 0, we may introduce deadlocks or unreachable vertices. In particular, collapsing the free edge $\tau = [(1, 3), (2, 3)]$ of the top blue square $\sigma =$

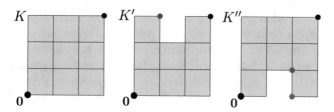

Fig. 5 Illustrating Example 7. On the left: the cubical complex K with initial vertex $\mathbf{0}$ and final vertex $(3, 3)$. In the center: The cubical complex K' which is a directed collapse of K. The deadlock in blue does not change the space of directed paths from $\mathbf{0}$ to any of the vertices between $\mathbf{0}$ and $(3, 3)$. On the right: the cubical complex K'' which is not a directed collapse of K. The space of directed paths into the unreachable red vertex, $(2, 0)$, becomes empty. The empty path space is reflected in the topology of the past link of the red vertex $(2, 1)$ (see Example 8)

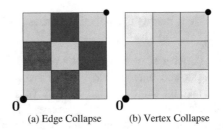

(a) Edge Collapse   (b) Vertex Collapse

Fig. 6 Illustrating Example 8. On the left: the collapsing of the free edge in the blue squares is an admitted directed collapse. The collapsing of the free edge in the red squares is not an admitted directed collapse. On the right: the collapsing of the free vertex in the yellow squares is an admitted directed collapse

$[(1, 2), (2, 3)]$ in Fig. 6 changes the space of directed paths $\overrightarrow{P}_{(1,3)}^{(3,3)}(K)$ from being trivial to empty in $K \backslash \{ \gamma \mid \tau \subseteq \gamma \subseteq \sigma \}$. Yet we care about preserving the space of directed paths from our designated start vertex $\mathbf{0}$ to any of the vertices (i, j) with $0 \leq i, j \leq 3$ since we ultimately are interested in preserving the path space $\overrightarrow{P}_{\mathbf{0}}^{(3,3)}(K)$. Because of this, such collapses should be allowed in our directed setting. Note that, in these cases, the past link of all vertices remains contractible. However, collapsing the free edge $\tau' = [(1, 0), (2, 0)]$ of the bottom red square $\sigma' = [(1, 0), (2, 1)]$ in Fig. 6 changes the path space $\overrightarrow{P}_{\mathbf{0}}^{(2,0)}(K)$ from being trivial to empty. This change is reflected in the non-contractible past link of $(2, 1)$ in $K \backslash \{ \gamma \mid \tau' \subseteq \gamma \subseteq \sigma' \}$ that consists of the two vertices $\mathbf{j} = (1, 0)$ and $\mathbf{j}' = (0, 1)$ but not the edge $\mathbf{j}'' = (1, 1)$ connecting them. Restricting our collapsing pairs to only include τ of dimension 0 allows for only two potential collapses, the corner vertices $(0, 3)$ and $(3, 0)$ into the yellow squares $[(0, 2), (1, 3)]$ and $[(2, 0), (3, 1)]$, respectively. Neither of these collapses create deadlocks or unreachable vertices and the contractibility of the past link at all vertices is preserved. Performing these corner vertex collapses exposes new free vertices that can be a part of subsequent collapses.

Lastly, we explain how the Swiss flag can be collapsed using a sequence of zero-collapses. The Swiss flag contains uncountably many paths between the initial and final vertex. After performing the sequence of zero-collapses as described in Example 9, the Swiss flag has only two paths up to reparametrization between the initial and final vertex. These two paths represent the two dihomotopy classes of paths that exists for the Swiss flag. Referring back to concurrent programming, we interpret the two paths as two inequivalent executions: either the first process holds a lock on the two resources then releases them so the other process can place a lock on the resources or vice versa.

Example 9 (0-Collapsing the Swiss Flag) The Swiss flag considered as a Euclidean cubical complex in the 5×5 grid has vertices with connected past links, except at $(4, 3)$ and $(3, 4)$. The vertex $(2, 2)$ and the cube $[1, 2] \times [1, 2]$ are a 0-collapsing pair. The vertex $(3, 3)$ and the cube $[3, 4] \times [3, 4]$ are not, since that collapse would produce a disconnected past link at $(4, 4)$. A sequence of 0-collapses preserving

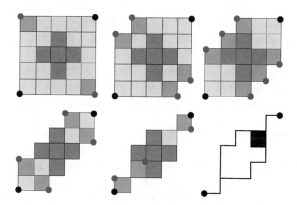

Fig. 7 Zero-collapsing the Swiss Flag. A sequence of zero-collapses is presented from the top left to bottom right. At each stage, the faces and vertices shaded in blue represent the zero-collapsing pairs. The result of the sequence is shown in the bottom right which is a one-dimensional Euclidean cubical complex and one two-cube

the initial and final point will give a one-dimensional Euclidean cubical complex and one 2-cube. Specifically, we get the edges $[0, 1] \times \{0\}$, $\{1\} \times [0, 1]$, $\{1\} \times [1, 3]$, $[1, 3] \times \{1\}$, $[1, 2] \times \{3\}$, $\{3\} \times [1, 2]$, $\{2\} \times [3, 4]$, $[3, 4] \times \{2\}$, $[2, 3] \times \{4\}$, $\{4\} \times [2, 3]$, the square $[3, 4] \times [3, 4]$, and lastly the edges $\{4\} \times [4, 5]$ and $[4, 5] \times \{5\}$ (Fig. 7).

6   Discussion

Directed topological spaces have a rich underlying structure and many interesting applications. The analysis of this structure requires tools that are not fully developed, and a further investigation into these methods will lead to a better understanding of directed spaces. In particular, the development of these notions, such as directed collapse, may lead to a better understanding of equivalence of directed spaces and their spaces of directed paths.

Interestingly, when comparing directed collapse with the notion of cubical collapse in the undirected case, two main contrasts arise. First, the notion of directed collapse is stronger than that of cubical collapse; any directed collapse is a cubical collapse, but not all cubical collapses satisfy the past link requirement of directed collapse. However, directed collapse is not related to existing notions of dihomotopy equivalence which involve continuous maps between topological spaces that preserve directed paths. Hence, directed collapse contrasts cubical collapse in the undirected case since any two spaces related by cubical collapses are homotopic. This contrast suggests the need for dihomotopy equivalence with respect to an initial point.

Directed collapse may not preserve dihomotopy equivalence, so we can collapse more than, e.g., Kahl. By Theorem 2, if K' is a directed collapse of K with respect to $\mathbf{v}$ and K' has trivial spaces of directed paths from $\mathbf{v}$, then so does K. Similarly, if all spaces of directed paths are connected in K', then all spaces of directed paths are connected in K. Hence, our definition of directed collapsibility preserves spaces of directed paths with an initial vertex of $\mathbf{0}$. Preserving spaces of directed paths allows us to study more types of concurrent programs and preserve notions of partial executions.

We plan to pursue many future avenues of research in the directed topological setting. First, we hope to find necessary and sufficient conditions for a pair of cubical cells (τ, σ) to be a collapsing pair. The key will be to have a better understanding of what removing a cubical cell does to the past link of a complex. Additionally, we would like to find directed conterparts to the various types of simplicial collapses. For example, is there a notion of strong directed collapse? As strong collapse also considers the link of a vertex, a consideration of how strong collapse extends to a directed setting seems natural.

Next, we would like to learn more about past link obstructions. We know that performing a directed collapse will not alter the space of directed paths of a Euclidean cubical complex; however, if we are unable to perform a directed collapse due to a past link obstruction, what happens to the space of directed paths? Theorem 3 is a start in understanding what happens to spaces of directed paths for 0 collapses. Another question may be, in what way are obstructions of type ∞ realized as non-contractible spaces of directed paths?

Another direction of research we hope to pursue is defining a way to compute a directed homology that is collapsing invariant. Even the two-dimensional setting (where the cubes are at most dimension two) has proved to be difficult, as adding one two-cell can have various effects, depending on the past links of the vertices involved. We would like to classify the spaces where such a dynamic programming approach would work.

Lastly, many computational questions arise on how to implement the collapse of a directed cubical complex. In [8], an example of collapsing a three-dimensional cubical complex is implemented in C++. This algorithm could be used as a model when handling the directed complex.

Many interesting theoretical and computational questions continue to emerge in the field of directed topology. We hope that our research excites others in studying cubical complexes in the directed setting.

Acknowledgements This research is a product of one of the working groups at the Women in Topology (WIT) workshop at MSRI in November 2017. This workshop was organized in partnership with MSRI and the Clay Mathematics Institute, and was partially supported by an AWM ADVANCE grant (NSF-HRD 1500481). In addition, LF and BTF further collaborated at the Hausdorff Research Institute for Mathematics during the Special Hausdorff Program on Applied and Computational Algebraic Topology (2017).

The authors also thank the generous support of NSF. RB is partially supported by the NSF GRFP (grant no. DGE 1649608). BTF is partially supported by NSF CCF 1618605. CR is partially supported by the NSF GRFP (grant no. DGE 1842165). SE is supported by the Swiss National Science Foundation (grant no. 200021-172636)

References

1. Jonathan Ariel Barmak and Elias Gabriel Minian. Strong homotopy types, nerves and collapses. *Discrete Comput. Geom.*, 47(2):301–328, 2012.
2. Edsger W. Dijkstra. Two starvation-free solutions of a general exclusion problem. 1977. Manuscript EWD625, from the archives of UT Austin, https://www.cs.utexas.edu/users/EWD/.
3. Lisbeth Fajstrup. Dipaths and dihomotopies in a cubical complex. *Advances in Applied Mathematics*, 35(2):188–206, 2005.
4. Lisbeth Fajstrup, Eric Goubault, Emmanuel Haucourt, Samuel Mimram, and Martin Raussen. *Directed Algebraic Topology and Concurrency*. Springer Publishing Company, Inc., 1st edition, 2016.
5. Lisbeth Fajstrup, Eric Goubault, and Martin Raussen. Algebraic topology and concurrency. *Theoretical Comuter Science*, pages 241–271, 2006.
6. Robin Forman. A user's guide to discrete Morse theory. *Sém. Lothar. Combin.*, 48, 12 2001.
7. Charles A.R. Hoare. Communicating sequential processes. *Commun. ACM*, 21(8):666–677, August 1978.
8. Jacques-Olivier Lachaud. Cubical complex collapse, 2017.
9. Martin Raussen. On the classification of dipaths in geometric models for concurrency. *Mathematical Structures in Computer Science*, 10(4):427–457, 2000.
10. Martin Raussen and Krzysztof Ziemiański. Homology of spaces of directed paths on euclidean cubical complexes. *Journal of Homotopy and Related Structures*, 9(1):67–84, Apr 2014.
11. John H.C. Whitehead. Simplicial spaces, nuclei and m-groups. *Proceedings of the London Mathematical Society*, s2-45(1):243–327.
12. Krzysztof Ziemiański. On execution spaces of PV-programs. *Theoretical Computer Science*, 619:87–98, 2016.

Contact Open Books and Symplectic Lefschetz Fibrations (Survey)

Bahar Acu

1 Introduction

A *contact manifold* (M, ξ) is a smooth $(2n + 1)$-dimensional manifold M equipped with a *maximally nonintegrable hyperplane distribution* $\xi \subset TM$. That is, locally $\xi = \ker \lambda$ whose defining 1-form λ on M satisfies

$$\lambda \wedge (d\lambda)^n \neq 0,$$

i.e. $\lambda \wedge (d\lambda)^n$ is a volume form on M. Then ξ is called a *contact structure* and the 1-form λ, which locally defines ξ, is called a *contact form* which locally defines ξ. Observe that the condition $\lambda \wedge (d\lambda)^n \neq 0$ is a property of the contact structure ξ, hence independent of the choice of the defining 1-form λ. A *symplectic manifold* (W, ω), on the other hand, is a $2n$-dimensional manifold equipped with a 2-form ω on W satisfying the non-degeneracy condition $\omega^n \neq 0$. The 2-form ω is then called a *symplectic structure*. Notice that if $\lambda \wedge (d\lambda)^n \neq 0$, then $d\lambda$ is a nondegenerate 2-form when restricted to ξ on a contact manifold M. That is to say, there is a strong formal link between symplectic and contact manifolds. For this reason, and many other fruitful relations, contact geometry is viewed as the odd-dimensional sibling of symplectic geometry.

Contact and symplectic manifolds have been central objects of a very active subject of study in which topology, geometry, and dynamics on manifolds mix and interact in several interesting ways. Over the last two decades, the global topology and geometry of symplectic and contact manifolds have undergone a vast expansion following the groundbreaking results of Giroux [17] who outlined

B. Acu (✉)
Department of Mathematics, Northwestern University, Evanston, IL, USA
e-mail: baharacu@northwestern.edu

© The Author(s) and the Association for Women in Mathematics 2020
B. Acu et al. (eds.), *Advances in Mathematical Sciences*, Association for
Women in Mathematics Series 21, https://doi.org/10.1007/978-3-030-42687-3_18

a program for characterizing contact manifolds in all dimensions in terms of *open book decompositions* with certain symplectic pages, of Gromov [19] who introduced *pseudoholomorphic curve* theory studying solutions to Cauchy-Riemann like PDEs and also of Donaldson [8] introducing *Lefschetz fibrations* on symplectic manifolds. The goal of this note is to survey these fundamental notions and introduce iterated planarity as in [2] to present some generalizations of the results in planar contact and symplectic topology. It also carries the, perhaps naive, optimism to provide some insight while exposing the already published results exploring the topology of contact and symplectic manifolds and their context to a broader audience in a reasonable number of pages.

2 Open Books and Lefschetz Fibrations

2.1 Open Book Decompositions

In this section, we will overview open book decompositions and some fundamental results concerning this topological machinery in contact geometry. For a more comprehensive discussion, we refer the reader to the lecture notes on open book decompositions by Etnyre [13].

An *abstract open book decomposition* is a pair (F, Φ), where

- F is a compact $2n$-dimensional manifold with boundary, called the *page* and
- $\Phi : F \to F$ is a diffeomorphism preserving ∂F, i.e. $\Phi|_{\partial F} = id$. This diffeomorphism is called the *monodromy*.

Given a compact oriented manifold M, an *open book decomposition* of M is a pair (B, π), where

- B is a codimension 2 submanifold of M with trivial normal bundle so that a tubular neighborhood of B looks like a product and
- the map $\pi : M - B \to S^1$ is a fiber bundle of the complement of B and the fiber bundle π restricted to a neighborhood of B agrees with the angular coordinate θ on the normal disk.

Define $F_\theta := \overline{\pi^{-1}(\theta)}$ and observe that $\partial F_\theta = B$ for all $\theta \in S^1$. We call the closure of the fiber $F = F_\theta$, for any θ, a *page* and B the *binding* of the open book. The monodromy of the fiber bundle π determines an isotopy class in the orientation preserving diffeomorphism group of a page F fixing its boundary, i.e., in $\mathrm{Diff}^+(F, \partial F)$ which we call the *monodromy* of the open book (Fig. 1).

By using this description, one can construct a closed oriented $(2n + 1)$-dimensional manifold M from an abstract open book with oriented pages in the following way:

Consider the mapping torus

Fig. 1 Local behavior of the map π in a neighborhood of the binding B

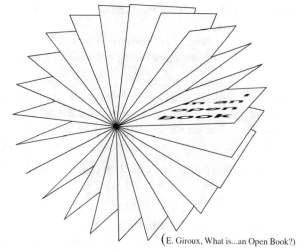

(E. Giroux, What is...an Open Book?)

$$F_\Phi = [0, 1] \times F/(0, \Phi(z)) \sim (1, z).$$

We set

$$M_{(F,\Phi)} = F_\Phi \cup_{\partial F_\Phi} \left(\partial F \times \mathbb{D}^2 \right),$$

by gluing $\partial(\partial F \times \mathbb{D}^2) = \partial F \times S^1$ to the boundary of the mapping torus F_Φ. Here the boundary of each disk $\{pt\} \times S^1$ in $\partial F \times \mathbb{D}^2$ gets glued to $S^1 \times \{pt\}$ in the mapping torus. Then (F, Φ) is an open book decomposition of a closed oriented $(2n + 1)$-dimensional manifold M if $M_{(F,\Phi)}$ is diffeomorphic to M. Note that the mapping torus F_Φ carries the structure of a smooth fibration $F_\Phi \to S^1$ whose fiber is the page F of the open book decomposition.

In this note (and in the literature), we often use abstract open books and (nonabstract) open book decompositions interchangeably. However, there is a basic difference between these two notions: For instance, observe that when studying regular open books, we discuss pages up to isotopy in M, whereas in abstract open books, we only discuss pages up to diffeomorphism.

2.2 Open Books vs. Contact Manifolds

By the works of J. W. Alexander [7] in dimension 3 (1923) and T. Lawson [22] in higher odd dimensions (1977), open book decompositions are known to exist in all odd dimensional manifolds. Hence, studying open books on contact manifolds is one good way to understand the topology of these manifolds by factoring them into lower dimensional pieces. Moreover, many different properties of contact manifolds such as symplectic fillability[1] can be read off from their associated open book decompositions.

A contact structure ξ on a manifold M is said to be *supported by an open book* (B, π) of M if it is the kernel of a contact form λ satisfying the following:

- $\lambda > 0$ on the binding[2] and
- $d\lambda$ is a positive symplectic form on the pages and the associated Liouville vector field on the pages is pointing outward along the binding. Equivalently, the orientation on the binding induced by the contact form λ agrees with the orientation on the pages induced by the symplectic form $d\lambda$ on the pages.

If these two conditions hold, then the open book (B, π) is called a *supporting open book* for the contact manifold (M, ξ) and the contact form λ is said to be *supported* by the open book (B, π).

Example 1 Consider the unit sphere (S^3, ξ_{std}) as the contact-type boundary of unit 4-ball in $\mathbb{C}^2$. Here the standard contact structure ξ_{std} is the set of complex tangents. That is,

$$\xi_{std} = TS^3 \cap i(TS^3)$$

which can also be described as

$$\xi_{std} = \ker(r_1^2 d\theta_1 + r_2^2 d\theta_2)$$

where $(z_1, z_2) = (r_1 e^{i\theta_1}, r_2 e^{i\theta_2})$ denote the complex coordinates on $\mathbb{C}^2$. Let the binding $B = \partial \mathbb{D}^2 \times \{0\} \subset S^3 \subset \mathbb{C}^2$. Notice that the map

$$\pi : S^3 - B \to S^1$$

$$(z_1, z_2) \mapsto \frac{z_2}{|z_2|}$$

defines an open book decomposition for S^3 with pages diffeomorphic to the 2-disk. We take the monodromy to be identity since all compactly supported diffeomorphisms of the open disk are isotopic in dimension two.

Alternatively,

Example 2 Consider the following open book for S^3 supporting the standard contact structure. Let $B \subset S^3$ be the Hopf link, i.e.

---

[1] Simply put, a symplectic filling is a cobordism between the empty set and a contact manifold. It comes in several flavors such as weak, strong, exact, and Stein with suitable compatibility conditions.

[2] Recall that the binding and the pages are oriented.

$$B = (\partial \mathbb{D}^2 \times \{0\}) \cup (\{0\} \times \partial \mathbb{D}^2).$$

Similarly, define

$$\pi : S^3 - B \to S^1$$

$$(z_1, z_2) \mapsto \frac{z_1 z_2}{|z_1 z_2|}.$$

This fibration defines another open book decomposition for S^3 with pages diffeomorphic to annulus and, this time, monodromy isotopic to a right-handed Dehn twist. We remark that the notion of monodromy is subtle to define and compute. For the purposes of the present note, we will not focus on the computation of monodromy.

As Examples 1 and 2 suggest, one can find two topologically different open book decompositions that support the same contact structure. Furthermore, any open book can be modified (by attaching a topological 1-handle to its page and modifying the monodromy) to a new one without changing the supported contact structure. This operation is called *positive stabilization.*

Thurston-Winkelnkemper [30] in dimension three and, 30 years later in 2002, Giroux [17] generalizing the construction in Thurston-Winkelnkemper's proof in higher dimensions showed that every open book decomposition gives rise to a supporting contact manifold. In a rather fancier language, to each triple (F^{2n}, λ, Φ), where $(F^{2n}, d\lambda)$ is an exact symplectic manifold with boundary and the monodromy $\Phi \in \text{Symp}(F, d\lambda, \partial F)^3$, we can associate a contact manifold (M^{2n+1}, ξ). In 2002, Emmanuel Giroux [17] announced the following groundbreaking result in three dimensional contact topology known as the Giroux correspondence:[4]

Theorem 1 (Giroux Correspondence, [17]) *Let M be a closed oriented 3-manifold. Then there exists a 1-1 correspondence between*

{*contact structures on M up to isotopy*}

⇕

{*abstract open book decompositions of M up to positive stabilization*}.

The question of whether there is a unique open book decomposition, up to positive stabilization, supporting a given contact structure is still open in higher dimensions.

---

[3]Here, $\text{Symp}(F, \omega, \partial F) \subset \text{Diff}^+(F, \partial F)$ is the symplectomorphism group of (F, ω) consisting of all orientation-preserving diffeomorphisms Φ on F that keep the boundary ∂F fixed and preserve the symplectic form ω.

[4]For an illuminating visual introduction to the Giroux correspondence, we refer the reader to [24].

Giroux correspondence plays a pivotal role in understanding Floer-theoretic invariants of contact structures, particularly those defined by Ozsváth and Szabó [27], and symplectic cobordisms of contact structures. Note that a filling of a contact manifold M is a special case of a cobordism (from empty set to M). A better understanding of filling properties of contact structures naturally leads to various important topological results in contact geometry such as results of Eliashberg [9], Etnyre [11, 12], Etnyre-Honda [14], and that of Gay [16].

Open books whose pages have zero genus, as in Example 1, play a particularly significant role in three dimensional contact topology. An open book decomposition on a 3-manifold is called *planar* if its pages have zero genus. A contact manifold is said to be *planar* if it is supported by a planar open book decomposition. Some other examples of planar contact manifolds are $S^1 \times S^2$ and the lens spaces $L(k, k - 1)$, see [34, Section 9.3].

While every contact structure is supported by some open book decomposition, not all contact structures are supported by planar open books. In 2004, Etnyre [11] proves that all overtwisted[5] contact 3-manifolds are planar. However, we know that there are examples of nonplanar contact manifolds: The simplest examples are the tori (T^3, ξ_k), see [34, Section 9.2].

2.3 Lefschetz Fibrations

Another natural way to characterize the topology of a contact manifold M is to look at Lefschetz fibration on its symplectic filling, if any.

A *Lefschetz fibration* on a $2n$-dimensional manifold W with boundary and corners is a surjective map $f : W \to \mathbb{D}^2$, where $\mathbb{D}^2$ is a 2-disk, with finitely many nondegenerate critical points all of which lie in the interior of W. Near each critical point, one can choose complex coordinates $(z_1, \ldots, z_n)$ such that in these coordinates

$$f(z_1, \ldots, z_n) = z_1^2 + \cdots + z_n^2.$$

Away from critical values, f is a trivial fibration. Note that this is a purely topological notion. However, we can adapt it to the case where the total space W is a symplectic manifold with boundary. Then we require that a generic fiber is a $(2n - 2)$-dimensional symplectic submanifold with boundary away from the critical points, while at the critical points the coordinates in which f looks locally like a complex Morse function can be chosen to be holomorphic for some compatible almost complex structure (Fig. 2).

---

[5]In the world of contact geometry, there is a fundamental dichotomy between overtwisted and tight contact structures, that is, those that do contain overtwisted disks (embedded 2-disk D whose boundary is tangent to ξ and interior is transversal to ξ everywhere except at one point) and those that do not, respectively.

Fig. 2 A Lefschetz fibration
with two singular fibers over
the critical values c_1 and c_2

W^4

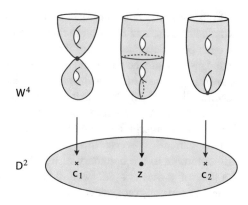

D^2

Lefschetz fibrations are known to exist on certain classes of exact symplectic manifolds. A *Weinstein domain* is a quadruple (W, ω, Z, ϕ) where (W, ω) is an exact symplectic manifold with nonempty boundary, Z is a Liouville vector field (that is $\mathcal{L}_Z\omega = \omega$) on W pointing outward transversely along the boundary of W, ϕ is a Morse function for which Z is gradient-like, and ∂W is a regular level set of ϕ. In 2016, Giroux and Pardon [18] showed that every Weinstein domain admits a Lefschetz fibration with fibers that are Weinstein domains.

2.3.1   What do Lefschetz Fibrations Have to Do with Open Book Decompositions?

Because all fibers and the base of the Lefschetz fibration f have boundary, ∂W naturally decomposes into two pieces which meet at a codimension two corner:

$$\partial W = \partial_v W \cup \partial_h W$$

where

- $\partial_v W$, vertical boundary, is the union of fibers over each point $z \in \partial\mathbb{D}^2$. That is, $\partial_v W = f^{-1}(\partial\mathbb{D}^2)$. Notice that $\partial_v W$ smoothly fibers over $\partial\mathbb{D}^2 = S^1$ since there are no critical values in $\partial\mathbb{D}^2$. In other words, $\partial_v W$ is (diffeomorphic to) the mapping torus for some monodromy map on a $(2n - 2)$-dimensional symplectic manifold with boundary.
- $\partial_h W$, horizontal boundary, is the union of boundaries of all, including singular, fibers. Namely, $\partial_h W = \cup_{z\in\mathbb{D}^2}\partial F_z$ where F_z is the fiber over $z \in \mathbb{D}^2$. Notice that $\partial_h W$ is diffeomorphic to a disjoint union of $\mathbb{D}^2$ family of the boundaries of fibers.[6] Hence, $\partial_h W$ naturally[7] fibers over $\mathbb{D}^2$.

---

[6]For instance, when $\dim_\mathbb{R} W = 4$, $\partial_h W$ is diffeomorphic to a disjoint union of solid tori since each fiber has boundary diffeomorphic to a disjoint union of S^1, and there exists a $\mathbb{D}^2$-family of these.

[7]Here we are imposing the condition that fibers meet the horizontal boundary transversely.

Looking at the boundary data above closely, one can observe that restricting a symplectic Lefschetz fibration to the smoothed[8] boundary of its total space induces an open book decomposition of the boundary. In that case, the fibers of the vertical boundary $\partial_v W$ are the pages and the boundary of the central fiber (i.e., central circles when $\dim_{\mathbb{R}} W = 4$) of the horizontal boundary is called the binding of the open book decomposition of ∂W. Note that the fibers of $\partial_v W$ are precisely the fibers of f over each point $z \in \partial \mathbb{D}^2$. Observe also that the pages of the open book are slightly larger than, but diffeomorphic to, the fibers of the vertical boundary.

One can also reverse this process and abstractly build a manifold that is diffeomorphic to the boundary of a Lefschetz fibration $f : W \to \mathbb{D}^2$. Let F_z be the fiber of f over $z \in \partial \mathbb{D}^2$ and Φ be the monodromy of the fibration $f|_{f^{-1}(\partial \mathbb{D}^2)}$. Then one can construct such a manifold diffeomorphic to ∂W as follows:

- Take $F_z \times [0, 1]$ and glue $F_z \times \{0\}$ and $F_z \times \{1\}$ by the monodromy Φ.
- Collapse the intervals on the boundary ∂F_z.

The most important topological fact about the relationship between open book decompositions and Lefschetz fibrations is that the monodromy of the open book decomposition depends on the critical points of the associated Lefschetz fibration. For instance, consider the case where Lefschetz fibration has no critical points, then the fibration is a regular fibration over the disk and thus trivial. Then the monodromy of the associated open book is also trivial. On the other hand, in the presence of critical points, the monodromy of the open book is nontrivial. See [26, Chapter 10] for a comprehensive discussion on this relationship.

A *planar Lefschetz fibration* is a Lefschetz fibration whose fibers are planar surfaces. In dimension three, there are several results studying the fillings of planar contact manifolds in the context of Lefschetz fibrations. One of the most remarkable ones reads as follows:

Theorem 2 (Wendl [33]) *Suppose that* $\pi : M - B \to S^1$ *is a planar open book decomposition supporting the contact structure on a contact manifold M. Then every strong[9] symplectic filling W of a planar contact manifold M admits a symplectic Lefschetz fibration f over the disk that induces the planar open book decomposition π on its boundary $f|_{\partial W}$.*

Equivalently, if M is a planar contact manifold and W is a strong filling of M, then (after possibly adding a symplectic cobordism) W admits a Lefschetz fibration over the disk inducing the given planar open book decomposition on M. An important consequence of Theorem 2 is that strongly fillable planar contact

---

[8]The corners of the total space can be rounded off to obtain a smooth manifold. See Lemma 7.6 in [28].

[9]A symplectic manifold (W, ω) is a *strong filling* of its contact boundary (M, ξ) if $\xi = \ker \iota_Z \omega$ for some vector field Z defined near M which points transversely outward at the boundary and satisfies $\mathcal{L}_Z \omega = \omega$.

manifolds are Stein fillable[10]. That is, when the contact manifold is planar, strongly fillable and Stein fillable are equivalent notions.

Theorem 2 also implies the first classification result in the study of symplectic fillings that reads as follows:

Theorem 3 (Gromov [19], Eliashberg [10]) $(\mathbb{D}^4, \omega_{std})$ *is the unique (weak) symplectic filling of* (S^3, ξ_{std}) *up to symplectic deformation equivalence and blowup.*

Let us see how Theorem 3 follows directly from Theorem 2:

Proof (Sketch) Let W be a symplectic filling of (S^3, ξ_{std}) and π be the open book decomposition of S^3 induced by the Lefschetz fibration on $(\mathbb{D}^4, \omega_{std})$ whose pages are annuli, binding is the Hopf link, and monodromy is a right-handed Dehn twist. By Theorem 2, W admits a symplectic Lefschetz fibration f over the disk inducing the open book on $\partial W = S^3$. In particular, regular fibers of the symplectic Lefschetz fibration f are annuli. It also has exactly one singular fiber. Notice that this is the same symplectic Lefschetz fibration carried by $(\mathbb{D}^4, \omega_{std})$. On the other hand, ∂W can be smoothed so that W becomes (up to symplectic deformation) a symplectic filling of the contact structure supported by the induced open book at the boundary [35, Theorem 5.5]. This implies that W is $(\mathbb{D}^4, \omega_{std})$ (up to deformation equivalence). $\qquad\square$

3 Iterated Planar Open Books and Lefschetz Fibrations

Planar open book decompositions and planar Lefschetz fibrations have been immensely studied to unlock several topological aspects of three dimensional contact geometry. However, generalizations of some of these results to higher dimensions are still mostly open due to absence of several consequences (such as automatic transversality, positivity of intersection, and filling properties) of these tools in higher dimensions. In what follows, we will define iterated planarity, introduced in [2], and provide a survey of some generalizations of these results to the case of iterated planar contact manifolds.

---

[10]A properly embedded complex submanifold of an affine space in $\mathbb{C}^n$ is called a *Stein manifold*. Note that such manifolds are necessarily noncompact. By intersecting with a sufficiently large ball in $\mathbb{C}^n$, we obtain a compact manifold W with boundary called a *Stein domain*. The symplectic form ω on W induces a contact structure ξ on the boundary ∂W. We then call (W, ω) a *Stein filling* of the boundary contact manifold $(\partial W, \xi)$.

3.1 Planarity in High Dimensions

There are natural generalizations of planarity in higher dimensions. For instance, one can consider standard open books in high dimensions, keeping S^1 as the base of the symplectic fibration, but imposing the condition that the fibers carry a suitable structure built inductively from a low-dimensional planar structure. In this vein, we can define the following notion as in [2]:

Definition 1 Given a $2n$-dimensional Weinstein domain (W^{2n}, ω), we say that W^{2n} admits an *iterated planar Lefschetz fibration* f_n if

- $\exists$ a sequence $f_i : W^{2i} \to \mathbb{D}^2$, for $i = 2, \ldots, n$, of symplectic Lefschetz fibrations where the regular fiber of f_i is the total space of f_{i-1}, and
- $f_2 : W^4 \to \mathbb{D}^2$ is a planar Lefschetz fibration.

Observe that when $n = 2$, an iterated planar Lefschetz fibration is a planar Lefschetz fibration. Iterated planar Lefschetz fibrations are seemingly powerful tools in studying high dimensional contact manifolds given by a symplectic manifold with convex boundary.

Example 3 Consider the unit disk bundle $W^{2n} = T^*S^n$ which can be symplectically identified in $\mathbb{C}^{n+1}$ with $\{z_1^2 + \cdots + z_{n+1}^2 = 1\}$ and take the Lefshetz fibration on W to be the projection on the last coordinate z_{n+1}. Now observe that the regular fiber of the Lefschetz fibration on T^*S^n is T^*S^{n-1} and the Lefschetz fibration on T^*S^2 is planar with fibers $T^*S^1 = [0, 1] \times S^1$. Hence, W^{2n} admits an iterated planar Lefschetz fibration.

Definition 2 An *iterated planar contact manifold* (M^{2n+1}, ξ) is a contact manifold supported by an open book whose pages admit an iterated planar Lefschetz fibration.

For instance, consider the overtwisted contact structure ξ_{OT} on S^5. (S^5, ξ_{OT}) is iterated planar since it is supported by the open book whose pages are T^*S^2 and T^*S^2 admits a planar Lefschetz fibration (as in Example 3) whose pages are annuli and monodromy is a left-handed Dehn twist.

3.2 Obstructions to Iterated Planarity

A contact 3-manifold (M^3, ξ) is called *weakly fillable* if it is the smooth boundary of a symplectic 4-manifold (W^4, ω), i.e. $\partial W = M$ as oriented manifolds, such that $\omega|_\xi > 0$. Namely, (W^4, ω) is a *weak filling* of (M^3, ξ). One can generalize this idea to higher dimensions, as in [25], by requiring that $\omega + \tau d\lambda|_\xi$ is symplectic for every $\tau \geq 0$, for one choice of contact form λ. The contact structure ξ on M is then called *weakly dominated* by ω and (W, ω) is called a *weak filling* of (M, ξ). In dimension three, this definition of weak filling reduces to the standard one.

We say that (M, ξ) is *weakly co-fillable* if there exists a connected weak filling (W, ω) whose boundary is the disjoint union of (M, ξ) with an arbitrary nonempty contact manifold. (M, ξ) is then said to admit a connected *semi-filling* with disconnected boundary. Massot, Niederkrüger, and Wendl [25] construct several examples of (exactly) co-fillable higher-dimensional contact manifolds. It is then natural to wonder which contact manifolds can fit into a symplectic co-filling. In dimension 3, we have the following result:

Theorem 4 (Etnyre [12]) *If W is a weak filling of a planar contact manifold M, then M is connected.*

That is, if M admits a semi-filling with disconnected boundary (i.e. M is weakly co-fillable), then it cannot be a planar contact manifold. The following result is a generalization of Etnyre's three dimensional result above to iterated planar contact manifolds which provides an obstruction to iterated planarity. The statement reads as follows:

Theorem 5 (Acu-Moreno [4]) *Iterated planar contact manifolds are not weakly co-fillable.*

Put another way, if a contact manifold is weakly co-fillable, then it cannot be iterated planar. A further obstruction to iterated planarity, generalizing a result by Albers-Bramham-Wendl [5], is the following:

Theorem 6 (Acu-Moreno [4]) *Iterated planar contact manifolds do not embed as nonseparating weak contact-type hypersurfaces in closed symplectic manifolds.*

This can be equivalently stated as follows: If an iterated planar contact manifold admits a weak contact-type embedding into a closed symplectic manifold, then it separates the latter into two disjoint pieces.

3.3 The Weinstein Conjecture

The Weinstein conjecture has been one of the major driving forces in the development of contact and symplectic geometry leading to many fruitful interactions between analysis, geometry and topology, hence deserves a separate section.

Given a contact form λ for a contact manifold (M, ξ), there exists a unique vector field called the *Reeb vector field* R, defined by $\iota_R d\lambda = 0$ and $\lambda(R) = 1$. Note that R is a contact vector field. The flow of this vector field preserves not only the contact structure ξ, but also the defining contact form, i.e. $\mathcal{L}_R \lambda = 0$. The conjecture, formulated by Alan Weinstein in 1978, poses one of most famous questions in the field of symplectic geometry in regard to the existence of closed orbits of such vector fields:

Conjecture 1 (Weinstein Conjecture, [32]) On a compact contact manifold, any Reeb vector field carries at least one periodic orbit.

For any given contact manifold, there exists a supporting open book decomposition whose binding is a closed Reeb orbit. That is, any contact form is isotopic to a form that admits a closed Reeb orbit. However, this does not automatically prove the Weinstein conjecture since the conjecture states that *every* contact form (not a form that is only isotopic to the given form) admits a closed Reeb orbit.

The conjecture was proven for all closed 3-dimensional manifolds by Taubes [29]. Despite an extensive literature due to Abbas, Albers, Etnyre, Floer, Hofer, Viterbo in [1, 6, 12, 15, 20, 21], and [31], it is still open in higher dimensions.

When M is a planar contact manifold, the Weinstein conjecture is known to be true by the work of Abbas, Cieliebak, and Hofer [1]. The generalization of this result to the case of iterated planar contact manifolds reads as follows:

Theorem 7 (Acu-Moreno [3, 4]) *Let* (M, ξ) *be a closed, oriented,* $(2n + 1)$-*dimensional iterated planar contact manifold. Then* (M, ξ) *satisfies the Weinstein conjecture.*

Notice that when $n = 1$, M is a planar contact manifold. The technical input for obtaining Theorems 5, 6, and 7 is a suitable symplectic handle attachment directly inspired by the handle attachments in [9] and [23].

Acknowledgements The author would like to thank Klaus Niederkrüger for revising an earlier draft of this note, the anonymous referee for careful reading and many helpful suggestions improving the exposition, and the Oberwolfach Research Institute for Mathematics, where part of the writing was carried out during the author's visit as a Leibniz Fellow, for their hospitality.

References

1. C. Abbas, K. Cieliebak, and H. Hofer, *The Weinstein conjecture for planar contact structures in dimension three.* Comment. Math. Helv. 80 (2005), no. 4, 771–793.
2. B. Acu, *On foliations of higher-dimensional symplectic manifolds and symplectic mapping class group relations*, PhD Thesis, 2017.
3. B. Acu, *The Weinstein conjecture for iterated planar contact structures.* Preprint arXiv:1710.07724.
4. B. Acu and A. Moreno, *Planarity in higher-dimensional contact manifolds.* Preprint arXiv:1810.11448.
5. P. Albers, B. Bramham, and C. Wendl, *On nonseparating contact hypersurfaces in symplectic 4-manifolds.* Algebr. Geom. Topol. 10 (2010), no. 2, 697–737.
6. P. Albers and H. Hofer, *On the Weinstein conjecture in higher dimensions.* Comment. Math. Helv. 84 (2009), no. 2, 429–436.
7. J. W. Alexander, *A lemma on systems of knotted curves.* Proc. Nat. Acad. Sci. U.S.A. 9 (1923), 93–95.
8. S. K. Donaldson, *Lefschetz pencils on symplectic manifolds.* J. Differential Geom. **53** (1999), no. 2, 205–236.
9. Y. Eliashberg, *A few remarks about symplectic filling.* Geom. Topol. 8 (2004), 277–293.
10. Y. Eliashberg, *Filling by holomorphic discs and its applications.* Geometry of low-dimensional manifolds, 2 (Durham, 1989), 45–67, London Math. Soc. Lecture Note Ser., 151, Cambridge Univ. Press, Cambridge, 1990.

11. J. B. Etnyre, *Planar open book decompositions and contact structures.* Int. Math. Res. Not. 2004, no. 79, 4255–4267.
12. J. B. Etnyre, *On symplectic fillings.* Algebr. Geom. Topol. 4 (2004), 73–80.
13. J. B. Etnyre, *Lectures on open book decompositions and contact structures.* Floer homology, gauge theory, and low-dimensional topology, 103–141, Clay Math. Proc., 5, Amer. Math. Soc., Providence, RI, 2006.
14. J. B. Etnyre and K. Honda, *On symplectic cobordisms.* Math. Ann. 323 (2002), no. 1, 31–39.
15. A. Floer, H. Hofer, and C. Viterbo, *The Weinstein conjecture in $P \times \mathbf{C}^l$.* Math. Z. 203 (1990), no. 3, 469–482.
16. D. Gay, *Explicit concave fillings of contact three-manifolds.* Proc. Cam. Phil. Soc.133 (2002), 431–441.
17. E. Giroux, *Géométrie de contact: de la dimension trois vers les dimensions supérieures.* (French) [Contact geometry: from dimension three to higher dimensions] Proceedings of the International Congress of Mathematicians, Vol. II (Beijing, 2002), 405–414, Higher Ed. Press, Beijing, 2002.
18. E. Giroux and J. Pardon, *Existence of Lefschetz fibrations on Stein and Weinstein domains.* Geom. Topol. 21 (2017), no. 2, 963–997.
19. M. Gromov, *Pseudo holomorphic curves in symplectic manifolds.* Invent. Math. 82 (1985), 307–347.
20. H. Hofer, *Pseudoholomorphic curves in symplectizations with applications to the Weinstein conjecture in dimension three.* Invent. Math. 114 (1993), no. 3, 515–563.
21. H. Hofer and C. Viterbo, *The Weinstein conjecture in the presence of holomorphic spheres.* Comm. Pure Appl. Math. 45 (1992), no. 5, 583–622.
22. T. Lawson, *Open book decompositions for odd dimensional manifolds.* Topology 17 (1978), no. 2, 189–192.
23. S. Lisi, J. Van Horn-Morris, and C. Wendl, *On symplectic fillings of spinal open book decompositions I: Geometric constructions.* Preprint arXiv:1810.12017.
24. P. Massot, *A visual introduction to the Giroux correspondence.* http://www.math.u-psud.fr/~pmassot/exposition/giroux_correspondance.html
25. P. Massot, K. Niederkrüger, and C. Wendl, *Weak and strong fillability of higher-dimensional contact manifolds.* Invent. Math. 192 (2013), no. 2, 287–373.
26. B. Ozbagci and A. I. Stipsicz, *Surgery on Contact 3-Manifolds and Stein Surfaces.* Bolyai Society Mathematical Studies, vol. 13 (Springer, Berlin, 2004).
27. P. Ozsváth and Z. Szabó, *Heegaard Floer homologies and contact structures.* Duke Math. J., 129(1):39–61, 2005.
28. P. Seidel, *Fukaya categories and Picard-Lefschetz theory,* Zurich Lectures in Advanced Mathematics, EMS, 2008.
29. C. Taubes, *The Seiberg-Witten equations and the Weinstein conjecture.* Geom. Topol. 11 (2007), 2117–2202.
30. W.P. Thurston and H.E. Winkelnkemper, *On the existence of contact forms.* Proc. Amer. Math. Soc. 52 (1975), 345–347.
31. C. Viterbo, *A proof of Weinstein's conjecture in $\mathbb{R}^{2n}$.* Ann. Inst. H. Poincaré Anal. Non Linéaire 4 (1987), no. 4, 337–356.
32. A. Weinstein, *On the hypotheses of Rabinowitz' periodic orbit theorems.* J. Differential Equations 33 (1979), no. 3, 353–358.
33. C. Wendl, *Strongly fillable contact manifolds and J-holomorphic foliations.* Duke Math. J. 151 (2010), no. 3, 337–384.
34. C. Wendl, *Holomorphic curves in low dimensions. From symplectic ruled surfaces to planar contact manifolds.* Lecture Notes in Mathematics, 2216. Springer, Cham, 2018. xiii+292 pp. ISBN: 978-3-319-91369-8; 978-3-319-91371-1.
35. C. Wendl, *Lectures on Contact 3-Manifolds, Holomorphic Curves and Intersection Theory* (Cambridge Tracts in Mathematics). Cambridge: Cambridge University Press. 2020. Preprint arXiv:1706.05540.

Part VII
Applied Mathematics

A Robust Preconditioner for High-Contrast Problems (Research)

Yuliya Gorb, Daria Kurzanova, and Yuri Kuznetsov

1 Introduction

In this paper, we consider an iterative solution of the linear system arising from the discretization of the diffusion problem

$$- \nabla \cdot [\sigma(x)\nabla u] = f, \quad x \in \Omega \tag{1}$$

with appropriate boundary conditions on $\Gamma = \partial\Omega$. We assume that Ω is a bounded domain $\Omega \subset \mathbb{R}^d$, $d \in \{2, 3\}$, that contains $m \geq 1$ polygonal or polyhedral subdomains $\mathcal{D}^i$, see Fig. 1. Also assume that the distance between the neighboring $\mathcal{D}^i$ and $\mathcal{D}^j$ is at least of order of the sizes of these subdomains, that is, bounded below by a multiple of their diameters. The main focus of this work is on the case when the coefficient function $\sigma(x) \in L^\infty(\Omega)$ varies largely within the domain Ω, that is, $\kappa = \dfrac{\sup_{x\in\Omega}\sigma(x)}{\inf_{x\in\Omega}\sigma(x)} \gg 1$. In this work, we assume that the domain Ω contains disjoint polygonal or polyhedral subdomains $\mathcal{D}^i$, $i \in \{1, \ldots, m\}$, where σ takes "large" values, e.g. of order $O(\kappa)$, but remains of $O(1)$ in the domain outside of $\mathcal{D} := \cup_{i=1}^m \mathcal{D}^i$.

The P1-FEM discretization of this problem results in a linear system

$$\mathcal{K}\overline{u} = \overline{F}, \tag{2}$$

with a large and sparse matrix $\mathcal{K}$. A major issue in numerical treatments of (1), with the coefficient σ discussed above, is that the high contrast leads to an ill-conditioned matrix $\mathcal{K}$ in (2). Indeed, if h is the discretization scale, then the condition number

Y. Gorb (✉) · D. Kurzanova · Y. Kuznetsov
Department of Mathematics, University of Houston, Houston, TX, USA
e-mail: gorb@math.uh.edu; yuri@math.uh.edu

© The Author(s) and the Association for Women in Mathematics 2020
B. Acu et al. (eds.), *Advances in Mathematical Sciences*, Association for
Women in Mathematics Series 21, https://doi.org/10.1007/978-3-030-42687-3_19

Fig. 1 The domain Ω with
highly conducting inclusions
$\mathcal{D}^i, i \in \{1, \dots, m\}$

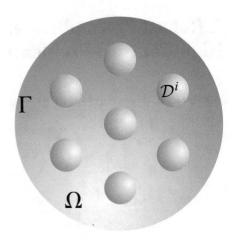

of the resulting stiffness matrix $\mathcal{K}$ grows proportionally to h^{-2} with coefficient of proportionality depending on κ. Because of that, the high contrast problems have been a subject of an active research recently, see e.g. [1, 2].

There is one more feature of the system (2) that we investigate in this paper. Recall that if $\mathcal{K}$ is symmetric and positive definite, then (2) is typically solved with the Conjugate Gradient (CG) method, if $\mathcal{K}$ is nonsymmetric then the most common solver for (2) is GMRES. In this paper, we introduce an additional variable that allows us to replace (2) with an equivalent formulation in terms of a linear system

$$\mathcal{A}x = \mathcal{F} \tag{3}$$

with a *saddle point matrix* $\mathcal{A}$ written in the block form:

$$\mathcal{A} = \begin{bmatrix} \mathbf{A} & \mathbf{B}^T \\ \mathbf{B} & -\mathbf{\Sigma} \end{bmatrix}, \tag{4}$$

where $\mathbf{A} \in \mathbb{R}^{n \times n}$ is symmetric and positive definite, $\mathbf{B} \in \mathbb{R}^{k \times n}$ is rank deficient, and $\mathbf{\Sigma} \in \mathbb{R}^{k \times k}$ is symmetric and positive semidefinite, so that the corresponding linear system (3) is singular but consistent. Unfortunately, Krylov space iterative methods tend to converge very slowly when applied to systems with saddle point matrices and preconditioners are needed to achieve faster convergence. The CG method that was mainly developed for the iterative solution of linear systems with symmetric and definite matrices is not, in general, robust for systems with indefinite matrices, [21]. The *Lanczos algorithm* of minimized iterations does not have such a restriction and has been utilized in this paper. Below, we introduce a construction of a robust preconditioner for solving (3) by the Lanczos iterative scheme [15], that is, whose convergence rate is independent of the contrast parameter $\kappa \gg 1$ and the discretization size $h > 0$.

Also, the special case of (3) with (4) considered in the Appendix of this paper is when $\boldsymbol{\Sigma} \equiv \mathbf{0}$. The problem of this type has received considerable attention over the years. But the most studied case is when $\mathcal{A}$ is *nonsingular*, in which case $\mathbf{B}$ must be of full rank, see e.g. [12, 16] and references therein. The main focus of this paper is on singular $\mathcal{A}$ with the rank deficient block $\mathbf{B}$. Below, we propose a block-diagonal preconditioner for the Lanczos method employed to solve the problem (3), and this preconditioner is also singular. We also rigorously justify its robustness with respect to h and κ. Our numerical experiments on simple test cases support our theoretical findings.

Finally, we point out that a robust numerical treatment of the described problem is crucial in developing the mutiscale strategies for models of composite materials with highly conducting particles. The latter find their application in particulate flows, subsurface flows in natural porous formations, electrical conduction in composite materials, and medical and geophysical imaging.

The rest of this paper is organized as follows. In Sect. 2 the mathematical problem formulation is presented and the main results are stated. Section 3 discusses proofs of the main results, and numerical results of the proposed procedure are given in Sect. 4. Conclusions are presented in Sect. 5. The proof of an auxiliary fact is given in Appendix.

2 Problem Formulation and Main Results

Consider an open, bounded domain $\Omega \subset \mathbb{R}^d$, $d \in \{2, 3\}$ with piece-wise smooth boundary Γ, that contains $m \geq 1$ subdomains $\mathcal{D}^i$, which are located at distances comparable to their sizes from one another, see Fig. 1. For simplicity, we assume that Ω and $\mathcal{D}^i$ are polygons if $d = 2$ or polyhedra if $d = 3$. The union of $\mathcal{D}^i$ is denoted by $\mathcal{D}$. In the domain Ω we consider the following elliptic problem

$$\begin{cases} -\nabla \cdot [\sigma(x)\nabla u] = f, \ x \in \Omega \\ \qquad\qquad u = 0, \ x \in \Gamma \end{cases} \tag{5}$$

with the coefficient σ that largely varies inside the domain Ω. For simplicity of the presentation, we focus on the case when σ is a piecewise constant function given by

$$\sigma(x) = \begin{cases} 1, & x \in \Omega \setminus \overline{\mathcal{D}} \\ 1 + \dfrac{1}{\varepsilon_i}, & x \in \mathcal{D}_i, \ i \in \{1, \dots, m\} \end{cases} \tag{6}$$

with $\varepsilon := \max_i \varepsilon_i \ll 1$. We also assume the source term in (5) is $f \in L^2(\Omega)$.

When performing a P1-FEM discretization of (5) with (6), we choose a FEM space $V_h \subset H_0^1(\Omega)$ to be the space of linear finite-element functions defined on a conforming quasi-uniform triangulation Ω_h of Ω of the size $h \ll 1$. For simplicity, we assume that $\partial \Omega_h = \Gamma$. With that, the classical FEM discretization results in the system of the type (2). We proceed differently and derive another discretized system of the saddle point type as shown below.

2.1 Derivation of a Singular Saddle Point Problem

If $\mathcal{D}_h^i = \Omega_h|_{\mathcal{D}^i}$ then we denote $V_h^i := V_h|_{\mathcal{D}_h^i}$ and $\mathcal{D}_h := \cup_{i=1}^m \mathcal{D}_h^i$. With that, we write the FEM formulation of (5)–(6) as

$$\text{Find} \quad u_h \in V_h \quad \text{and} \quad \lambda_h = (\lambda_h^1, \ldots, \lambda_h^m) \quad \text{with} \quad \lambda_h^i \in V_h^i \quad \text{such that}$$

$$\int_{\Omega_h} \nabla u_h \cdot \nabla v_h \, dx + \int_{\mathcal{D}_h} \nabla \lambda_h \cdot \nabla v_h \, dx = \int_{\Omega_h} f v_h \, dx, \quad \forall v_h \in V_h, \tag{7}$$

provided

$$u_h = \varepsilon_i \lambda_h^i + c_i \quad \text{in} \quad \mathcal{D}_h^i, \quad i \in \{1, \ldots, m\}, \tag{8}$$

where c_i is an arbitrary constant. First, we turn out attention to the FEM discretization of (7) that yields a system of linear equations

$$\mathbf{A}\bar{u} + \mathbf{B}^T \bar{\lambda} = \bar{F}, \tag{9}$$

and then discuss implications of (8).

To provide the comprehensive description of all elements of the system (9), we introduce the following notations for the number of degrees of freedom in different parts of Ω_h. Let N be the total number of nodes in Ω_h, and n be the number of nodes in $\overline{\mathcal{D}}_h$ so that $n = \sum_{i=1}^m n_i$, where n_i denotes the number of degrees of freedom in $\overline{\mathcal{D}}_h^i$, and, finally, n_0 is the number of nodes in $\Omega_h \setminus \overline{\mathcal{D}}_h$, so that we have $N = n_0 + n = n_0 + \sum_{i=1}^m n_i$. Then in (9), the vector $\bar{u} \in \mathbb{R}^N$ has entries $u_i = u_h(x_i)$ with $x_i \in \overline{\Omega}_h$. We count the entries of $\bar{u}$ in such a way that its first n elements correspond to the nodes of $\overline{\mathcal{D}}_h$, and the remaining n_0 entries correspond to the nodes of $\overline{\Omega}_h \setminus \overline{\mathcal{D}}_h$. Similarly, the vector $\bar{\lambda} \in \mathbb{R}^n$ has entries $\lambda_i = \lambda_h(x_i)$ where $x_i \in \overline{\mathcal{D}}_h$.

The symmetric positive definite matrix $\mathbf{A} \in \mathbb{R}^{N \times N}$ of (9) is the stiffness matrix that arises from the discretization of the Laplace operator with the homogeneous Dirichlet boundary conditions on Γ. Entries of $\mathbf{A}$ are defined by

$$(\mathbf{A}\bar{u}, \bar{v}) = \int_{\Omega_h} \nabla u_h \cdot \nabla v_h \, dx, \quad \text{where} \quad \bar{u}, \bar{v} \in \mathbb{R}^N, \quad u_h, v_h \in V_h, \tag{10}$$

where $(\cdot, \cdot)$ is the standard dot-product of vectors. This matrix can also be partitioned into

$$\mathbf{A} = \begin{bmatrix} A_{\mathcal{D}\mathcal{D}} & A_{\mathcal{D}0} \\ A_{0\mathcal{D}} & A_{00} \end{bmatrix}, \tag{11}$$

where the block $A_{\mathcal{D}\mathcal{D}} \in \mathbb{R}^{n \times n}$ is the stiffness matrix corresponding to the highly conducting inclusions $\overline{\mathcal{D}}_h^i$, $i \in \{1, \ldots, m\}$, the block $A_{00} \in \mathbb{R}^{n_0 \times n_0}$ corresponds to the region outside of $\overline{\mathcal{D}}_h$, and the entries of $A_{\mathcal{D}0} \in \mathbb{R}^{n \times n_0}$ and $A_{0\mathcal{D}} = A_{\mathcal{D}0}^T$ are assembled from contributions both from finite elements in $\overline{\mathcal{D}}_h$ and $\overline{\Omega}_h \setminus \overline{\mathcal{D}}_h$.

The matrix $\mathbf{B} \in \mathbb{R}^{n \times N}$ of (9) is also written in the block form as

$$\mathbf{B} = \begin{bmatrix} \mathcal{B}_{\mathcal{D}} & 0 \end{bmatrix} \tag{12}$$

with zero-matrix $0 \in \mathbb{R}^{n \times n_0}$ and $\mathcal{B}_{\mathcal{D}} \in \mathbb{R}^{n \times n}$ that corresponds to the highly conducting inclusions. In its turn, $\mathcal{B}_{\mathcal{D}}$ is written in the block form as $\mathcal{B}_{\mathcal{D}} = \mathrm{diag}\,(B_1, \ldots, B_m)$, with matrices $B_i \in \mathbb{R}^{n_i \times n_i}$, whose entries are similarly defined by

$$(B_i \overline{u}, \overline{v}) = \int_{\mathcal{D}_h^i} \nabla u_h \cdot \nabla v_h \, dx, \quad \text{where} \quad \overline{u}, \overline{v} \in \mathbb{R}^{n_i}, \quad u_h, v_h \in V_h^i. \tag{13}$$

The matrix B_i is the stiffness matrix in the discretization of the Laplace operator in the domain $\overline{\mathcal{D}}_h^i$ with the Neumann boundary conditions on $\partial \mathcal{D}_h^i$. We remark that each B_i is positive semidefinite with $\ker B_i = \mathrm{span} \left\{ \begin{bmatrix} 1 \\ \vdots \\ 1 \end{bmatrix} \right\}$.

Finally, the vector $\overline{F} \in \mathbb{R}^N$ of (9) is defined in a similar way by

$$(\overline{F}, \overline{v}) = \int_{\Omega_h} f v_h \, dx, \quad \text{where} \quad \overline{v} \in \mathbb{R}^N, \quad v_h \in V_h.$$

To complete the derivation of the linear system corresponding to (7)–(8), we rewrite (8) in the weak form that is as follows:

$$\int_{\mathcal{D}_h^i} \nabla u_h \cdot \nabla v_h^i \, dx - \varepsilon_i \int_{\mathcal{D}_h^i} \nabla \lambda_h^i \cdot \nabla v_h^i \, dx = 0, \quad \text{for all } v_h^i \in V_h^i, \tag{14}$$

for $i \in \{1, \ldots, m\}$, and add the discrete analog of (14) to the system (9). For that, denote $\Sigma_\varepsilon = \mathrm{diag}\,(\varepsilon_1 B_1, \ldots, \varepsilon_m B_m)$, then (14) implies

$$\mathbf{B}\overline{u} - \Sigma_\varepsilon \overline{\lambda} = \overline{0}. \tag{15}$$

This together with (9) yields

$$
\begin{cases} A\overline{u} + B^T \overline{\lambda} = \overline{F}, \\ B\overline{u} - \Sigma_\varepsilon \overline{\lambda} = \overline{0}, \end{cases} \quad \overline{u} \in \mathbb{R}^N, \quad \overline{\lambda} \in \mathbb{R}^n, \ \overline{\lambda} \in \operatorname{Im} \mathcal{B}_\mathcal{D}, \tag{16}
$$

$$
\text{or} \quad \mathcal{A}_\varepsilon \mathbf{x}_\varepsilon = \overline{\mathcal{F}}, \tag{17}
$$

$$
\text{where} \quad \mathcal{A}_\varepsilon = \begin{bmatrix} A & B^T \\ B & -\Sigma_\varepsilon \end{bmatrix} = \begin{bmatrix} A_{\mathcal{DD}} & A_{\mathcal{D}0} & \mathcal{B}_\mathcal{D} \\ A_{0\mathcal{D}} & A_{00} & \mathbf{0}^T \\ \mathcal{B}_\mathcal{D} & \mathbf{0} & -\Sigma_\varepsilon \end{bmatrix}, \quad \mathbf{x}_\varepsilon = \begin{bmatrix} \overline{u} \\ \overline{\lambda} \end{bmatrix}, \quad \overline{\mathcal{F}} = \begin{bmatrix} F \\ 0 \end{bmatrix}. \tag{18}
$$

This saddle point formulation (17)–(18) for the PDE (5)–(6) was first proposed in [14]. Clearly, there exists a unique solution $\overline{u} \in \mathbb{R}^N$ and $\overline{\lambda} \in \mathbb{R}^n$, $\overline{\lambda} \in \operatorname{Im} \mathcal{B}_\mathcal{D}$ of (17)–(18).

It is important to point out that the main feature of the problem (16) is in rank deficiency of the matrix **B**. This would lead to the introduction in the next Sect. 2.3 of a *singular* block-diagonal preconditioner for the Lanczos method employed to solve the problem (18). Independence of the convergence of the employed Lanczos method on the discretization size $h > 0$ follows from the spectral properties of the constructed preconditioner that are independent of h due to the norm preserving extension theorem of [20]. Independence on contrast parameters ε_i follows from the closeness of spectral properties of the matrices of the original system (18) and the limiting one (51), also demonstrated in Appendix. Our numerical experiments below also show independence of the iterative procedure on the number of different contrasts ε_i, $i \in \{1, \ldots, m\}$, in the inclusions $\mathcal{D}^i$.

2.2 Preconditioned System and Its Implementation

Lanczos Method
In principal, we could have used the CG method that was mainly developed for the iterative solution of linear systems with symmetric and *definite* matrices, and apply it to the square of the matrix of the preconditioned system. However, the *Lanczos method* of minimized iterations is not restricted to the definite matrices, and, since it has the same arithmetic cost as CG, is employed in this paper. A symmetric and positive semidefinite block-diagonal preconditioner of the form

$$
\mathcal{P} = \begin{bmatrix} \mathcal{P}_A & 0 \\ 0 & \mathcal{P}_B \end{bmatrix}, \tag{19}
$$

for this method is also proposed in this section, where the role of the blocks $\mathcal{P}_A$ and $\mathcal{P}_B$ will be explained below. But, first, for the completeness of presentation, we describe the Lanczos algorithm.

For a symmetric and positive semidefinite matrix $\mathcal{H}$ that later will be defined as the Moore-Penrose pseudo inverse[1] $\mathcal{P}^\dagger$, introduce a new scalar product $(\overline{x}, \overline{y})_\mathcal{H} := (\mathcal{H}\overline{x}, \overline{y})$, for all $\overline{x}, \overline{y} \in \mathbb{R}^{N+n}$, $\overline{x}, \overline{y} \perp \ker \mathcal{H}$, and consider the preconditioned Lanczos iterations, see [15], $\overline{z}^k = \begin{bmatrix} \overline{u}^k \\ \overline{\lambda}^k \end{bmatrix} \in \mathbb{R}^{N+n}$, $k \geq 1$: $\overline{z}^k = \overline{z}^{k-1} - \beta_k \overline{y}_k$, where

$$\beta_k = \frac{(\mathcal{A}_\varepsilon \overline{z}^{k-1} - \overline{\mathcal{F}}, \mathcal{A}_\varepsilon \overline{y}_k)_\mathcal{H}}{(\mathcal{A}_\varepsilon \overline{y}_k, \mathcal{A}_\varepsilon \overline{y}_k)_\mathcal{H}}, \quad y_k = \begin{cases} \mathcal{H}(\mathcal{A}_\varepsilon \overline{z}^0 - \overline{\mathcal{F}}), & k = 1 \\ \mathcal{H}\mathcal{A}_\varepsilon \overline{y}_1 - \alpha_2 \overline{y}_1, & k = 2 \\ \mathcal{H}\mathcal{A}_\varepsilon \overline{y}_{k-1} - \alpha_k \overline{y}_{k-1} - \gamma_k \overline{y}_{k-2}, & k > 2, \end{cases}$$

with $\alpha_k = \dfrac{(\mathcal{A}_\varepsilon \mathcal{H}\mathcal{A}_\varepsilon \overline{y}_{k-1}, \mathcal{A}_\varepsilon \overline{y}_{k-1})_\mathcal{H}}{(\mathcal{A}_\varepsilon \overline{y}_{k-1}, \mathcal{A}_\varepsilon \overline{y}_{k-1})_\mathcal{H}}$, and $\gamma_k = \dfrac{(\mathcal{A}_\varepsilon \mathcal{H}\mathcal{A}_\varepsilon \overline{y}_{k-1}, \mathcal{A}_\varepsilon \overline{y}_{k-1})_\mathcal{H}}{(\mathcal{A}_\varepsilon \overline{y}_{k-2}, \mathcal{A}_\varepsilon \overline{y}_{k-2})_\mathcal{H}}$.

Proposed Preconditioner

It was previously observed, see e.g. [13, 14], that the following matrix

$$\mathbf{P} = \begin{bmatrix} \mathbf{A} & \mathbf{0} \\ \mathbf{0} & \mathbf{BA}^{-1}\mathbf{B}^T \end{bmatrix}, \tag{20}$$

is the best choice for a block-diagonal preconditioner of $\mathcal{A}_\varepsilon$. This is because the eigenvalues of the generalized eigenvalue problem

$$\mathcal{A}_\varepsilon x = \mu \mathbf{P}x, \quad \overline{u} \in \mathbb{R}^N, \quad \overline{\lambda} \in \operatorname{Im} \mathcal{B}_D, \tag{21}$$

belong to the union of $[c_1, c_2] \cup [c_3, c_4]$ with $c_1 \leq c_2 < 0$ and $0 < c_3 \leq c_4$, with numbers c_i being independent of both h, and ε_i, see [11, 13, 14]. For the reader's convenience, the proof of this statement is also shown in the Appendix below (see Lemma 6).

The preconditioner $\mathbf{P}$ of (20) is of limited practical use and is a subject of primarily theoretical interest. To construct a preconditioner that one can actually use in practice, we will find a matrix $\mathcal{P}$ such that there exist constants α, β independent on the mesh size h and that

$$\alpha(\mathbf{P}x, x) \leq (\mathcal{P}x, x) \leq \beta(\mathbf{P}x, x) \quad \text{for all } x \in \mathbb{R}^{N+n}. \tag{22}$$

---

[1] $M^\dagger$ is the Moore-Penrose pseudo inverse of M if and only if it satisfies the following Moore-Penrose equations, see e.g. [3]:

(i) $M^\dagger M M^\dagger = M^\dagger$,   (ii) $M M^\dagger M = M$,   (iii) $M M^\dagger$ and $M^\dagger M$ are symmetric.

This property (22) is hereafter referred to as *spectral equivalence* of $\mathcal{P}$ to $\mathbf{P}$ of (20). Obviously, the matrix $\mathcal{P}$ of the form (19) has to be such that the block $\mathcal{P}_A$ is spectrally equivalent to $\mathbf{A}$, whereas $\mathcal{P}_B$ is spectrally equivalent to $\mathbf{BA}^{-1}\mathbf{B}^T$, see also [11, 13, 14].

For the block $\mathcal{P}_A$, one can use any existing symmetric and positive definite preconditioner devised for the discrete Laplace operator on quasi-uniform and regular meshes. Note that for a regular hierarchical mesh, the best preconditioner for $\mathbf{A}$ would be the BPX preconditioner, see [4]. However, to extend our results to the hierarchical meshes, one needs the corresponding norm preserving extension theorem as in [20]. Hence, this paper is not investigating the effect of the choice $\mathcal{P}_A$, and our primary aim is to propose a preconditioner $\mathcal{P}_B$ that could be effectively used in solving (16).

To that end, for our Lanczos method of minimized iterations, we use the following block-diagonal preconditioner:

$$\mathcal{P} = \begin{bmatrix} \mathcal{P}_A & 0 \\ 0 & \mathcal{B}_D \end{bmatrix}, \tag{23}$$

and in the numerical experiments below, we will simply take $\mathcal{P}_A = \mathbf{A}$. Finally, we define

$$\mathcal{H} = \mathcal{P}^\dagger = \begin{bmatrix} \mathcal{P}_A^{-1} & 0 \\ 0 & [\mathcal{B}_D]^\dagger \end{bmatrix}, \tag{24}$$

and remark that even though the matrix $\mathcal{B}_D$ is singular, as evident from the Lanczos algorithm above, one actually never needs to use its pseudo inverse at all. Indeed, this is due to the block-diagonal structure (24) of $\mathcal{H}$, and the block form (18) of the original matrix $\mathcal{A}_\varepsilon$.

2.3 Main Result: Block-Diagonal Preconditioner

The main theoretical result of this paper establishes a robust preconditioner for solving (16) or, equivalently (17), and is given in the following theorem.

Theorem 1 *Let the triangulation Ω_h for (5)–(6) be conforming and quasi-uniform. Then the matrix $\mathcal{B}_D$ is spectrally equivalent to the matrix $\mathbf{BA}^{-1}\mathbf{B}^T$, that is, there exist constants $\mu_\star, \mu^\star > 0$ independent of h and such that*

$$\mu_\star \leq \frac{(\mathcal{B}_D\overline{\psi}, \overline{\psi})}{(\mathbf{BA}^{-1}\mathbf{B}^T\overline{\psi}, \overline{\psi})} \leq \mu^\star, \quad \text{for all} \quad 0 \neq \overline{\psi} \in \mathbb{R}^n, \ \overline{\psi} \in \text{Im}\,\mathcal{B}_D. \tag{25}$$

This theorem asserts that the nonzero eigenvalues of the generalized eigenproblem

$$\mathbf{BA}^{-1}\mathbf{B}^T\overline{\psi} = \mu\mathcal{B}_{\mathcal{D}}\overline{\psi}, \quad \overline{\psi} \in \mathbb{R}^n, \; \overline{\psi} \in \mathrm{Im}\,\mathcal{B}_{\mathcal{D}}, \tag{26}$$

are bounded. Hence, its proof is based on the construction of the **upper** and **lower** bounds for μ in (26) and is comprised of the following facts many of which are proven in the next section.

Lemma 1 *The following equality of matrices holds*

$$\mathbf{BA}^{-1}\mathbf{B}^T = \mathcal{B}_{\mathcal{D}}S_{00}^{-1}\mathcal{B}_{\mathcal{D}}, \tag{27}$$

where $S_{00} = A_{\mathcal{DD}} - A_{\mathcal{D}0}A_{00}^{-1}A_{0\mathcal{D}}$, is the Schur complement to the block A_{00} of the matrix $\mathbf{A}$ of (51).

This fact is straightforward and comes from the block structure of matrices $\mathbf{A}$ of (11) and $\mathbf{B}$ of (12). Indeed, using this, the generalized eigenproblem (26) can be rewritten as

$$\mathcal{B}_{\mathcal{D}}S_{00}^{-1}\mathcal{B}_{\mathcal{D}}\overline{\psi} = \mu\mathcal{B}_{\mathcal{D}}\overline{\psi}, \quad \overline{\psi} \in \mathbb{R}^n, \; \overline{\psi} \in \mathrm{Im}\,\mathcal{B}_{\mathcal{D}}. \tag{28}$$

Introduce a matrix $B_{\mathcal{D}}^{1/2}$ via $\mathcal{B}_{\mathcal{D}} = B_{\mathcal{D}}^{1/2}B_{\mathcal{D}}^{1/2}$ and note that $\ker\mathcal{B}_{\mathcal{D}} = \ker B_{\mathcal{D}}^{1/2}$.

Lemma 2 *The generalized eigenvalue problem (28) is equivalent to*

$$\mathcal{B}_{\mathcal{D}}^{1/2}S_{00}^{-1}\mathcal{B}_{\mathcal{D}}^{1/2}\overline{\varphi} = \mu\,\overline{\varphi}, \quad \overline{\varphi} \in \mathbb{R}^n, \; \overline{\varphi} \in \mathrm{Im}\,\mathcal{B}_{\mathcal{D}}, \tag{29}$$

in the sense that they both have the same eigenvalues μ's, and the corresponding eigenvectors are related via $\overline{\varphi} = \mathcal{B}_{\mathcal{D}}^{1/2}\overline{\psi} \in \mathrm{Im}\,\mathcal{B}_{\mathcal{D}}$.

Lemma 3 *The generalized eigenvalue problem (29) is equivalent to*

$$\mathcal{B}_{\mathcal{D}}\overline{u}_{\mathcal{D}} = \mu S_{00}\overline{u}_{\mathcal{D}}, \quad \overline{u}_{\mathcal{D}} \in \mathbb{R}^n, \quad \overline{u}_{\mathcal{D}} \in \mathrm{Im}\,(S_{00}^{-1}\mathcal{B}_{\mathcal{D}}), \tag{30}$$

in the sense that both problems have the same eigenvalues μ's, and the corresponding eigenvectors are related via $\overline{u}_{\mathcal{D}} = S_{00}^{-1}\mathcal{B}_{\mathcal{D}}^{1/2}\overline{\varphi} \in \mathrm{Im}\,(S_{00}^{-1}\mathcal{B}_{\mathcal{D}})$.

This result is also straightforward and can be obtained multiplying (29) by $S_{00}^{-1}\mathcal{B}_{\mathcal{D}}^{1/2}$.

To that end, establishing the upper and lower bounds for the eigenvalues of (30) and due to equivalence of (30) with (29), and hence (28), we obtain that eigenvalues of (26) are bounded. We are interested in nonzero eigenvalues of (30) for which the following result holds.

Lemma 4 *Let the triangulation Ω_h for (5)–(6) be conforming and quasi-uniform. Then there exists $\hat{\mu}_\star > 0$ independent of the mesh size $h > 0$ such that*

$$\hat{\mu}_\star \leq \frac{(\mathcal{B}_\mathcal{D} \bar{u}_\mathcal{D}, \bar{u}_\mathcal{D})}{(S_{00} \bar{u}_\mathcal{D}, \bar{u}_\mathcal{D})} \leq 1, \quad for\ all \quad 0 \neq \bar{u}_\mathcal{D} \in \mathrm{Im}\,(S_{00}^{-1}\mathcal{B}_\mathcal{D}). \tag{31}$$

3 Proofs of Statements in Sect. 2.3

Harmonic Extensions

Hereafter, we will use the index $\mathcal{D}$ to indicate vectors or functions associated with the domain $\mathcal{D}$ that is the union of all inclusions, and index 0 to indicate quantities that are associated with the domain outside the inclusions $\Omega \setminus \overline{\mathcal{D}}$.

Now we recall some classical results from the theory of elliptic PDEs. Suppose a function $u^\mathcal{D} \in H^1(\mathcal{D})$, then consider its harmonic extension $u^0 \in H^1(\Omega \setminus \overline{\mathcal{D}})$ that satisfies

$$\begin{cases} -\triangle u^0 = 0, & \text{in } \Omega \setminus \overline{\mathcal{D}}, \\ u^0 = u^\mathcal{D}, & \text{on } \partial \mathcal{D}, \\ u^0 = 0, & \text{on } \Gamma. \end{cases} \tag{32}$$

For such functions the following holds true:

$$\int_\Omega |\nabla u|^2 \, dx = \min_{v \in H_0^1(\Omega)} \int_\Omega |\nabla v|^2 \, dx, \tag{33}$$

where $u = \begin{cases} u^\mathcal{D}, & \text{in } \mathcal{D} \\ u^0, & \text{in } \Omega \setminus \overline{\mathcal{D}} \end{cases}$ and $v = \begin{cases} u^\mathcal{D}, & \text{in } \mathcal{D} \\ v^0, & \text{in } \Omega \setminus \overline{\mathcal{D}} \end{cases}$ with the function $v^0 \in$ $H^1(\Omega \setminus \overline{\mathcal{D}})$ such that $v^0|_\Gamma = 0$, and

$$\|u\|_{H_0^1(\Omega)} \leq C \|u^\mathcal{D}\|_{H^1(\mathcal{D})} \quad \text{with the constant } C \text{ independent of } u^\mathcal{D}, \tag{34}$$

where $\| \cdot \|_{H^1(\Omega)}$ denotes the standard norm of $H^1(\Omega)$:

$$\|v\|_{H^1(\Omega)}^2 = \int_\Omega |\nabla v|^2 dx + \int_\Omega v^2 dx, \tag{35}$$

and $\|v\|_{H_0^1(\Omega)}^2 = \int_\Omega |\nabla v|^2 dx$.

In view of (33), the function u^0 of (33) is the *best extension* of $u^\mathcal{D} \in H^1(\mathcal{D})$ among all $H^1(\Omega \setminus \overline{\mathcal{D}})$ functions that vanish on Γ. The algebraic linear system that corresponds to (33) satisfies a similar property. Namely, if the vector $\bar{u}_0 \in \mathbb{R}^{n_0}$ is a FEM discretization of the function $u^0 \in H_0^1(\Omega \setminus \overline{\mathcal{D}})$ of (32), then for a given $\bar{u}_\mathcal{D} \in \mathbb{R}^n$, the best extension $\bar{u}_0 \in \mathbb{R}^{n_0}$ would satisfy

$$A_{0\mathcal{D}}\, \bar{u}_\mathcal{D} + A_{00}\, \bar{u}_0 = 0, \tag{36}$$

and

$$\left(\mathbf{A} \begin{bmatrix} \overline{u}_{\mathcal{D}} \\ \overline{u}_0 \end{bmatrix}, \begin{bmatrix} \overline{u}_{\mathcal{D}} \\ \overline{u}_0 \end{bmatrix} \right) = \min_{\overline{v}_0 \in \mathbb{R}^{n_0}} \left(\mathbf{A} \begin{bmatrix} \overline{u}_{\mathcal{D}} \\ \overline{v}_0 \end{bmatrix}, \begin{bmatrix} \overline{u}_{\mathcal{D}} \\ \overline{v}_0 \end{bmatrix} \right). \tag{37}$$

Proof of Lemma 2

Consider generalized eigenvalue problem (28) and replace $\mathcal{B}_{\mathcal{D}}$ with $\mathcal{B}_{\mathcal{D}}^{1/2}\mathcal{B}_{\mathcal{D}}^{1/2}$ there, then $\mathcal{B}_{\mathcal{D}}^{1/2}\mathcal{B}_{\mathcal{D}}^{1/2}S_{00}^{-1}\mathcal{B}_{\mathcal{D}}^{1/2}\mathcal{B}_{\mathcal{D}}^{1/2}\overline{\psi} = \mu\mathcal{B}_{\mathcal{D}}^{1/2}\mathcal{B}_{\mathcal{D}}^{1/2}\overline{\psi}$. Now multiply both sides by the Moore-Penrose pseudo inverse $\left[\mathcal{B}_{\mathcal{D}}^{1/2}\right]^{\dagger}$:

$$\left[\mathcal{B}_{\mathcal{D}}^{1/2}\right]^{\dagger}\mathcal{B}_{\mathcal{D}}^{1/2}\mathcal{B}_{\mathcal{D}}^{1/2}S_{00}^{-1}\mathcal{B}_{\mathcal{D}}^{1/2}\mathcal{B}_{\mathcal{D}}^{1/2}\overline{\psi} = \mu\left[\mathcal{B}_{\mathcal{D}}^{1/2}\right]^{\dagger}\mathcal{B}_{\mathcal{D}}^{1/2}\mathcal{B}_{\mathcal{D}}^{1/2}\overline{\psi}.$$

This pseudo inverse has the property that $\left[\mathcal{B}_{\mathcal{D}}^{1/2}\right]^{\dagger}\mathcal{B}_{\mathcal{D}}^{1/2} = P_{im}$, where P_{im} is an orthogonal projector onto the image $\mathcal{B}_{\mathcal{D}}^{1/2}$, hence, $P_{im}\mathcal{B}_{\mathcal{D}}^{1/2} = \mathcal{B}_{\mathcal{D}}^{1/2}$ and therefore, $\mathcal{B}_{\mathcal{D}}^{1/2}S_{00}^{-1}\mathcal{B}_{\mathcal{D}}^{1/2}\overline{\varphi} = \mu\overline{\varphi}$, where $\overline{\varphi} = \mathcal{B}_{\mathcal{D}}^{1/2}\overline{\psi}$.

Conversely, consider the eigenvalue problem (29), and multiply its both sides by $\mathcal{B}_{\mathcal{D}}^{1/2}$. Then $\mathcal{B}_{\mathcal{D}}^{1/2}\mathcal{B}_{\mathcal{D}}^{1/2}S_{00}^{-1}\mathcal{B}_{\mathcal{D}}^{1/2}\overline{\varphi} = \mu\mathcal{B}_{\mathcal{D}}^{1/2}\overline{\varphi}$, where we replace $\overline{\varphi}$ by $\mathcal{B}_{\mathcal{D}}^{1/2}\overline{\psi}$: $\mathcal{B}_{\mathcal{D}}^{1/2}\mathcal{B}_{\mathcal{D}}^{1/2}S_{00}^{-1}\mathcal{B}_{\mathcal{D}}^{1/2}\mathcal{B}_{\mathcal{D}}^{1/2}\overline{\psi} = \mu\mathcal{B}_{\mathcal{D}}^{1/2}\mathcal{B}_{\mathcal{D}}^{1/2}\overline{\psi}$ to obtain (28). $\qquad\square$

Proof of Lemma 4

I. Upper Bound for the Generalized Eigenvalues of (26)

Consider $\overline{u} = \begin{bmatrix} \overline{u}_{\mathcal{D}} \\ \overline{u}_0 \end{bmatrix} \in \mathbb{R}^N$ with $\overline{u}_{\mathcal{D}} \in \text{Im}\,(S_{00}^{-1}\mathcal{B}_{\mathcal{D}})$, satisfying (36), then

$$(S_{00}\overline{u}_{\mathcal{D}}, \overline{u}_{\mathcal{D}}) = (\mathbf{A}\overline{u}, \overline{u}). \tag{38}$$

Using (10) and (13) we obtain from (38):

$$\mu = \frac{(\mathcal{B}_{\mathcal{D}}\overline{u}_{\mathcal{D}}, \overline{u}_{\mathcal{D}})}{(S_{00}\,\overline{u}_{\mathcal{D}}, \overline{u}_{\mathcal{D}})} = \frac{(\mathcal{B}_{\mathcal{D}}\overline{u}_{\mathcal{D}}, \overline{u}_{\mathcal{D}})}{(\mathbf{A}\overline{u}, \overline{u})} = \frac{\displaystyle\int_{\mathcal{D}_h} |\nabla u_h^{\mathcal{D}}|^2 \, dx}{\displaystyle\int_{\Omega_h} |\nabla u_h|^2 \, dx} \le 1, \tag{39}$$

$$\text{with} \qquad u_h = \begin{cases} u_h^{\mathcal{D}}, & \text{in } \mathcal{D}_h \\ u_h^0, & \text{in } \Omega \setminus \overline{\mathcal{D}}_h \end{cases} \tag{40}$$

where u_h^0 is the harmonic extension of $u_h^{\mathcal{D}}$ into $\Omega_h \setminus \overline{\mathcal{D}}_h$ in the sense (32). $\qquad\square$

II. Lower Bound for the Generalized Eigenvalues of (26)

Before providing the proofs, we introduce one more construction to simplify our consideration below. Because all inclusions are located at distances that are comparable to their sizes, we construct new domains $\hat{\mathcal{D}}^i$, $i \in \{1, \ldots, m\}$, see Fig. 2, centered at the centers of the original inclusions $\mathcal{D}^i$, $i \in \{1, \ldots, m\}$, but of sizes much larger of those of $\mathcal{D}^i$ and such that $\hat{\mathcal{D}}^i \cap \hat{\mathcal{D}}^j = \emptyset$, for $i \neq j$. This yields that the problem (5)–(6) might be partitioned into m independent subproblems. Hence, without loss of generality, in this part of the construction, we assume that there is only one inclusion, that is, $m = 1$.

We also recall a few important results from classical PDE theory analogs of which will be used below. Namely, for a given $v \in H^1(\mathcal{D})$ there exists an extension v_0 of v to $\Omega \setminus \overline{\mathcal{D}}$ so that

$$\|v_0\|_{H^1(\Omega \setminus \mathcal{D})} \leq C \|v\|_{H^1(\mathcal{D})}, \quad \text{with} \quad C = C(d, \mathcal{D}, \Omega). \tag{41}$$

One can also introduce a number of norms equivalent to (35), and, in particular, below we will use

$$\|v\|_{\mathcal{D}}^2 := \int_{\mathcal{D}} |\nabla v|^2 dx + \frac{1}{R^2} \int_{\mathcal{D}} v^2 dx, \tag{42}$$

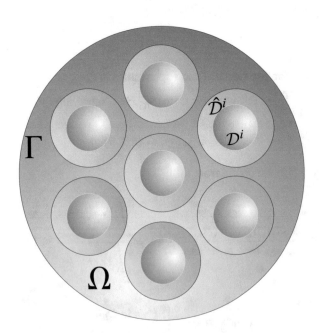

Fig. 2 New domains $\hat{\mathcal{D}}^i$ for our construction of the lower bound of μ

where R is the radius of the particle $\mathcal{D} = \mathcal{D}_1$. The scaling factor $1/R^2$ is needed for transforming the classical results from a reference (i.e. unit) disk to the disk of radius $R \neq 1$.

We note that the FEM analog of the extension result of (41) for a regular grid was shown in [20], from which it also follows that the constant C of (41) is independent of the mesh size $h > 0$. We utilize this observation in our construction below. Consider $u_h \in V_h$ given by (40). Introduce a space $\hat{V}_h = \left\{ v_h \in V_h : v_h = 0 \text{ in } \Omega_h \setminus \widehat{\mathcal{D}}_h \right\}$. Similarly to (40), define

$$\hat{V}_h \ni \hat{u}_h = \begin{cases} u_h^{\mathcal{D}}, & \text{in } \mathcal{D}_h \\ \hat{u}_h^0, & \text{in } \Omega_h \setminus \overline{\mathcal{D}}_h \end{cases}, \tag{43}$$

where $\hat{u}_h^0$ is the harmonic extension of $u_h^{\mathcal{D}}$ into $\widehat{\mathcal{D}}_h \setminus \overline{\mathcal{D}}_h$ in the sense (32) and $\hat{u}_h^0 = 0$ on $\partial \widehat{\mathcal{D}}_h$. Also, by (33) we have $\displaystyle\int_{\Omega_h \setminus \mathcal{D}_h} |\nabla u_h^0|^2 dx \leq \int_{\Omega_h \setminus \mathcal{D}_h} |\nabla \hat{u}_h^0|^2 dx$. Define the

matrix $\hat{\mathbf{A}} := \begin{bmatrix} \hat{A}_{\mathcal{D}\mathcal{D}} & \hat{A}_{\mathcal{D}0} \\ \hat{A}_{0\mathcal{D}} & \hat{A}_{00} \end{bmatrix}$ by $\left(\hat{\mathbf{A}}\overline{v}, \overline{w} \right) = \displaystyle\int_{\Omega_h} \nabla v_h \cdot \nabla w_h dx$, where $\overline{v}, \overline{w} \in \mathbb{R}^N$,

$v_h, w_h \in \hat{V}_h$. As before, introduce the Schur complement to the block $\hat{A}_{00}$ of $\hat{\mathbf{A}}$: $\hat{S}_{00} = A_{\mathcal{D}\mathcal{D}} - \hat{A}_{\mathcal{D}0}\hat{A}_{00}^{-1}\hat{A}_{0\mathcal{D}}$, and consider a new generalized eigenvalue problem $\mathcal{B}_{\mathcal{D}}\overline{u}_{\mathcal{D}} = \hat{\mu}\hat{S}_{00}\overline{u}_{\mathcal{D}}$ with $\overline{u}_{\mathcal{D}} \in \text{Im}\,(S_{00}^{-1}\mathcal{B}_{\mathcal{D}})$. By (37) and (38) we have

$$(S_{00}\overline{u}_{\mathcal{D}}, \overline{u}_{\mathcal{D}}) \leq \left(\hat{S}_{00}\overline{u}_{\mathcal{D}}, \overline{u}_{\mathcal{D}} \right) \quad \text{for all} \quad \overline{u}_{\mathcal{D}} \in \text{Im}\,(S_{00}^{-1}\mathcal{B}_{\mathcal{D}}). \tag{44}$$

Now, we consider a new generalized eigenvalue problem similar to one in (29), namely,

$$\mathcal{B}_{\mathcal{D}}^{1/2}\hat{S}_{00}^{-1}\mathcal{B}_{\mathcal{D}}^{1/2}\overline{\varphi} = \hat{\mu}\,\overline{\varphi}, \quad \overline{\varphi} \in \text{Im}\,\mathcal{B}_{\mathcal{D}}. \tag{45}$$

We plan to replace $\mathcal{B}_{\mathcal{D}}^{1/2}$ in (45) with a new symmetric positive-definite matrix $\hat{\mathcal{B}}_{\mathcal{D}}^{1/2}$, given below in (47), so that

$$\mathcal{B}_{\mathcal{D}}^{1/2}\mathcal{B}_{\mathcal{D}}^{1/2}\overline{\xi} = \mathcal{B}_{\mathcal{D}}^{1/2}\hat{\mathcal{B}}_{\mathcal{D}}^{1/2}\overline{\xi} = \hat{\mathcal{B}}_{\mathcal{D}}^{1/2}\mathcal{B}_{\mathcal{D}}^{1/2}\overline{\xi} \quad \text{for all} \quad \overline{\xi} \in \text{Im}\,\mathcal{B}_{\mathcal{D}}, \tag{46}$$

with what (45) has the same nonzero eigenvalues as the problem $\hat{\mathcal{B}}_{\mathcal{D}}^{1/2}\hat{S}_{00}^{-1}\hat{\mathcal{B}}_{\mathcal{D}}^{1/2}\overline{\varphi} = \hat{\mu}\,\overline{\varphi}$, $\overline{\varphi} \in \text{Im}\,\mathcal{B}_{\mathcal{D}}$. For this purpose, we consider the decomposition: $\mathcal{B}_{\mathcal{D}} = W\Lambda W^T$, where $W \in \mathbb{R}^{n \times n}$ is an orthogonal matrix composed of eigenvectors $\overline{w}_i$, $i \in \{1, \ldots, n\}$, of $\mathcal{B}_{\mathcal{D}}\overline{w} = v\overline{w}$, $\overline{w} \in \mathbb{R}^n$, and $\Lambda = \text{diag}\,[v_1, v_2, \ldots, v_n]$. Then

$\overline{w}_1$ is an eigenvector of $\mathcal{B}_D$ corresponding to $\nu_1 = 0$ and $\overline{w}_1 = \dfrac{1}{\sqrt{n}} \begin{bmatrix} 1 \\ \vdots \\ 1 \end{bmatrix}$. To that

end, we choose

$$\hat{\mathcal{B}}_D = \mathcal{B}_D + \beta\,\overline{w}_1 \otimes \overline{w}_1 = \mathcal{B}_D + \beta\,\overline{w}_1 \overline{w}_1^T, \tag{47}$$

where $\beta > 0$ is some constant parameter chosen below. Note that the matrix $\hat{\mathcal{B}}_D$ is symmetric and positive-definite, and satisfies (46). It is trivial to show that $\hat{\mathcal{B}}_D$ given by (47) is spectrally equivalent to $\mathcal{B}_D + \beta\mathrm{I}$ for any $\beta > 0$. Also, for quasi-uniform grids, the matrix $h^2\mathrm{I}$ (in 3-dim case, $h^3\mathrm{I}$) is spectrally equivalent to the mass matrix M_D given by $(M_D\overline{u}, \overline{v}) = \displaystyle\int_{\mathcal{D}_h^1} u_h v_h\, dx$, where $\overline{u}, \overline{v} \in \mathbb{R}^{n_1}$, $u_h, v_h \in V_h^1$, see e.g. [19]. This implies there exists a constant $C > 0$ independent of h, such that

$$\left(\hat{\mathcal{B}}_D\overline{u}_D, \overline{u}_D\right) \geq C\left(\left(\mathcal{B}_D + \frac{1}{R^2}M_D\right)\overline{u}_D, \overline{u}_D\right), \quad \text{with} \quad \beta = \frac{h^2}{R^2}. \tag{48}$$

The choice of the matrix $\mathcal{B}_D + \frac{1}{R^2}M_D$ for the spectral equivalence was motivated by the fact that the right hand side of (48) describes $\|\cdot\|_{\mathcal{D}_h}$-norm (42) of the FEM function $u_h^{\mathcal{D}} \in V_h^1$ that corresponds to the vector $\overline{u}_D \in \mathbb{R}^n$.

Now consider $\overline{u} = \begin{bmatrix} \overline{u}_D \\ \overline{u}_0 \end{bmatrix} \in \mathbb{R}^N$ with $\overline{u}_D \in \mathbb{R}^n$, $\overline{u}_D \in \mathrm{Im}\,(S_{00}^{-1}\mathcal{B}_D)$, and $\overline{u}_0 \in$

$\mathbb{R}^{n_0}$ satisfying (36), and similarly choose $\overline{\overline{u}} = \begin{bmatrix} \overline{u}_D \\ \hat{\overline{u}}_0 \end{bmatrix} \in \mathbb{R}^N$ with $\hat{\overline{u}}_0 \in \mathbb{R}^{n_0}$ satisfying

$\hat{A}_{0D}\,\overline{u}_D + \hat{A}_{00}\,\hat{\overline{u}}_0 = 0$, which implies

$$\left(\hat{S}_{00}\overline{u}_D, \overline{u}_D\right) = \left(\hat{A}\overline{\overline{u}}, \overline{\overline{u}}\right). \tag{49}$$

Then $\left(\hat{A}\overline{\overline{u}}, \overline{\overline{u}}\right) = \displaystyle\int_{\Omega_h} |\nabla\hat{u}_h|^2 dx$

$$= \int_{\hat{\mathcal{D}}_h \setminus \mathcal{D}_h} |\nabla\hat{u}_h^0|^2 dx + \int_{\mathcal{D}_h} |\nabla u_h^{\mathcal{D}}|^2 dx \leq (C^* + 1)\|u_h^{\mathcal{D}}\|_{\mathcal{D}_h}^2, \tag{50}$$

where $\hat{u}_h \in \hat{V}_h$ is the same extension of $u_h^{\mathcal{D}}$ from $\overline{\mathcal{D}}_h$ to $\Omega_h \setminus \overline{\mathcal{D}}_h$ as defined in (43). For the inequality of (50), we applied the FEM analog of the extension result of (41) by [20], that yields that the constant C^* in (50) is independent of h.

With all the above, we have the following chain of inequalities:

$$\frac{(\mathcal{B}_{\mathcal{D}}\bar{u}_{\mathcal{D}}, \bar{u}_{\mathcal{D}})}{(\mathsf{S}_{00}\bar{u}_{\mathcal{D}}, \bar{u}_{\mathcal{D}})} \underset{(46),(47)}{=} \frac{((\mathcal{B}_{\mathcal{D}} + \beta\,\overline{w}_1 \otimes \overline{w}_1)\,\bar{u}_{\mathcal{D}}, \bar{u}_{\mathcal{D}})}{(\mathsf{S}_{00}\bar{u}_{\mathcal{D}}, \bar{u}_{\mathcal{D}})}$$

$$\underset{(44)}{\geq} \frac{((\mathcal{B}_{\mathcal{D}} + \beta\,\overline{w}_1 \otimes \overline{w}_1)\,\bar{u}_{\mathcal{D}}, \bar{u}_{\mathcal{D}})}{(\hat{\mathsf{S}}_{00}\bar{u}_{\mathcal{D}}, \bar{u}_{\mathcal{D}})} \underset{(49),(48)}{\geq} C\frac{\left(\left(\mathcal{B}_{\mathcal{D}} + \frac{1}{R^2}\mathsf{M}_{\mathcal{D}}\right)\bar{u}_{\mathcal{D}}, \bar{u}_{\mathcal{D}}\right)}{\left(\hat{\mathbf{A}}\bar{\hat{u}}, \bar{\hat{u}}\right)}$$

$$\underset{(50)}{\geq} \frac{C\|u_h^{\mathcal{D}}\|_{\mathcal{D}_h}^2}{(C^* + 1)\|u_h^{\mathcal{D}}\|_{\mathcal{D}_h}^2} = \frac{C}{(C^* + 1)} =: \hat{\mu}_\star,$$

with $\beta = \dfrac{h^2}{R^2}$. Clearly, $\mu_\star$ is independent of $h > 0$.

From the obtained above bounds, we have (31). □

4   Numerical Results

In this section, we use four examples to show the numerical advantages of the Lanczos iterative scheme with the preconditioner $\mathcal{P}$ defined in (23) over the existing preconditioned conjugate gradient method.

Our numerical experiments are performed by implementing the described above Lanczos algorithm for the problem (5)–(6), where the domain Ω is chosen to be a disk of radius 5 with $m = 37$ identical circular inclusions $\mathcal{D}^i$, $i \in \{1, \ldots, m\}$. Inclusions are equally spaced. The function f of the right hand side of (5) is chosen to be a constant, $f = 50$.

In the first set of experiments the values of ε_i's of (6) are going to be identical in all inclusions and vary from 10^{-1} to 10^{-8}. In the second set of experiments we consider four groups of particles with the same values of ε in each group that vary from 10^{-4} to 10^{-7}. In the third set of experiments we consider the case when all inclusions have different values of ε_i's that vary from 10^{-1} to 10^{-9}. Finally, in the fourth set of experiments we decrease the distance between neighboring inclusions.

The initial guess z^0 is a random vector that was fixed for all experiments. The stopping criteria is the Euclidian norm of the relative residual $(\mathcal{A}_\varepsilon \bar{z}^k - \overline{\mathcal{F}})/\overline{\mathcal{F}}$ being less than a fixed tolerance constant.

We test our results agains standard $\texttt{pcg}$ function of MATLAB® with $\mathcal{P}_A = \mathbf{A}$. The same matrix is also used in the implementation of the described above Lanczos algorithm. In the following tables **PCG** stands for preconditioned conjugate gradient method by MATLAB® and **PL** stands for preconditioned Lanczos method of this paper.

Experiment 1 For the first set of experiments we consider particles $\mathcal{D}^i$ of radius $R = 0.45$ in the disk Ω. This choice makes distance d between neighboring inclusions approximately equal to the radius R of inclusions. The triangular mesh Ω_h has $N = 32,567$ nodes. Tolerance is chosen to be equal to 10^{-4}. This experiment concerns the described problem with parameter ε being the same in each inclusion. Table 1 shows the number of iterations corresponding to the different values of ε. Based on these results, we first observe that our **PL** method requires less iterations as ε goes less than 10^{-4}. We also notice that number of iterations in the Lanczos algorithm does not depend on ε.

Experiment 2 In this experiment we leave radii of the inclusions to be the same, namely, $R = 0.45$. Tolerance is chosen to be 10^{-6}. We now distinguish four groups of particles of different ε's. The first group consists of one inclusion—in the center—with the coefficient $\varepsilon = \varepsilon_1$, whereas the second, third and fourth groups are comprised of the disks in the second, third and fourth circular layers of inclusions with coefficients ε_2, ε_3 and ε_4, respectively, see Fig. 3 (inclusions of the same group are located on the same concentric circle of particles indicated by the same shade of grey). We perform this type of experiments for three different triangular meshes with the total number of nodes $N = 5249$, $N = 12, 189$ and $N = 32, 567$. Tables 2, 3, and 4 below show the number of iterations corresponding to three meshes respectively.

These results yield that **PL** requires much less iterations than the corresponding **PCG** with the number of iterations still being independent of both the contrast ε and the mesh size h for **PL**.

Table 1 Number of iterations in *Experiment 1*, $N = 32, 567$

ε	10^{-1}	10^{-2}	10^{-3}	10^{-4}	10^{-5}	10^{-6}	10^{-7}	10^{-8}
PCG	10	20	32	40	56	183	302	776
PL	33	37	37	37	37	37	37	37

Fig. 3 The domain Ω with highly conducting inclusions $\mathcal{D}^i$ of fours groups

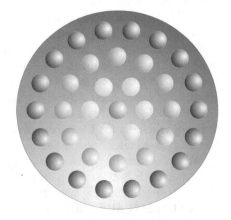

Table 2 Number of iterations in *Experiment 2*, $N = 5249$

ε_1	ε_2	ε_3	ε_4	**PCG**	**PL**
10^{-5}	10^{-5}	10^{-4}	10^{-4}	217	39
10^{-5}	10^{-5}	10^{-4}	10^{-3}	208	39
10^{-6}	10^{-5}	10^{-4}	10^{-3}	716	39
10^{-7}	10^{-6}	10^{-5}	10^{-4}	571	39

Table 3 Number of iterations in *Experiment 2*, $N = 12, 189$

ε_1	ε_2	ε_3	ε_4	**PCG**	**PL**
10^{-5}	10^{-5}	10^{-4}	10^{-4}	116	39
10^{-5}	10^{-5}	10^{-4}	10^{-3}	208	39
10^{-6}	10^{-5}	10^{-4}	10^{-3}	457	39
10^{-7}	10^{-6}	10^{-5}	10^{-4}	454	39

Table 4 Number of iterations in *Experiment 2*, $N = 32, 567$

ε_1	ε_2	ε_3	ε_4	**PCG**	**PL**
10^{-5}	10^{-5}	10^{-4}	10^{-4}	311	35
10^{-5}	10^{-5}	10^{-4}	10^{-3}	311	35
10^{-6}	10^{-5}	10^{-4}	10^{-3}	697	35
10^{-7}	10^{-6}	10^{-5}	10^{-4}	693	35

Table 5 Number of iterations in *Experiment 3*, $N = 12, 189$

Range of ε	PL
10^{-1}–10^{-8}	53
10^{-1}–10^{-3}	53
10^{-7}–10^{-9}	39

Experiment 3 The next point of interest is to assign different value of ε for each of 37 inclusions. The geometrical setup is the same as in *Experiment 2*. The value of ε_i, $i \in \{1, \ldots, 37\}$, is randomly assigned to each particle and is chosen from the range of ε's reported in Table 5 above. The tolerance is 10^{-6} as above. The triangular mesh Ω_h has 12, 189 nodes. We run **ten** tests for each range of contrasts and obtain the **same number of iterations** in every case, and that number is being reported in Table 5. We also observe that as the contrast between conductivities in the background domain $\Omega \setminus \overline{\mathcal{D}}$ and the one inside particles $\mathcal{D}_i$, $i \in \{1, \ldots, 37\}$, becomes larger our preconditioner demonstrates better convergence, as the third row of Table 5 reports. This is expected since the preconditioner constructed above was chosen for the case of absolutely conductive particles. These sets of tests are not compared against the **PCG** due to the large number of considered contrasts that prevent this test to converge in a reasonable amount of time.

Experiment 4 In the next set of experiments we intend to test how well our algorithm performs if the distance between particles decreases. Recall that the assumption made for our procedure to work is that the interparticle distance d is of order of the particles' radius R. With that, we take the same setup as in *Experiment 2* and decrease the distance between particles by making radius of each disk larger. We set $R = 0.56$ obtaining that the radius of each inclusion is now twice larger than the distance d, and also consider $R = 0.59$ so that the radius of an inclusion is three times larger than d. The triangular mesh Ω_h has $N = 6329$ and $N = 6497$ nodes, respectively. The tolerance is chosen to be 10^{-6}. Tables 6 and 7 show the number of iterations in each case. Here we observe that number of iterations increases for both **PCG** and **PL**, while this number still remains independent of ε for **PL**.

Table 6 Number of iterations in *Experiment 4*, $R = 0.56$, $N = 6329$

ε_1	ε_2	ε_3	ε_4	**PCG**	**PL**
10^{-5}	10^{-5}	10^{-4}	10^{-4}	799	61
10^{-7}	10^{-6}	10^{-5}	10^{-4}	859	61

Table 7 Number of iterations in *Experiment 4*, $R = 0.59$, $N = 6497$

ε_1	ε_2	ε_3	ε_4	**PCG**	**PL**
10^{-5}	10^{-5}	10^{-4}	10^{-4}	832	73
10^{-7}	10^{-6}	10^{-5}	10^{-4}	890	73

We then continue to decrease the distance d, and set $R = 0.62$ that is approximately four times larger than the distance between two neighboring inclusions d. Choose the same tolerance 10^{-6} as above, and the triangular mesh Ω_h of $N = 6699$ nodes, and we observed that our **PL** method does not reach the desired tolerance in 1128 iterations, that confirms our expectations. Further research is needed to develop novel techniques for the case of closely spaced particles that the authors intend to pursue in future.

5 Conclusions

This paper focuses on a construction of the robust preconditioner (23) for the Lanczos iterative scheme that can be used in order to solve PDEs with *high-contrast* coefficients of the type (5)–(6). A typical FEM discretization yields an ill-conditioning matrix when the contrast in σ becomes high (i.e. $\varepsilon \ll 1$). We propose an alternative saddle point formulation (18) of the given problem with the symmetric and indefinite matrix and propose a preconditioner for the employed Lanczos method for solving (18). The main feature of this novel approach is that we precondition the given linear system with a symmetric and positive semidefinite matrix. The key theoretical outcome is the that the condition number of the constructed preconditioned system is of $O(1)$, which makes the proposed methodology more beneficial for high-contrast problems' application than existing iterative substructuring methods [7, 8, 17–19]. Finally, our numerical results based on simple test scenarios confirm theoretical findings of this paper, and demonstrate convergence of the constructed **PL** scheme to be independent of the contrast ε, mesh size h, and also on the number of different contrasts ε_i, $i \in \{1, \ldots, m\}$ in the inclusions. In the future, we plan to employ the proposed preconditioner to other types of problems to fully exploit its feature of the independence on contrast and mesh size.

Appendix

Discussions About System (16)

Along with the problem (17)–(18) and its solution $\mathbf{x}_\varepsilon$ by (18), we consider an auxiliary linear system

$$\mathcal{A}_o \mathbf{x}_o = \begin{bmatrix} \mathbf{A} & \mathbf{B}^T \\ \mathbf{B} & \mathbf{0} \end{bmatrix} \begin{bmatrix} \bar{u}_o \\ \bar{\lambda}_o \end{bmatrix} = \begin{bmatrix} \mathbf{F} \\ \mathbf{0} \end{bmatrix}, \tag{51}$$

$$\text{or} \quad \begin{cases} \mathbf{A}\bar{u}_o + \mathbf{B}^T \bar{\lambda}_o = \bar{\mathbf{F}}, \\ \mathbf{B}\bar{u}_o = \bar{0}. \end{cases} \tag{52}$$

where matrices $\mathbf{A}$, $\mathbf{B}$ and the vector $\bar{\mathbf{F}}$ are the same as above. The linear system (51) or, equivalently (52), emerges in a FEM discretization of the diffusion problem posed in the domain Ω whose inclusions are *infinitely conducting*, that is, when $\varepsilon = 0$ in (6). The corresponding PDE formulation for problem (52) might be as follows (see e.g. [5])

$$\begin{cases} \Delta u = f, & x \in \Omega \setminus \overline{\mathcal{D}} \\ u = \text{const}, & x \in \partial \mathcal{D}^i, \ i \in \{1, \dots, m\} \\ \int_{\partial \mathcal{D}^i} \nabla u \cdot \mathbf{n}_i \, ds = 0, & i \in \{1, \dots, m\} \\ u = 0, & x \in \Gamma \end{cases} \tag{53}$$

where $\mathbf{n}_i$ is the outer unit normal to the surface $\partial \mathcal{D}^i$. If $u \in H_0^1(\Omega \setminus \overline{\mathcal{D}})$ is an electric potential then it attains constant values on the inclusions $\mathcal{D}^i$ and these constants are not known a priori so that they are unknowns of the problem (53) together with u.

Formulation (51) or (52) also arises in constrained quadratic optimization problem and solving the Stokes equations for an incompressible fluid [6], and solving elliptic problems using methods combining fictitious domain and distributed Lagrange multiplier techniques to force boundary conditions [9].

Then the following relation between solutions of systems (16) and (52) holds true.

Lemma 5 *Let* $\mathbf{x}_\varepsilon = \begin{bmatrix} \bar{u}_\varepsilon \\ \bar{\lambda}_\varepsilon \end{bmatrix} \in \mathbb{R}^{N+n}$ *the solution of* (16), *and* $\mathbf{x}_o = \begin{bmatrix} \bar{u}_o \\ \bar{\lambda}_o \end{bmatrix} \in \mathbb{R}^{N+n}$ *be the solution of the linear system* (52). *Then* $\bar{u} \to \bar{u}_o$ *as* $\varepsilon := \max_{i \in \{1, \dots, m\}} \varepsilon_i \to 0$.

This lemma asserts that the discrete approximation for the problem (5)–(6) converges to the discrete approximation of the solution of (53) as $\varepsilon \to 0$. We also note that the continuum version of this fact was shown in [10].

Proof Hereafter, we denote by C a positive constant independent of ε.

Subtract (52) from (16):

$$\begin{cases} \mathbf{A}(\overline{u}_\varepsilon - \overline{u}_o) + \mathbf{B}^T (\overline{\lambda}_\varepsilon - \overline{\lambda}_o) = \overline{0}, \\ \mathbf{B}(\overline{u}_\varepsilon - \overline{u}_o) - \Sigma_\varepsilon (\overline{\lambda}_\varepsilon - \overline{\lambda}_o) = \Sigma_\varepsilon \overline{\lambda}_o. \end{cases} \tag{54}$$

and multiply by $\overline{u}_\varepsilon - \overline{u}_o$ to obtain

$$(\mathbf{A}(\overline{u}_\varepsilon - \overline{u}_o, \overline{u}_\varepsilon - \overline{u}_o)) + \left(\mathbf{B}^T (\overline{\lambda}_\varepsilon - \overline{\lambda}_o), \overline{u}_\varepsilon - \overline{u}_o \right) = \overline{0}.$$

Denote $\| \cdot \|_{\mathbf{D}} := (\mathbf{D} \cdot, \cdot)$ for any symmetric and positive-definite matrix $\mathbf{D}$ (and the same notation for the semi-norm associated with a symmetric, positive semi-definite $\mathbf{D}$), hence,

$$\| \overline{u}_\varepsilon - \overline{u}_o \|_{\mathbf{A}} \leq \| \overline{\lambda}_\varepsilon - \overline{\lambda}_o \|_{\mathcal{B}_D}, \tag{55}$$

where we have used that $\| \overline{u}_\varepsilon - \overline{u}_o \|_{\mathcal{B}_D} \leq \| \overline{\lambda}_\varepsilon - \overline{\lambda}_o \|_{\mathcal{B}_D}$. Now eliminate $(\overline{u}_\varepsilon - \overline{u}_o)$ from (54) and denote $\overline{\xi} := \overline{\lambda}_\varepsilon - \overline{\lambda}_o \in \mathbb{R}^n$ to have

$$(\mathbf{B}\mathbf{A}^{-1}\mathbf{B}^T \overline{\xi}, \overline{\xi}) + (\Sigma_\varepsilon \overline{\xi}, \overline{\xi}) = -(\Sigma_\varepsilon \overline{\lambda}_o, \overline{\xi}). \tag{56}$$

It was shown in (39) that $(\mathcal{B}_D \overline{\xi}, \overline{\xi}) \leq (\mathbf{B}\mathbf{A}^{-1}\mathbf{B}^T \overline{\xi}, \overline{\xi})$. It is also obvious that

$$(\Sigma_\varepsilon \overline{\xi}, \overline{\xi}) \geq \underline{\varepsilon}(\mathcal{B}_D \overline{\xi}, \overline{\xi}), \quad \text{where} \quad \underline{\varepsilon} := \min_{i \in \{1,\dots,m\}} \varepsilon_i$$

Thus, from (56), we have $(1 + \underline{\varepsilon})\|\overline{\xi}\|_{\mathcal{B}_D} \leq \varepsilon \|\overline{\lambda}_o\|_{\mathcal{B}_D}$. From this inequality, norm equivalence, and (55), we obtain $\|\overline{u}_\varepsilon - \overline{u}_o\| \to 0$ as $\varepsilon \to 0$. $\quad\square$

It was also previously observed, see e.g. [11, 14, 19], that the matrix (20) is the best choice for a preconditioner of $\mathcal{A}_o$. This is because there are exactly three eigenvalues of $\mathcal{A}_o$ associated with the following generalized eigenvalue problem (see, e.g. [11, 19])

$$\mathcal{A}_o \begin{bmatrix} u \\ \lambda \end{bmatrix} = \mu \mathbf{P} \begin{bmatrix} u \\ \lambda \end{bmatrix}, \quad \overline{u} \in \mathbb{R}^N, \quad \overline{\lambda} \in \operatorname{Im} \mathcal{B}_D, \tag{57}$$

and they are: $\mu_1 < 0$, $\mu_2 = 1$ and $\mu_3 > 1$, and, hence, a Krylov subspace iteration method applied for a preconditioned system for solving (57) with (20) *converges to the exact solution in three iterations*.

Now, we turn back to the problem (16). Then the following statement about the generalized eigenvalue problem (21) holds true.

Lemma 6 *There exist constants $c_1 \leq c_2 < 0 < c_3 \leq c_4$ independent of the discretization scale $h > 0$ or the contrast parameters ε_i, $i \in \{1, \dots, m\}$, such that the eigenvalues of the generalized eigenvalue problem*

$$\mathcal{A}_\varepsilon x = \nu \mathbf{P} x, \quad \bar{u} \in \mathbb{R}^N, \; \bar{\lambda} \in \operatorname{Im} \mathcal{B}_\mathcal{D},$$

belong to $[c_1, c_2] \cup [c_3, c_4]$.

Remark that the endpoints c_i of the eigenvalues' intervals might depend on eigenvalues of (57).

Proof Without loss of generality, here we also assume that all ε_i, $i \in \{1, \ldots, m\}$, are the same and equal to ε, that is, $\mathbf{\Sigma}_\varepsilon = \varepsilon \mathcal{B}_\mathcal{D}$. Write the given eigenvalue problem as: $\begin{bmatrix} \mathbf{A} & \mathbf{B}^T \\ \mathbf{B} & -\varepsilon \mathcal{B}_\mathcal{D} \end{bmatrix} \begin{bmatrix} \bar{u} \\ \bar{\lambda} \end{bmatrix} = \nu \begin{bmatrix} \mathbf{A} & \mathbf{0} \\ \mathbf{0} & \mathbf{B}\mathbf{A}^{-1}\mathbf{B}^T \end{bmatrix} \begin{bmatrix} \bar{u} \\ \bar{\lambda} \end{bmatrix}$, $\bar{u} \in \mathbb{R}^N, \bar{\lambda} \in \operatorname{Im} \mathcal{B}_\mathcal{D}$, which leads to the equation for ν, which is as follows

$$\nu - \frac{1}{\nu - 1} = -\varepsilon \left[\frac{(\mathcal{B}_\mathcal{D} \bar{\lambda}, \bar{\lambda})}{(\mathbf{B}\mathbf{A}^{-1}\mathbf{B}^T \bar{\lambda}, \bar{\lambda})} \right], \quad \bar{\lambda} \in \operatorname{Im} \mathcal{B}_\mathcal{D}. \tag{58}$$

The fraction of the right-hand side of the above equation, that we denote by μ, has been estimated in Theorem 1: $\mu_\star \leq \mu \leq 1$, where $\mu_\star$ is independent of the discretization size $h > 0$ due to the norm-preserving extension theorem, [20]. From (58), we obtain that the eigenvalues ν of (21) that differ from one, $\nu \neq 1$, are $\nu^\pm = \dfrac{1 - \varepsilon\mu \pm \sqrt{5 + 2\varepsilon\mu + \varepsilon^2\mu^2}}{2}$, and as $\varepsilon \to 0$, we have $0 > \nu^- \to \frac{1-\sqrt{5}}{2}$ and $0 < \nu^+ \to \frac{1+\sqrt{5}}{2}$.

Finally, using the bounds for μ by (25), we have from (58) that the endpoints of the intervals $[c_1, c_2] \ni \nu^-$ and $[c_3, c_4] \ni \nu^+$ are independent of both h and ε. In particular, for $0 < \varepsilon \ll 1$ and $\mu_\star \leq \mu \leq 1$, we have that $\nu^- \in \left[-\sqrt{2}, \frac{1-\sqrt{5}}{2} \right]$ and $\nu^+ \in \left[\frac{\sqrt{5}}{2}, 2 \right]$.

If we one has variable ε_i then it yields a sum over $i \in \{1, \ldots, m\}$ in the right hand side of (58). This can be estimated by taking maximal and minimal values of ε_i. $\square$

This lemma demonstrates that (20) is the best (theoretical) preconditioner for $\mathcal{A}_\varepsilon$ as well as for $\mathcal{A}_o$.

Acknowledgements First two authors were supported by the NSF grant DMS-1350248.

References

1. B. Aksoylu, I. G. Graham, H. Klie, and R. Scheichl, "Towards a rigorously justified algebraic preconditioner for high-contrast diffusion problems", *Computing and Visualization in Science*, **11:4-6**, 2008, pp. 319–331
2. J. Aarnes, and T. Y. Hou, "Multiscale domain decomposition methods for elliptic problems with high aspect ratios", *Acta Mathematicae Applicatae Sinica. English Series*, **18:1**, 2002, pp. 63–76

3. O. Axelsson, Iterative Solution Methods, Cambridge University Press, 1994
4. J. H. Bramble, J. E. Pasciak, and J. Xu, "Parallel multilevel preconditioners", *Math. Comp.*, **55**, 1990, pp.1–22
5. V. M. Calo, Y. Efendiev, and J. Galvis, "Asymptotic expansions for high-contrast elliptic equations", *Mathematical Models and Methods in Applied Sciences*, **24:3**, 2014, pp. 465–494
6. H. C. Elman, D. J. Silvester, and A. J. Wathen, "Finite elements and fast iterative solvers: with applications in incompressible fluid dynamics", in *Numerical Mathematics and Scientific Computation*, Oxford University Press, New York, 2005
7. C. Farhat, M. Lesoinne, P. LeTallec, K. Pierson, and D. Rixen, "FETI-DP: a dual-primal unified FETI method. Part I. A faster alternative to the two-level FETI method", *International Journal for Numerical Methods in Engineering*, **50:7**, 2001, pp. 1523–1544
8. C. Farhat, F.X. Roux, "A Method of Finite Element Tearing and Interconnecting and its Parallel Solution Algorithm", *International Journal for Numerical Methods in Engineering*, **32**, 1991, pp.1205–1227.
9. R. Glowinski, and Yu. Kuznetsov, "On the solution of the Dirichlet problem for linear elliptic operators by a distributed Lagrange multiplier method", *Comptes Rendus de l'Académie des Sciences. Série I. Mathématique*, **327:7**, 1998, pp. 693–698
10. Y. Gorb, and Yu. Kuznetsov, "Asymptotic Expansions for High-Contrast Scalar and Vectorial PDEs", *preprint*.
11. Yu. Iliash, T. Rossi, and J. Toivanen, "Two iterative methods to solve the Stokes problem", *Technical Report No. 2. Lab. Sci. Comp.*, Dept. Mathematics, University of Jyväskylä. Jyväskylä, Finland, 1993
12. C. Keller, N. I. M. Gould, and A. J. Wathen, "Constraint preconditioning for indefinite linear systems", *SIAM Journal on Matrix Analysis and Applications*, **21:4**, 2000, pp. 1300–1317
13. Yu. Kuznetsov, "Efficient iterative solvers for elliptic finite element problems on nonmatching grids", *Russian Journal of Numerical Analysis and Mathematical Modelling*, **10:3**, 1995, pp. 187–211
14. Yu. Kuznetsov, "Preconditioned iterative methods for algebraic saddle-point problems", *Journal of Numerical Mathematics*, **17:1**, 2009, pp. 67–75
15. Yu. Kuznetsov, and G. Marchuk, "Iterative methods and quadratic functionals". In *Méthodes de l'Informatique–4*, eds. J.-L. Lions and G. Marchuk, pp. 3–132, Paris, 1974 (In French)
16. L. Lukšan, and J. Vlček, "Indefinitely preconditioned inexact Newton method for large sparse equality constrained non-linear programming problems", *Numerical Linear Algebra with Applications*, **5:3**, 1998, pp. 219–247
17. J. Mandel, "Balancing domain decomposition", *Comm. Numer. Methods Engrg.*, **9**, 1993, pp. 233–241
18. J. Mandel and R. Tezaur, "On the convergence of a dual-primal substructuring method", *Numerische Mathematik*, **88**, 2001, pp. 543–558
19. A. Toselli, and O. Widlund, "Domain decomposition methods – algorithms and theory", *Springer Series in Computational Mathematics*, **34**, Springer-Verlag, Berlin, 2005
20. O. B. Widlund, "An Extension Theorem for Finite Element Spaces with Three Applications", Chapter *Numerical Techniques in Continuum Mechanics* in *Notes on Numerical Fluid Mechanics*, **16**, 1987, pp. 110–122
21. X. Wu, B. P. B. Silva, and J. Y. Yuan, "Conjugate gradient method for rank deficient saddle point problems", *Numerical Algorithms*, **35:2–4**, 2004, pp. 139–154

On the Dimension Reduction in Prestrained Elasticity (Survey)

Silvia Jiménez Bolaños

1 Introduction

This review discusses the analytical and geometrical questions coming from the study of elastic materials that exhibit residual stress at free equilibria. There is an abundance of applications where such structures and their actuations are present, for example: growing tissues, plastically strained sheets, specifically engineered swelling or shrinking gels, atomically thin graphene layers; just to mention a few. These and other phenomena can be studied through a variational model, pertaining to the non-Euclidean version of nonlinear elasticity, which postulates the formation of a target Riemannian metric, resulting in the morphogenesis of the film attaining an orientation-preserving configuration closest to being the isometric immersion of the metric. Shape formation driven by internal prestrain has also been studied by means of formal methods, numerics, and analytical arguments [6, 9–12, 20].

In [8], Gero Friesecke, Richard D. James, and Stefan Müller used an asymptotic framework to understand which theories of thin objects (plates, shells) are predicted by the three dimensional nonlinear theory in the classical elasticity. They derived a hierarchy of the limiting theories as Γ-limits (see Sect. 2.1) of the rescaled versions of three dimensional energies; these theories are unlike each other by the qualitatively different responses to external forces and boundary conditions of the film. In the context of the prestrain-driven response, the parallel theories are differentiated by the embeddability properties of the target metrics and, a-posteriori, by the emergence of isometry constraints on deformations with low regularity. In turn, results on thin limit models have consequences for the three dimensional original model in terms of energy scaling laws, understanding of the

S. Jiménez Bolaños (✉)
Department of Mathematics, Colgate University, Hamilton, NY, USA
e-mail: sjimenez@colgate.edu

© The Author(s) and the Association for Women in Mathematics 2020
B. Acu et al. (eds.), *Advances in Mathematical Sciences*, Association for
Women in Mathematics Series 21, https://doi.org/10.1007/978-3-030-42687-3_20

role of curvature in determining the mechanical properties of the material, and in the effects of the symmetry and the symmetry breaking in the solutions to the resulting Euler-Lagrange equations.

This review paper is organized as follows. In Sect. 2, we give the background on thin films and dimension reduction in prestrained (incompatible) elasticity and briefly explain the Γ-convergence formalism. In Sect. 3, we account for some relevant literature on the subject and current developments. In Sect. 4, we propose a new result in the described direction, anticipating the work done by the author in collaboration with Anna Zemlyanova in [3]. In Sect. 5, we discuss a more general result developed in collaboration with Marta Lewicka and Anna Zemlyanova. Finally, conclusions are presented in Sect. 6.

2 Background on Thin Films and Dimension Reduction in Prestrained Elasticity

We model a thin film as the Cartesian product:

$$\Omega^h = \omega \times (-h/2, h/2),$$

with midplate ω and small thickness $0 < h \ll 1$. The midplate ω is an open bounded subset of $\mathbb{R}^2$. A typical point in Ω^h is denoted by $x = (x_1, x_2, x_3) = (x', x_3)$, where $x' \in \omega$ and $|x_3| < h/2$.

Let $G^h : \Omega^h \to \mathbb{R}^{3\times 3}_{sym, pos}$ be a smooth Riemannian metric on Ω^h. It is a well known fact that the manifold (Ω^h, G^h) can be isometrically immersed in $\mathbb{R}^3$ if and only if the Riemann curvature tensor of G^h vanishes in Ω^h, i.e.: $\mathrm{Riem}(G^h) \equiv 0$. When this happens, there exists a smooth *deformation* u^h of Ω^h into $\mathbb{R}^3$ which is an isometric embedding of G^h:

$$\nabla u^h(x)^T \nabla u^h(x) = G^h(x), \quad \text{for all } x \in \Omega^h. \tag{1}$$

If $\mathrm{Riem}(G^h) \not\equiv 0$, one looks for an orientation-preserving deformation u^h that minimizes the difference between the tensor fields in the right and the left hand sides of (1). It is then the goal to study a variational model called *the prestrained elasticity*:

$$I(u^h) = \int_{\Omega^h} \mathrm{dist}^2 \left(\nabla u^h(x)(G^h(x))^{-1/2}, SO(3)\right) dx, \tag{2}$$

describing its critical points, minimizers / almost minimizers, particularly in the singular limit where $h \to 0$. Observe that $I(u^h) = 0$ iff $\nabla u^h(x) \in SO(3)(G^h(x))^{1/2}$ for a.e. x; which is, by the polar decomposition theorem, in turn equivalent to $(\nabla u^h)^T (\nabla u^h) = G^h$ and $\det \nabla u^h > 0$. The energy functional in (2) is defined for all $u^h \in W^{1,2}(\Omega^h, \mathbb{R}^3)$, where $SO(3)$ denotes the special orthogonal group of proper rotations, and dist $(B, SO(3))$ stands for the minimal distance:

$$|B - R| = \left(\text{Trace}(B - R)^T (B - R)\right)^{1/2}$$

of a given $B \in \mathbb{R}^{3 \times 3}$ from all $R \in SO(3)$.

More precisely, we are interested in analyzing the scaling of the infimum energy and the corresponding structure of minimizers to the energy functional:

$$I_W^h(u^h) = \frac{1}{h} \int_{\Omega^h} W(\nabla u^h (G^h)^{-1/2}) dx \quad \forall u^h \in W^{1,2}(\Omega^h, \mathbb{R}^3), \tag{3}$$

in terms of powers of the vanishing thickness parameter h. The stored energy density function $W : \mathbb{R}^{3 \times 3} \to \mathbb{R}_+$ is assumed to satisfy the standard conditions of normalization, frame indifference with respect to the special orthogonal group $SO(3)$ of proper rotations in $\mathbb{R}^3$, and second order nondegeneracy, given by:

$$\exists c > 0 \; \forall F \in \mathbb{R}^{3 \times 3} \; \forall R \in SO(3) \quad W(R) = 0, \quad W(RF) = W(F),$$

$$W(F) \geq c \, \text{dist}^2(F, SO(3)). \tag{4}$$

Note that, as a consequence, both W and its first derivative DW vanish on the energy well $SO(3)$.

2.1 Γ-Convergence

The above mentioned properties of W contradict the possibility of imposing suitable convexity assumptions and, thus, preclude applying the direct methods of Calculus of Variations in studying (3). Instead, minimizing sequences of the family of problems (3) are studied through asymptotic analysis, exploiting the small thickness regime. The first step in this approach is to establish compactness for sequences of approximate minimizers of I_W^h as $h \to 0$. These will naturally vary among different ranges of the scaling exponent β in $\inf I_W^h \sim h^\beta$, that is in turn induced by the prestrain encoded in the curvatures of G^h. The second step is to look for suitable "dimensionally reduced" energies defined on effective domains $\mathcal{A}_\beta$ consisting of admissible limiting deformations u^h, that carry the structure of I_W^h. The variational method used in this context is the Γ-convergence.

The notion of Γ-convergence, introduced by Ennio De Giorgi in a series of papers published between 1975 and 1983 [5], has become the standard notion of convergence for variational problems [4]. There are by now many applications of this tool to a variety of asymptotic problems, yielding results in the theory of partial differential equations.

In the present set-up for thin films, proving Γ-convergence of the scaled energies $\frac{1}{h^\beta} I_W^h$ consists of deriving two inequalities, after fixing a metric topology on the space of deformations u^h:

(1) The first inequality establishes a lower bound:

$$\mathcal{I}_\beta(u) \geq \liminf_{h \to 0} \frac{1}{h^\beta} I_W^h(u^h),$$

for any sequence u^h converging to $u \in \mathcal{A}_\beta$.

(2) The second inequality serves to prove that the previous bound is optimal in the sense that, for any given admissible $u \in \mathcal{A}_\beta$ there holds:

$$\mathcal{I}_\beta(u) = \limsup_{h \to 0} \frac{1}{h^\beta} I_W^h(u^h)$$

for some "recovery sequence" u^h converging to u.

We say that $\frac{1}{h^\beta} I_W^h(u^h)$ Γ-converges to the residual energy $\mathcal{I}_\beta(u)$, provided that (1) and (2) are satisfied.

The importance of Γ-convergence is that it implies, under quite mild compactness assumptions, that the limits as $h \to 0$ of any converging sequence of approximate minimizers to I_W^h coincide with the minimizers of $\mathcal{I}_\beta$. Hence, identifying the governing variational principle for the asymptotic behaviour of (3) is accomplished by deriving the Γ-limit of $\frac{1}{h^\beta} I_W^h(u^h)$.

2.2 Notation

For a matrix F, its $n \times m$ principal minor is denoted by $F_{n \times m}$. When $m = n$ then the symmetric part of a square matrix F is: $\mathrm{sym} F = 1/2(F + F^T)$. The superscript T refers to the transpose of a matrix or an operator. The operator $\mathrm{curl}^T \mathrm{curl}$ acts on 2×2 square matrix fields F by taking first curl of each row and then taking the curl of the resulting two dimensional vector, so that: $\mathrm{curl}^T \mathrm{curl}\, F = \partial_{11}^2 F_{22} - \partial_{12}^2 (F_{12} + F_{21}) + \partial_{22}^2 F_{11}$. In particular, we see that: $\mathrm{curl}^T \mathrm{curl}\, F = \mathrm{curl}^T \mathrm{curl}(\mathrm{sym} F)$.

By ∇_{tan} we denote taking derivatives ∂_1 and ∂_2 in the in-plate directions $e_1 = (1, 0, 0)^T$ and $e_2 = (0, 1, 0)^T$. The derivative ∂_3 is taken in the out-of-plate direction $e_3 = (0, 0, 1)^T$.

3 Relevant Literature and Current Developments

The study of dimensionally reduced models helps the understanding of the role of the curvature tensor in the stress distribution within a three dimensional prestrained body, and eventually leads to an adequate mathematical description of the morphogenesis phenomena.

In the case when the prestrain $G^h = G$ is independent of the thickness variable h, the recent developments in [13] completed the analysis in [1, 18, 19]. It contains the derivation of all the Γ-limits of the rescaled non-Euclidean elastic energies in the scaling regime h^β, $\beta \geq 2$. These correspond to the even scaling powers $\beta = 2k$, which are the only scalings possible and the regimes of their validity can be identified, in terms of the Riemann curvatures. In paper [14], a more general class of incompatibilities is treated, where the transversal dependence of the lower order terms is not necessarily linear, extending the previous results to arbitrary metrics and higher order scalings.

On the other hand, for the prestrain metrics G^h that are a perturbation of the flat I_3 metric, we quote papers [16, 17, 21]. In particular, in [17] the authors derived a new variational model consisting of minimizing a biharmonic energy $\int_\omega |\nabla^2 v|^2 \, dx'$ of the out-of plane displacement $v \in W^{2,2}(\omega, \mathbb{R})$, satisfying the Monge-Ampère constraint:

$$\det \nabla^2 v = f,$$

where $f = -\mathrm{curl}^T \mathrm{curl}\, S_{2\times2}$ is the linearized Gauss curvature of the Riemannian metrics in:

$$G^h(x', x_3) = I_3 + 2h^\gamma S(x') + 2h^{\gamma/2} x_3 B(x'). \tag{5}$$

This work was done in the parameter range $0 < \gamma < 2$, whereas the case $\gamma = 2$ has been treated in [15], leading to the derivation of the Föppl-von Kármán equations accounting for presence of the prestrain. In [3], the authors carried out the analysis for the parameter range $\gamma > 2$ and the results will be presented in Sect. 4. In [2], the authors looked at a more general version of the Riemannian metric (5), given by:

$$G^h(x', x_3) = I_3 + 2h^\alpha S(x') + 2h^{\gamma/2} x_3 B(x'),$$

and identified the asymptotic behavior of the minimizers of I_W^h (see (3)) as $h \to 0$, through deriving the Γ-limit of the rescaled energies $\frac{1}{h^{\delta+2}} I_W^h$. Different models occur depending on the relationship between the parameters α, γ, and δ. The analysis recovers previous results by using the appropriate values of α, γ, and δ:

- If $\alpha = \gamma = \delta = 2$, we recover the results from [15].
- If $\alpha = \gamma = \delta \in (0, 2)$, we recover the results from [17].
- If $\alpha = \gamma = \delta > 2$, we recover the results from [3].

These results are presented in Sect. 5.

4 Relative Bending Energy for Weakly Prestrained Shells

In connection with (5), we write: $(G^h)^{1/2} = A^h + h.o.t$, where $A^h = [A^h_{ij}] : \overline{\Omega^h} \rightarrow \mathbb{R}^{3\times3}$, $\det A^h > 0$, are given by:

$$A^h(x', x_3) = I_3 + h^\gamma S(x') + h^{\gamma/2}x_3 B(x'), \qquad \gamma > 2 \qquad (6)$$

via the smooth tensor fields S ("stretching") and B ("bending"); where $S, B : \overline{\omega} \rightarrow \mathbb{R}^{3\times3}$. For a deformation $u^h : \Omega^h \rightarrow \mathbb{R}^3$, the elastic energy $I^h_W(u^h)$, given in (3), is then written in terms of the elastic tensor $\nabla u^h (A^h)^{-1}$ accounting for the reorganization of Ω^h in response to A^h in:

$$I^h_W(u^h) = \frac{1}{h} \int_{\Omega^h} W((\nabla u^h)(A^h)^{-1}) \, dx.$$

At a technical level, we also assume that there exists a monotone non-negative function $v : [0, +\infty] \rightarrow [0, +\infty]$ which converges to zero at 0, and a quadratic form Q_3 on $\mathbb{R}^{3\times3}$, with:

$$\forall F \in \mathbb{R}^{3\times3} \quad \left| W(I_3 + F) - \frac{1}{2}Q_3(F) \right| \leq v(|F|)|F|^2. \qquad (7)$$

If W is C^2 regular in a neighborhood of $SO(3)$, then $Q_3 = D^2 W(I_3)$. Note that (7) implies that Q_3 is nonnegative, is positive definite on symmetric matrices, and $Q_3(F) = Q_3(\text{sym } F)$ for all $F \in \mathbb{R}^{3\times3}$.

As discussed in Sect. 2, $I^h_W(u^h) = 0$ is equivalent to:

$$(\nabla u^h)^T \nabla u^h = (A^h)^T (A^h) = G^h \quad \text{and} \quad \det \nabla u^h > 0.$$

Therefore, the quantity:

$$e_h = \inf \left\{ I^h_W(u^h); \ u^h \in W^{1,2}(\Omega^h, \mathbb{R}^3) \right\}$$

measures the residual energy at free equilibria of the configuration Ω^h.

Expanding the energy to the deformation:

$$u^h(x', x_3) = (x', 0)^T + h^{\gamma/2} V(x') + x_3 N_h(x'),$$

where $N_h(x')$ is the unit normal to the midplate and $V : \omega \to \mathbb{R}^3$, we obtain:

$$I_W^h(u^h) = \frac{h^{2\gamma}}{8} \int_\omega \mathcal{Q}_3(-2\mathrm{sym}S + \nabla V^T \nabla V) dx'$$

$$+ \frac{h^{\gamma+2}}{24} \int_\omega \mathcal{Q}_3(-\mathrm{sym}B - \nabla^2 V_3 + h^{\frac{\gamma}{2}}\mathcal{D}) dx' + h.o.t.,$$

where the matrix D is given by:

$$\mathcal{D} = \begin{pmatrix} V_{3,1}V_{1,11} + V_{3,2}V_{2,11} & V_{3,1}V_{1,12} + V_{3,2}V_{2,12} \\ V_{3,1}V_{1,21} + V_{3,2}V_{2,21} & V_{3,1}V_{1,22} + V_{3,2}V_{2,22} \end{pmatrix}.$$

Since $\gamma > 2$, we observe that $I_W^h(u^h) \approx C h^{\gamma+2}$ and hence, we expect the Γ-limit of $\frac{1}{h^{\gamma+2}} I_W^h(u^h)$ to be only the first order change in the linear bending energy:

$$\frac{1}{24} \int_\omega \mathcal{Q}_3(B + \nabla^2 V_3) \, dx'.$$

The goal of the analysis whose results we announce below, is to identify the asymptotic behavior of the minimizers of I_W^h as $h \to 0$. We have:

Theorem 1 ([3]) *Let A^h be given as in (6). Assume that a sequence of deformations $u^h \in W^{1,2}(\Omega^h, \mathbb{R}^3)$ satisfies:*

$$I_W^h(u^h) \le Ch^{\gamma+2},$$

where W fulfills (4) and (7). Then, there exist rotations $\bar{R}^h \in SO(3)$ and translations $c^h \in \mathbb{R}^3$ such that, for the normalized deformations:

$$y^h \in W^{1,2}(\Omega^1, \mathbb{R}^3), \qquad y^h(x', x_3) = (\bar{R}^h)^T u^h(x', hx_3) - c^h,$$

the following hold (up to a subsequence that we do not relabel):

(1) $y^h(x', x_3) \to x'$ in $W^{1,2}(\Omega^1, \mathbb{R}^3)$.
(2) The scaled displacements:

$$V^h(x') = \frac{1}{h^{\gamma/2}} \int_{-1/2}^{1/2} y^h(x', t) - x' dt$$

converge to a vector field V of the form $V = (0, 0, V_3)^T$. This convergence is strong in $W^{1,2}(\omega, \mathbb{R}^3)$. The only non-zero out-of-plane scalar component V_3 of V belongs to $W^{2,2}(\omega, \mathbb{R})$.

(3) Moreover:

$$\liminf_{h \to 0} \frac{1}{h^{\gamma+2}} I_W^h(u^h) \geq \mathcal{I}_\gamma(V_3),$$

where $\mathcal{I}_\gamma : W^{2,2}(\omega) \to \bar{\mathbb{R}}_+$ is given by:

$$\mathcal{I}_\gamma(V_3) = \frac{1}{24} \int_\omega \mathcal{Q}_2(\nabla^2 V_3 + (\mathrm{sym}B(x'))_{2\times 2}) \, dx' \tag{8}$$

and the quadratic non-degenerate form $\mathcal{Q}_2$ is:

$$\mathcal{Q}_2(F) = \min \left\{ \mathcal{Q}_3(\tilde{F}) : \tilde{F} \in \mathbb{R}^{3\times 3}, \ \tilde{F}_{2\times 2} = F \right\} \qquad \forall F \in \mathbb{R}^{2\times 2}. \tag{9}$$

Theorem 2 ([3]) *Assume (6) and that W satisfies (4) and (7). If ω is simply connected then, for every $V_3 \in W^{2,2}(\omega, \mathbb{R})$, there exists a sequence of deformations $u^h \in W^{1,2}(\Omega^h, \mathbb{R}^3)$ such that the following hold:*

(1) The sequence $y^h(x', x_3) = u^h(x', hx_3)$ converges in $W^{1,2}(\Omega^1, \mathbb{R}^3)$ to x'.
(2) The displacements:

$$V^h(x') = \frac{1}{h^{\gamma/2}} \int_{-h/2}^{h/2} \left(u^h(x', t) - x' \right) dt$$

converge in $W^{1,2}(\omega, \mathbb{R}^3)$ to $(0, 0, V_3)^T$.
(3) Recalling (8) one has:

$$\lim_{h \to 0} \frac{1}{h^{\gamma+2}} I_W^h(u^h) = \mathcal{I}_\gamma(V_3).$$

As a result of Theorems 1 and 2 we have:

Corollary 1 *Assume (6), (4), and (7). Moreover, assume that ω is simply connected and that $\gamma > 2$. Then there exist a uniform constant $C \geq 0$ such that:*

$$e_h = \inf I_W^h \leq C h^{\gamma+2}.$$

Under this condition, for any minimizing sequence $u^h \in W^{1,2}(\Omega^h, \mathbb{R}^3)$ for I_W^h, i.e. when:

$$\lim_{h \to 0} \frac{1}{h^{\gamma+2}} \left(I_W^h(u^h) - \inf I_W^h \right) = 0, \tag{10}$$

the convergences (1), (2) of Theorem 1 hold up to a subsequence, and the limit V_3 is a minimizer of the functional I_γ defined as in (8).

Moreover, for any (global) minimizer V_3 of I_γ, there exists a minimizing sequence u^h, satisfying (10) together with (1), (2), and (3) of Theorem 2.

5 General Scaling Result

In the same fashion as in the previous section, we are interested in understanding the asymptotic behavior of the minimizers of I_W^h (see (3)) as $h \to 0$, through deriving the Γ-limit of the rescaled energies $\frac{1}{h^{\delta+2}} I_W^h(u^h)$, where $(G^h)^{1/2} = A^h + h.o.t$, and A^h are given by:

$$A^h(x', x_3) = I_3 + h^\alpha S(x') + h^{\gamma/2} x_3 B(x'). \tag{11}$$

Here, we work with $\delta \geq 2$, $\delta = \min\{\delta, 2\alpha, \gamma\}$, and we obtain different results, depending on the relation between the three parameters.

First, we quote the following approximation result, which can be directly obtained from the geometric rigidity estimate found in Theorem 1.6 of [7], in view of the following bounds:

$$Var(A^h) = \left\| \nabla_{tan}(A^h\big|_{x_3=0}) \right\|_{L^\infty(\omega)} + \left\| \partial_3 A^h \right\|_{L^\infty(\Omega^h)} \leq Ch^{\alpha \wedge \frac{\gamma}{2}},$$

$$\left\| A^h \right\|_{L^\infty(\Omega^h)} + \left\| (A^h)^{-1} \right\|_{L^\infty(\Omega^h)} \leq C.$$

Theorem 3 ([15]) *Assume $I_W^h(u^h) \leq Ch^{2+\delta}$. Then, there exist matrix fields $R^h \in W^{1,2}(\omega, \mathbb{R}^{3\times3})$, such that $R^h(x') \in SO(3)$ for a.e. $x' \in \omega$, and:*

$$\frac{1}{h} \int_{\Omega^h} \left| \nabla u^h(x) - R^h(x')A^h(x) \right|^2 dx \leq Ch^{2+\min\{\delta, 2\alpha, \gamma\}},$$

$$\int_\omega \left| \nabla R^h \right|^2 \leq Ch^{\min\{\delta, 2\alpha, \gamma\}}.$$

Let us define the averaged rotations: $\tilde{R}^h = \mathbb{P}_{SO(3)} \fint_\omega R^h \, dx'$, which allow us to define: $\hat{R}^h = \mathbb{P}_{SO(3)} \fint_{\Omega^h} (\tilde{R}^h)^T \nabla u^h \, dx$, as well as: $\bar{R}^h = \tilde{R}^h \hat{R}^h$.

Different cases occur depending on the relationship between the parameters α, γ, and δ. The first thing to notice is that, in order to characterize the Γ-convergence of the rescaled energies, we need to have:

$$\frac{1}{h^{2+\delta}} W((R^h)^T \nabla u^h (A^h)^{-1}) = \frac{1}{h^{2+\delta}} W(Id + h^{1+\delta/2} P^h) = \frac{1}{2} Q_3(P^h) + error,$$

where the rescaled strains $P^h \in L^2(\Omega^1, \mathbb{R}^{3\times3})$ are given by:

$$P^h(x', x_3) = \frac{1}{h^{\delta/2+1}} \left(\left(R^h(x')\right)^T \nabla u^h(x', hx_3) \left(A^h(x', hx_3)\right)^{-1} - Id \right),$$

and satisfy that $P^h \to P$ weakly in $L^2(\Omega^1, \mathbb{R}^{3\times3})$ as $h \to 0$. From this, the first constraint that appears is that $\delta = \min\{\delta, 2\alpha, \gamma\}$.

For the limiting strain P, we obtain that:

$$P(x', x_3)_{3\times2} = P(x', 0)_{3\times2} - x_3 \begin{cases} 0 & \text{when } \delta < \gamma \\ B(x')_{3\times2} & \text{when } \delta = \gamma, \end{cases}$$

$$- x_3 \left(\nabla^2 V_3(x')\right)_{3\times2} + 2x_3 \begin{cases} 0 & \text{if } \delta < 2\alpha \\ (\nabla_{tan} Se_3)_{3\times2} & \text{if } \delta = 2\alpha \end{cases},$$

where $V_3 \in W^{2,2}(\omega, \mathbb{R})$ is given in part (2) of Theorem 4 and satisfies:

$$(-D_{3,1}, -D_{3,2})^T = (\partial_1 V_3, \partial_2 V_3)^T - \begin{cases} 0 & \text{when } \delta < 2\alpha \\ (S_{3,1}, S_{3,2})^T & \text{when } \delta = 2\alpha \end{cases},$$

where D is the $L^2(\omega, \mathbb{R}^{3\times3})$-limit of $\frac{1}{h^{\delta/2}} \left((\bar{R}^h)^T R^h - Id\right)$ as $h \to 0$, and where $P(x', 0)_{3\times2} \in L^2(\omega, \mathbb{R}^{3\times2})$ satisfies:

$$\text{sym}P(x', 0)_{2\times2} =$$

$$\begin{cases} \text{sym}\nabla w & \text{if } \delta > 2 \text{ and } (S_{2\times2} = 0 \text{ or } \alpha > 1 + \delta/2) \\ \text{sym}\nabla w - S_{2\times2} & \text{if } \delta > 2 \text{ and } \alpha = 1 + \delta/2 \\ \text{sym}\nabla w - \frac{1}{2}(D^2)_{2\times2} & \text{if } \delta = 2 \text{ and } (S_{2\times2} = 0 \text{ or } \alpha > 1 + \delta/2 = 2) \\ \text{sym}\nabla w - \frac{1}{2}(D^2)_{2\times2} - S_{2\times2} & \text{if } \delta = 2 \text{ and } \alpha = 1 + \delta/2 = 2 \end{cases},$$

with w given in part (3) of Theorem 4.

If $\delta \in (0, 2)$, the stretching term $\text{sym}P(x', 0)_{2\times2}$ is discarded, and we have that:

- When $\delta < \alpha$ or $S_{2\times2} = 0$: then $\text{sym}\nabla w = -1/2\nabla V_3 \otimes \nabla V_3$, which is equivalent to $\det\nabla^2 V_3 = 0$.
- When $\delta = \alpha$: then $\text{sym}\nabla w = -1/2\nabla V_3 \otimes \nabla V_3 + S_{2\times2}$, which is equivalent to $\det\nabla^2 V_3 = -\text{curl}^T \text{curl } S_{2\times2}$.

Theorem 4 ([2]) *Let A^h be given as in (11). Assume that a sequence of deformations $u^h \in W^{1,2}(\Omega^h, \mathbb{R}^3)$ satisfies:*

$$I_W^h(u^h) \leq Ch^{\delta+2},$$

where W fulfills (4) and (7). Then, there exist rotations $\bar{R}^h \in SO(3)$ and translations $c^h \in \mathbb{R}^3$ such that, for the normalized deformations:

$$y^h \in W^{1,2}(\Omega^1, \mathbb{R}^3), \qquad y^h(x', x_3) = (\bar{R}^h)^T u^h(x', hx_3) - c^h,$$

the following hold (up to a subsequence that we do not relabel):

(1) $y^h(x', x_3) \to x'$ in $W^{1,2}(\Omega^1, \mathbb{R}^3)$.
(2) The scaled displacements:

$$V^h(x') = \frac{1}{h^{\delta/2}} \int_{-1/2}^{1/2} y^h(x', t) - x' dt$$

converge to a vector field V of the form $V = (0, 0, V_3)^T$. This convergence is strong in $W^{1,2}(\omega, \mathbb{R}^3)$. The only non-zero out-of-plane scalar component V_3 of V belongs to $W^{2,2}(\omega, \mathbb{R})$.

(3) The scaled in-plane displacements $\frac{1}{h} V_{tan}^h$ converge (up to a subsequence), weakly in $W^{1,2}(\omega, \mathbb{R}^2)$ to an in-plane displacement field $w \in W^{1,2}(\omega, \mathbb{R}^2)$.
(4) Moreover:

$$\liminf_{h \to 0} \frac{1}{h^{\delta+2}} I_W^h(u^h) \geq \mathcal{I}(w, V_3),$$

where:

$$\mathcal{I}(w, V_3) = \frac{1}{2} \int_\omega \mathcal{Q}_2(\mathrm{sym}P(x', 0)_{2\times2}) \, dx' + \frac{1}{24} \int_\omega \mathcal{Q}_2(\mathcal{S}) \, dx', \qquad (12)$$

and $\mathcal{S}$ is given by:

$$\mathcal{S} = \nabla^2 V_3 - \begin{cases} 0 & \text{when } \delta < \min\{\gamma, 2\alpha\} \\ -B_{2\times2} & \text{when } \delta = \gamma < 2\alpha \\ 2\nabla_{tan}(S_{3,1}, S_{3,2}) & \text{if } \delta = 2\alpha < \gamma \\ 2\nabla_{tan}(S_{3,1}, S_{3,2}) - B_{2\times2} & \text{if } \delta = 2\alpha = \gamma \end{cases}$$

and the quadratic non-degenerate form $\mathcal{Q}_2$ is given by (9).

Theorem 5 ([2]) *Assume (11) and that W satisfies (4) and (7). If ω is simply connected then, for every $V_3 \in W^{2,2}(\omega, \mathbb{R})$ and $w \in W^{1,2}(\omega, \mathbb{R}^3)$, there exists a sequence of deformations $u^h \in W^{1,2}(\Omega^h, \mathbb{R}^3)$ such that the following hold:*

(1) *The sequence* $y^h(x', x_3) = u^h(x', hx_3)$ *converges in* $W^{1,2}(\Omega^1, \mathbb{R}^3)$ *to* x'.

(2) *The displacements* $V^h(x') = \dfrac{1}{h^{\gamma/2}} \displaystyle\int_{-h/2}^{h/2} \left(u^h(x', t) - x' \right) dt$ *converge in*

$W^{1,2}(\omega, \mathbb{R}^3)$ *to* $(0, 0, V_3)^T$.

(3) $\dfrac{1}{h} V_{tan}^h$ *converge in* $W^{1,2}(\omega, \mathbb{R}^2)$ *to* w.

(4) *Recalling (12) one has:*

$$\lim_{h \to 0} \frac{1}{h^{\delta+2}} I_W^h(u^h) = \mathcal{I}(w, V_3).$$

6   Conclusions

In this note, recent developments on the analysis and derivation of thin film models for prestrained structures were showcased. We also announced the result of [3] in which the dimensionally reduced model for the weak prestrain in (6) is derived, completing thus the prior analysis of [17]; as well as the results of [2], in which the more general prestrain (11) is considered.

As part of the work in [2], we were able to identify the Γ-limits as the sum of norms of $\mathrm{curl}^T \mathrm{curl}\, S_{2\times2}$ and/or $\mathrm{curl}\, B_{2\times2}$, corresponding to the stretching and bending terms. For example, in the case when $\delta > 2$ and ($\delta = 2\alpha - 2$ and $S_{2\times2} \neq 0$), by Theorems 4 and 5, the Γ-convergence is done with respect to w and V_3, and the Γ-limit is given by:

$$\frac{1}{2} \int_\omega \mathcal{Q}_2(\mathrm{sym}\nabla w - S_{2\times2})\, dx'$$

$$+ \frac{1}{24} \int_\omega \mathcal{Q}_2\left(\nabla^2 V_3 - \begin{cases} 0 & \text{when } \delta < \gamma \\ -B_{2\times2} & \text{when } \delta = \gamma \end{cases}\right) dx'.$$

However, we may also bring down the Γ-convergence to be with respect to V_3 only; where the first term above (stretching term) is then replaced by the following constant quantity:

$$\frac{1}{2} \mathrm{dist}_{\mathcal{Q}_2}^2 \left(S_{2\times2}, \left\{ \mathrm{sym}\nabla w;\ w \in W^{1,2}(\omega, \mathbb{R}^2) \right\} \right).$$

This quantity may, in fact, be viewed as the following norm:

$$\left\| \mathrm{curl}^T \mathrm{curl}\, S_{2\times2} \right\|_{H^{-2}(\omega)}^2.$$

Similarly, when $\delta = \gamma$, the second (bending) term may be replaced by the norm:

$$\left\|\operatorname{curl} B_{2 \times 2}\right\|_{H^{-1}(\omega)}^{2}.$$

We thus see that:

$$\liminf_{h \to 0} \frac{1}{h^{\delta+2}} I_W^h(u^h) \sim \left\|\operatorname{curl}^T \operatorname{curl} S_{2 \times 2}\right\|_{H^{-2}(\omega)}^{2} + \left\|\operatorname{curl} B_{2 \times 2}\right\|_{H^{-1}(\omega)}^{2},$$

where the symbol $\sim$ indicates equality up to a certain universal constant.

References

1. Kaushik Bhattacharya, Marta Lewicka, and Mathias Schäffner. Plates with incompatible prestrain. *Archive for Rational Mechanics and Analysis*, 221(1):143–181, Jul 2016.
2. Silvia Jiménez Bolaños and Marta Lewicka. Scalings and limiting models for prestrain metrics G^h that are a perturbation of the flat I_3 metric. *Preprint*, 2020.
3. Silvia Jiménez Bolaños and Anna Zemlyanova. Relative bending energy for weakly prestrained shells. Rocky Mountains J. Math. *Preprint*, 2019.
4. Andrea Braides. Γ-convergence for beginners. In *Oxford Lecture Series in Mathematics and its Applications*. Oxford University Press, 2002.
5. Ennio De Giorgi. G-operators and Γ-convergence. *Proceedings of the International Congress of Mathematicians*, 1,2:1175–1191, 1984.
6. Efi Efrati, Eran Sharon, and Raz Kupferman. Elastic theory of unconstrained non-Euclidean plates. *Journal of the Mechanics and Physics of Solids*, 57(4):762 – 775, 2009.
7. Gero Friesecke, Richard D. James, and Stefan Müller. A theorem on geometric rigidity and the derivation of nonlinear plate theory from three-dimensional elasticity. *Communications on Pure and Applied Mathematics*, 55(11):1461–1506, 2002.
8. Gero Friesecke, Richard D. James, and Stefan Müller. A hierarchy of plate models derived from nonlinear elasticity by Gamma-convergence. *Archive for Rational Mechanics and Analysis*, 180(2):183–236, May 2006.
9. Gareth W. Jones and L. Mahadevan. Optimal control of plates using incompatible strains. *Nonlinearity*, 28(9):3153–3174, aug 2015.
10. Yael Klein, Efi Efrati, and Eran Sharon. Shaping of elastic sheets by prescription of non-Euclidean metrics. *Science*, 315(5815):1116–1120, 2007.
11. Raz Kupferman and Cy Maor. A Riemannian approach to the membrane limit in non-Euclidean elasticity. *Communications in Contemporary Mathematics*, 16, 10 2014.
12. Raz Kupferman and Jake P. Solomon. A Riemannian approach to reduced plate, shell, and rod theories. *Journal of Functional Analysis*, 266(5):2989 – 3039, 2014.
13. Marta Lewicka. Quantitative immersability of Riemann metrics and the infinite hierarchy of prestrained shell models. *eprint arXiv:1812.09850*.
14. Marta Lewicka and Danka Lucic. Dimension reduction for thin films with transversally varying prestrain: the oscillatory and the non-oscillatory case. *Communications on Pure and Applied Mathematics*, 2019.
15. Marta Lewicka, L. Mahadevan, and Mohammad Reza Pakzad. The Föppl-von Kármán equations for plates with incompatible strains. *Proceedings of the Royal Society A: Mathematical, Physical and Engineering Sciences*, 467(2126):402–426, 2011.
16. Marta Lewicka, L. Mahadevan, and Mohammad Reza Pakzad. Models for elastic shells with incompatible strains. *Proceedings of the Royal Society A: Mathematical, Physical and Engineering Sciences*, 470(2165):20130604, 2014.

17. Marta Lewicka, Pablo Ochoa, and Mohammad Reza Pakzad. Variational models for prestrained plates with Monge-Ampère constraint. *Differential Integral Equations*, 28(9/10):861–898, 09 2015.
18. Marta Lewicka, Annie Raoult, and Diego Ricciotti. Plates with incompatible prestrain of high order. *Annales de l'Institut Henri Poincaré C, Analyse non linéaire*, 34(7):1883 – 1912, 2017.
19. Marta Lewicka and Mohammad Reza Pakzad. Scaling laws for non-Euclidean plates and the $W^{2,2}$ isometric immersions of Riemannian metrics. *ESAIM: Control, Optimisation and Calculus of Variations*, 17(4):1158–1173, 2011.
20. Emily Rodriguez, Anne Hoger, and Andrew D. McCulloch. Stress-dependent finite growth in soft elastic tissues. *Journal of biomechanics*, 27 4:455–67, 1994.
21. Bernd Schmidt. Plate theory for stressed heterogeneous multilayers of finite bending energy. *Journal de Mathématiques Pures et Appliquées, v.88, 107–122 (2007)*, 88, 07 2007.

Machine Learning in Crowd Flow Exit Data (Survey)

F. Patricia Medina

1 Introduction

This paper is based on the work published by the author with her co-authors in [12] where the results of exploratory analysis of black box simulation data modeling crowds exiting different configurations of a one story building are described. Here we focused in the machine learning approach given in section 8 in [12], where we proposed a methodology to generate features inspired by a similar technique used in computer vision.

The main problem is to predict the exit times of 100 agents in a given room configuration using simulated data. Given the small size of the simulated data, we present a technique for generating more features and data points (the "sliding window" technique described in Sect. 2.1). We also explore the difference between unsupervised linear dimension reduction using Principal Component Analysis and unsupervised non-linear dimension reduction using auto-encoders.

Crowd dynamics, or pedestrian dynamics, is an area of research that covers a wide range of approaches related to understanding and modeling crowd behavior. The most practical motivation for understanding crowd dynamics is to improve human safety in real-world crowd situations.

We mention three main applications motivating the study of crowd dynamics. First, gaining a better understanding and modeling of crowd flow can improve evacuation strategies from buildings, stadiums, airline terminals and other public spaces. For example, in 2017 there were multiple crowd stampedes resulting in injury and death (e.g. [3, 17]). Second, research in crowd flow in real world crowd

F. P. Medina (✉)
Yeshiva University, New York, NY, USA
e-mail: patricia.medina@yu.edu

© The Author(s) and the Association for Women in Mathematics 2020
B. Acu et al. (eds.), *Advances in Mathematical Sciences*, Association for
Women in Mathematics Series 21, https://doi.org/10.1007/978-3-030-42687-3_21

flow data, can improve the tracking of people's movements. A third application is producing computer-generated crowds in video games.

Most of the time, crowd behaviors are based in agent-based modeling [6]. In an agent-based model of crowd dynamics, individual people or groups of people are represented by agents, which are given a set of rules and properties, possibly all the same or differing by agent, for how the agent should interact with its environment and the other agents. Methods for formulating the rules to steer pedestrians include: ego-centric fields [16] and social force models [14].

The development of software to execute a given agent-based modeling system is another problem in crowd dynamics. One example is the program Menge [8], which was created for simulating pedestrian movement in a crowd. In our work, we used data generated by the recently developed SteerSuite platform, designed to be an open framework for developing, evaluating and sharing steering algorithms [14–16, 25].

Another agent-based modeling approach is based on fluid dynamics by treating the crowd as a continuous flow. For example, models based on optimal transport [20], and mean field games [18], as well as in modeling crowd emotions [7, 28].

One of the problems in computer vision focus on the detection of abnormal behavior and extraction of useful features from video streams [23, 26]. The most frequently used features for crowd abnormal behavior include global flow-based features and local spatiotemporal based features (see [23, 26, 27]). The simulated data we used for our experiments includes the position of the agents and a list of features for each agent. We focus in the inference of crowd flow exit-dynamics.

Each building configuration had three rooms on the north side, two rooms on the east side, and two rooms on the south side. For a detail description of the room configurations, see [12]. We show one of the building configurations (scenario 2) in Fig. 1.

The data for each run consisted of 23-dimensional vectors for each agent at each time step. The 23 features were: agent id, time, x and y coordinates for position and velocity, goal and final target (the exit), the radius, acceleration, personal space threshold, agent repulsion importance, query radius, body force, agent body force, sliding friction force, maximum speed, two other features for nearby agents and two wall parameters (see Table 1 for feature description as in SteerSuite).

2 Methods for Estimating Exit Times

We follow closely the exposition presented in [12]. The idea is to show a general machine learning framework involving feature generation, dimension reduction and feeding a neural network for supervised multi-output regression obtained from the given unsupervised dimensionality reduction method. The final output of the latter neural network are the predicted exit times.

Fig. 1 Building configuration 2. Figure extracted from [12], p. 244

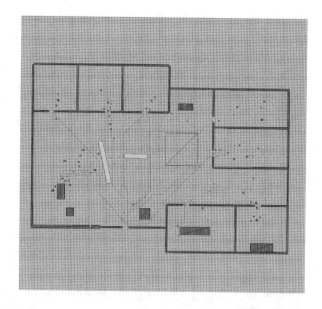

Accordingly, the outline of our main experiments is as follows (see Fig. 2):

1. Perform feature engineering to generate a new data set using a subset of the original data.
2. Perform dimensionality reduction using either PCA (for a linear projection) or a 3-layer auto-encoder (for a non-linear projection).

    (a) If using PCA, then use the projected features as the predictors for our supervised learning.
    (b) If using an auto-encoder, then use the hidden layer as the predictors for our supervised learning.

3. Provide a training sample of our projected data to a 2-layer feed-forward neural network (from Sect. 3.1) to make predictions of the agents' exit times.

In Sect. 2.1, we describe the feature engineering step consisting on a "sliding window" approach. In Sect. 2.2, we give the general idea of how to used auto-encoders to perform dimensionality reduction of the data.

We emphasize that option 2a (PCA) tries to find a linear map that maximizes certain cost function (the solution space for the associated optimization problem is convex), while option 2b (deep auto-encoders) tries to find a nonlinear map and belongs to the class of non-convex techniques [19].

Table 1 Features and their description for agents and their trajectories. Table extracted from [12], p. 243

Trajectory features	Description of features
ID	Agent's ID
Time	Timestamp (seconds)
Position x	x coordinate of current position
Position y	y coordinate of current position
Velocity x	x coordinate of current velocity
Velocity y	y coordinate of current position
Target x	x coordinate of final target
Target y	y coordinate of final target
Agent features	Description of features
Radius	Radius of the agent
Acceleration	The inertia related to mass
Personal space threshold	The distance between a wall and an agent within which a repulsive force begins to act
Agent repulsion importance	The factor which decides how much the penetration depth affects both the repulsive force and frictional force between two agents
Query radius	Defines the area, in which all objects act force on the subject agent.
Body force	Factor of repulsive force between an agent and a wall
Agent body force	Factor of repulsive force between two agents
Sliding friction force	Factor of frictional force
agent_b	The proximity force between two agents is agent_a $* \mathrm{EXP}(-d*$ agent_b), where d is the closest distance between two agents' outlines
agent_a	agent_b $* \mathrm{EXP}(-d*$ agent_a)
wall_b	The proximity force between an agent and a wall defined by wall_a $* \mathrm{EXP}(-d*$ wall_b), where d is the closest distance between two agent's outline and a wall
wall_a	wall_b $* \mathrm{EXP}(-d*$ wall_a)
Maximum speed	The maximum speed of an agent

2.1 Feature Engineering: Sliding Window Approach

Typical use of feed-forward networks which employ a sliding window approach are market predictions, meteorological and network traffic forecasting [5, 9–11]. In the context of computer vision, a "sliding window" is rectangular region of fixed width and height that "slides" across an image, [4]. In our context, the sliding window would be moving across time slices.

Our "sliding window" approach is used to generate a new data set using the original data set. In order to simplify the explanation of a feature generation technique, we divide our explanation in three main steps. First, we create a "window" of a given size for a fixed agent by combining original agent features at different slices at times t_0, t_1, t_2, ..., as shown in Fig. 3. Second, we further create

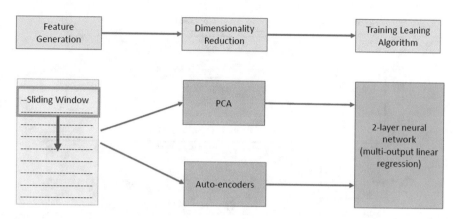

Fig. 2 Diagram describing the three stage process to estimate the agents exit times

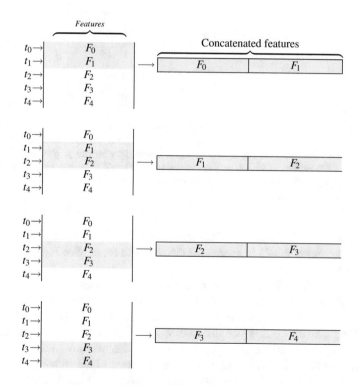

Fig. 3 First stage of the "sliding window technique" (agent and run are fixed). The figure is an illustration of the "sliding window" technique for a fixed agent and run. The number of generated new data points is proportional to the number of windows, and the dimension is proportional to the size of the window. In this case, we have 4 windows of size 2 each. Figure extracted from [12], p. 264

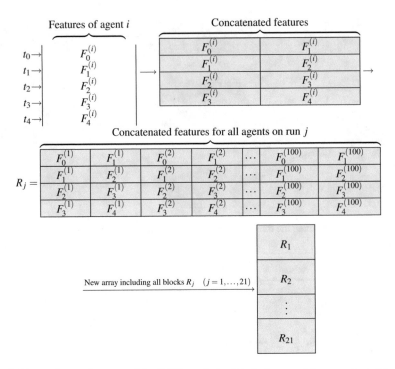

Fig. 4 First and second stages of the "sliding window" technique used to generate a new data set. Considering the features for agent i, we obtain a new array (as described in Fig. 3). Then concatenate the features for all 100 agents to obtain a new array for run j, R_j. Note that in this case, we have 4 windows of size 2, so the dimension of the new data is $p = d \times s_w \times N = 21 \times 2 \times 100 = 4200$. The number of points is $n_w \times$ number of runs $= 4 \times 21 = 84$. Figure extracted from [12], p. 265

another data frame by putting all 100 agents data frames together. Finally, we use the available *r*uns to create instances for the new data frame. See Fig. 4 to visualize the complete technique.

For example, for a fixed agent, consider the set of features F_0 at time t_0, and the set of feature F_1 at time t_1. Note that F_0 and F_1 are rows of the same original data frame. Since we started by just considering two *r*ows of the original data frame, we are working with a "window" of size two. We now repeat the previous process but now with the set of features $\{F_1, F_2\}$. At this point, the "window" just *s*lid. We stop at time t_4. See Fig. 3 for an illustration of this first stage of the technique.

Note that, so far, we have generated a new data frame for fixed agent i. Now, for a fixed run j, generate new data frames by repeating the same process just described for all 100 agents, and create a new data frame R_j as shown in Fig. 4. Finally, we create a new data set by putting all arrays R_j together ($j = 1, \ldots, 21$).

Observe that our new data set has dimension $p = d \times s_w \times N$, where d is the dimension of the original data, s_w is the size of the window, and N is the number of agents. Moreover, the number of instances is now $n_w \times$ Total number of runs, where n_w is the number of windows.

2.2 Auto-encoders for Dimensionality Reduction

Auto-encoders are feed-forward neural networks with an odd number of hidden layers and shared weights between the left and right layers. The input data X (input layer) and the output data $\hat{X}$ (output layer) have $d^{(0)}$ nodes. For a more detailed description on neural networks, see [2, 13] and [22]. More precisely, auto-encoders learn a non-linear map from the input to itself through a pair of encoding and decoding phases [29, 30]

$$\hat{X} = D(E(X)), \tag{1}$$

where E maps the input layer X to the "most" hidden layer (*encodes* the input data) in a non-linear fashion, D is a non-linear map from the "most" hidden layer to the output layer (decodes the "most" hidden layer), and $\hat{X}$ is the recovered version of the input data. An auto-encoder therefore solves the optimization problem:

$$\underset{E,\,D}{\arg\min} \, \|X - D(E(X))\|_2^2, \tag{2}$$

We are motivated to include multi-layer auto-encoders in our exploratory analysis in crowd flow data, since they have demonstrated to be effective for discovering non-linear features across problem domains.

We use an auto-encoder with three inner layers to perform dimension reduction in our experiments. We aim to find functions E and D which are solutions to the corresponding optimization problem (2), with first hidden layer (leftmost hidden layer) $S_l \in \mathbb{R}^{d^{(1)}}$, "deepest hidden layer" $Z \in \mathbb{R}^{d^{(3)}}$ and "rightmost inner layer" $S_r \in \mathbb{R}^{d^{(3)}}$.

In the next section, we perform dimensionality reduction by applying the encoder E to the input layer X. We then feed the neural network with $Z = E(X)$.

3 Estimation of Exit Times

We focus on the class of *feed forward* neural networks, which means that there are no backward pointing arrows and no jumps to other layers. In this paper, we consider a multi-output regression in the final stage of the process of estimating stopping times $T_1, \ldots, T_{100}$.

Recall that the input data is the new data frame produced by our "sliding window" technique approach for feature generation described on Sect. 2.1. Next, we reduce the dimension of the data by using PCA (linear and unsupervised) or an auto-encoder (nonlinear and unsupervised). We feed the multi-output neural network with new predictors obtained from the dimensionality reduction stage.

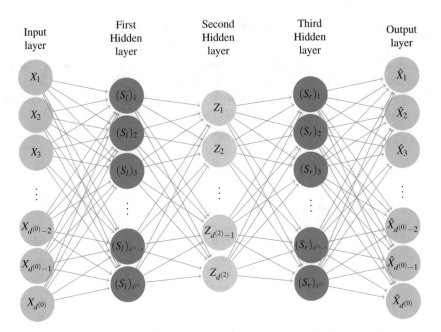

Fig. 5 3-Layer auto-encoder diagram. The input layer has dimension $d^{(0)}$, the three inner layers S_l, Z, S_r have dimensions $d^{(1)}$, $d^{(2)}$ and $d^{(3)}$, respectively. The dimension of the outer layer $\hat{X}$ has dimension $d^{(0)}$ since this is an auto-encoder. Figure extracted from [12], p. 269

More precisely, the input for the neural network has dimension $d^{(0)} = 20$ as the explained variance for PCA is more than 95% when using 20 components, and the output has dimension $d^{(L)} = 100$ which are exit times for each agent. We have chosen the dimension of the inner layer of the auto-encoder based on the number of components of the PCA for the sake of comparison (Fig. 5).

Since this is a multi-output regression, we computed the R^2-scores or coefficient of determination:

$$R^2 = 1 - \frac{\|h(X) - T\|_2^2}{\|T - \bar{T}\|_2^2} \tag{3}$$

where $T = (T_1, T_2, \ldots, T_N)$, $\bar{T}$ is the N dimensional vector with all entries $\frac{1}{N} \sum_{i=1}^{N} T_i$ and h is the function learned by the machine learning algorithm. In our case, $N = 100$.

An R^2 score near 1, means that the model is predicting the data (stopping time) very well, a score close to zero means that the model is predicting the *mean* of the stopping times, and the score can be arbitrarily negative indicating bad predictions by the model.

In all experiments we split the initial input data. We train the learning algorithm with 80% of the data and test it with the remaining 20% of the data (we do appropriate normalization of the data, z-scores). Also, we have considered a multi-linear regression for the learning algorithm to produce stopping times for $N = 100$ agents at once.

3.1 Experiments: PCA vs Auto-encoder with Neural Networks

The main goal of this section is to study the differences between *linear* dimensionality reduction using Principal Component Analysis (PCA) and using an auto-encoder to do non-linear dimensionality reduction. In both cases, after the new lower-dimensional features are produced, a feed-forward neural network is used to make predictions of the agent exit times. Our proposed methodology promises a number of advantages. First, since the dimensionality reduction is performed without access to the measured exit times of the agents, at least that part of our procedure is safe from over-fitting. Second, as the problem of interest is quite complex, there is likely a non-linear relationship between our measured features (e.g., the initial position of the agents) and their final exit time.

We present two main sets of experiments. The first set involves feature engineering using PCA to produce new features to feed a forward neural network. For the second set, we train a 3-layer auto-encoder instead of using PCA.

In all experiments, the neural network architecture consists of an input layer made of 20 inputs $(Z_1, \ldots, Z_{20})$ obtained after dimensionality reduction, two hidden layers (first hidden layer has 50 units, second hidden layer has 70 units). The output layer consists of 100 units representing exit time for each of the 100 agents.

We had little data for training, so we use a "sliding window" technique (see Fig. 3 for illustration) in order to produce more data points and improve the performance of the neural network. Overlapping windows and non-overlapping windows are considered. Note that the input data that gives the least number of data points corresponds to window size 1, which includes the original data together with all runs for a fixed scenario at one single time step.

We have used TensorFlow (an open source software library for numerical computation using data flow graphs, see [1]) to build the auto-encoder. The rest of the scripts are in Python using Sci-kit Learn [24] and Pandas [21] libraries.

Recall that our main steps for the proposed algorithm are as follows:

1. Generate new data X from the original data by using the sliding window approach (as described in Sect. 2.1).

Table 2 R^2 scores using
PCA for dimensionality
reduction with overlapping
windows for scenario 19.
Table extracted from [12], p.
273

Window size	# windows	R^2-score
2	1	−8.45903
	2	−0.34136
	3	−0.19203
	4	0.99402
3	1	−7.702292
	2	−2.15774
	3	−0.54476
	4	0.99563

Table 3 R^2 scores for
encoder (from 3-layer
auto-encoder) with
overlapping windows for
scenario 19. Table extracted
from [12], p. 274

Window size	# windows	R^2-score
2	1	−0.54033
	2	−0.03344
	3	−0.19686
	4	0.04300
3	1	−0.86243
	2	0.05440
	3	0.23429
	4	0.20910

2. Reduce the dimension of the new generated data by either using PCA or using
   an auto-encoder:

   (a) Perform PCA on the new input data X to obtain a new input layer Z.
   (b) Construct a auto-encoder with three inner layers and extract the encoder E.
       Apply the encoder E to the initial data X to reduce the dimension to 20.

3. Finally, estimate exit times $\{T_i\}$ with a neural network:

   (a) Feed the original 2-layer neural network with $Z = XM \in \mathbb{R}^{20}$, where M is
       the matrix of the principal components associated to the covariance matrix
       of the input data. See R^2 score in Table 2.
   (b) Feed the original 2-layer neural network with $Z = E(X) \in \mathbb{R}^{20}$. See R^2
       score in Table 3.

We reduce the dimension of the space performing PCA. The cumulative variance
is computed to get an estimate for the number of components. We end up choosing
20 components as the cumulative variance is more than 95%.

After augmenting the dimension of the input layer, we perform PCA to reduce the
dimension to 20, then we train a neural network with two hidden layers ($L = 4$). The
input layer has dimension $d^{(0)} = 20$ and the output, $d^{(3)} = 100$, so the output layer
contains T the stopping times for each agent. The hidden layers have dimensions
$d^{(1)} = 50$ and $d^{(2)} = 70$. The time step used in all experiments is 0.05 s.

A big improvements is observed whenever we use the sliding window technique
as seen in Tables 2, 3 when the window size is equal or greater than 2. In the sample

tables, we can see that the 3-layer auto-encoder performs better than PCA whenever a single window, two windows and three widows of size three are used. We have positive scores for windows of size 3 when the number of windows is 2 or greater if using an auto-encoder (Table 3), and when using PCA with overlapping windows in Table 2 scores are all negative except for the last one (0.99563).

We emphasize to the reader that there are two large jumps in accuracy seen in experiments involving PCA. We attribute this large jumps to the fact that studying the agent for longer time provides substantially more information that leads to more accurate predictions. The non-linear properties of auto-encoders allow for more accurate predictions from a small number of features. However, we don't have enough instances to train the auto-encoder and obtain a better performance.

Notice that in the "sliding window" technique a consecutive combinations of time steps is used to generate the new data set (input for the dimension reduction method). Instead of having information of the time step at a single time, we are combining either two or more time steps at once. Also, observe that the sliding window technique combines all 100 agents and runs in a single data set.

4 Summary and Future Research Directions

A "sliding window" approach was used for feature generation and two types of dimensionality reduction were used: PCA and encoder-decoder. The results were evaluated using the R^2 score. Exit times were successfully predicted for certain combinations of the methods. The "sliding window" technique resulted in a big improvement in exit time prediction and use of the encoder resulted in a small improvement over PCA dimensionality reduction.

Future work includes analyzing real-world tracking data from crowds at public events. Also, performing "sliding window" technique by including early time steps and compare the exit time prediction with a set of windows using later time steps, or by considering a fixed agent and included the features of the neighbor agents. In other words, a comparison study of predictions using different "windows".

Also, we can perform similar studies in real exit data and configurations such as the data set from the Asiana flight crash in San Francisco. We would like to give credit to Andrea Bertozzi for suggesting experimentation on the Asiana flight data as part of our future work in this topic.

Acknowledgements This research started at the *Women in Data Science and Mathematics Research Collaboration Workshop (WiSDM)*, July 17-21, 2017, at the *Institute for Computational and Experimental Research in Mathematics (ICERM)*. The workshop was partially supported by grant number NSF-HRD 1500481-AWM ADVANCE and co-sponsored by Brown's Data Science Initiative. Many thanks to our project group leader, Dr. Linda Ness, for advising the submission of this paper. We gratefully acknowledge Mubbasir Kapadia for permitting use of the Steer-Suite platform [25] to provide the data for our research effort and Weining Lu for designing and executing the simulation scenarios and for providing and documenting the data. This Machine Learning approach to this topic would not have been possible without the help of Dr. Randy C. Paffenroth (Worcester Polytechnic Institute) who brought his valuable expertise to the project.

References

1. M. ABADI, A. AGARWAL, P. BARHAM, E. BREVDO, Z. CHEN, C. CITRO, G. S. CORRADO, A. DAVIS, J. DEAN, M. DEVIN, S. GHEMAWAT, I. GOODFELLOW, A. HARP, G. IRVING, M. ISARD, Y. JIA, R. JOZEFOWICZ, L. KAISER, M. KUDLUR, J. LEVENBERG, D. MANÉ, R. MONGA, S. MOORE, D. MURRAY, C. OLAH, M. SCHUSTER, J. SHLENS, B. STEINER, I. SUTSKEVER, K. TALWAR, P. TUCKER, V. VANHOUCKE, V. VASUDEVAN, F. VIÉGAS, O. VINYALS, P. WARDEN, M. WATTENBERG, M. WICKE, Y. YU, AND X. ZHENG, *Tensor-Flow: Large-scale machine learning on heterogeneous systems*, 2015. Software available from tensorflow.org.
2. Y. S. ABU-MOSTAFA, M. MAGDON-ISMAIL, AND H.-T. LIN, *Learning From Data (e-chapter)*, AMLBook, 2012.
3. A. ALAMI, *Morocco food stampede leaves 15 dead and a country shaken*, The New York Times, (2017). Available: https://www.nytimes.com/2017/11/19/world/africa/morocco-stampede.html. Last accessed: 1 Jan. 2018.
4. Y. AMIT AND P. F. FELZENSZWALB, *Object detection*, in Computer Vision, A Reference Guide, 2014, pp. 537–542.
5. S. BENGIO, F. FESSANT, AND D. COLLOBERT, *A connectionist system for medium-term horizon time series prediction*, in IN PROC. INTL. WORKSHOP APPLICATION NEURAL NETWORKS TO TELECOMS, 1995, pp. 308–315.
6. E. BONABEAU, *Agent-based modeling: Methods and techniques for simulating human systems*, PNAS, 99 (suppl3) (2002), pp. 7280–7287.
7. T. BOSSE, R. DUELL, Z. A. MEMON, J. TREUR, AND C. N. VAN DER WAL, *Multi-agent model for mutual absorption of emotions*, ECMS, 2009 (2009), pp. 212–218.
8. S. CURTIS, A. BEST, AND D. MANOCHA, *Menge: A modular framework for simulating crowd movement*, Collective Dynamics, 1 (2016), pp. 1–40.
9. G. DORFFNER, *Neural networks for time series processing*, Neural Network World, 6 (1996), pp. 447–468.
10. T. EDWARDS, D. S. W. TANSLEY, R. J. FRANK, N. DAVEY, AND N. T. (NORTEL LIMITED, *Traffic trends analysis using neural networks*, in Proceedings of the International Workshop on Applications of Neural Networks to Telecommuncations, 1997, pp. 157–164.
11. R. J. FRANK, N. DAVEY, AND S. P. HUNT, *Time series prediction and neural networks*, J. Intell. Robotics Syst., 31 (2001), pp. 91–103.
12. A. GRIM, B. ISKRA, N. JU, A. KRYSHCHENKO, F. P. MEDINA, L. NESS, M. NGAMINI, M. OWEN, R. PAFFENROTH, AND S. TANG, *Analysis of simulated crowd flow exit data: Visualization, panic detection and exit time convergence, attribution and estimation*, in Research in Data Science, Associations for Women in Mathematics Series, E. Gasparovic and C. Domeniconi, eds., Springer, Switzerland, 2019, pp. 239–281.
13. T. HASTIE, R. TIBSHIRANI, AND J. FRIEDMAN, *The elements of statistical learning: data mining, inference and prediction*, Springer, 2 ed., 2009.
14. D. HELBING AND P. MOLNÁR, *Social force model for pedestrian dynamics*, Phys. Rev. E, 51 (1995), pp. 4282–4286.
15. M. KAPADIA, N. PELECHANO, J. ALLBECK, AND N. BADLER, *Virtual crowds: Steps toward behavioral realism*, Synthesis Lectures on Visual Computing, 7 (2015), pp. 1–270.
16. M. KAPADIA, S. SINGH, W. HEWLETT, AND P. FALOUTSOS, *Egocentric affordance fields in pedestrian steering*, in Symposium on Interactive 3D graphics and games, I3D, ACM, 2009, pp. 215–223.
17. H. KUMAR, *Stampede at Mumbai railway station kills at least 22*, The New York Times, (2017). Available: https://www.nytimes.com/2017/09/29/world/asia/mumbai-railway-stampede-elphinstone.html. Last accessed: 1 Jan. 2018.
18. A. LACHAPELLE AND M.-T. WOLFRAM, *On a mean field game approach modeling congestion and aversion in pedestrian crowds*, Transportation research part B: methodological, 45 (2011), pp. 1572–1589.

19. E. P. LAURENS VAN DER MAATEN AND J. VAN DEN HERIK, *Dimensionality reduction: A comparative review*, tech. report, TiCC, Tilburg University, 01 2009.
20. B. MAURY, A. ROUDNEFF-CHUPIN, F. SANTAMBROGIO, AND J. VENEL, *Handling congestion in crowd motion modeling*, Net. Het. Media, 6 (2011), pp. 485–519.
21. W. MCKINNEY, *Data structures for statistical computing in python*, in Proceedings of the 9th Python in Science Conference, S. van der Walt and J. Millman, eds., 2010, pp. 51 – 56.
22. T. M. MITCHELL, *Machine Learning*, McGraw-Hill, New York, 1997.
23. C. L. MUMFORD, *Computational intelligence: collaboration, fusion and emergence*, vol. 1, Springer Science & Business Media, 2009.
24. F. PEDREGOSA, G. VAROQUAUX, A. GRAMFORT, V. MICHEL, B. THIRION, O. GRISEL, M. BLONDEL, P. PRETTENHOFER, R. WEISS, V. DUBOURG, J. VANDERPLAS, A. PASSOS, D. COURNAPEAU, M. BRUCHER, M. PERROT, AND E. DUCHESNAY, *Scikit-learn: Machine learning in Python*, Journal of Machine Learning Research, 12 (2011), pp. 2825–2830.
25. S. SINGH, M. KAPADIA, P. FALOUTSOS, AND G. REINMAN, *An open framework for developing, evaluating, and sharing steering algorithms*, in Proceedings of the 2nd International Workshop on Motion in Games, MIG '09, Berlin, Heidelberg, 2009, Springer-Verlag, pp. 158–169.
26. N. SJARIF, S. SHAMSUDDIN, AND S. HASHIM, *Detection of abnormal behaviors in crowd scene: a review*, Int. J. Advance. Soft Comput. Appl, 4 (2012), pp. 1–33.
27. H. SWATHI, G. SHIVAKUMAR, AND H. MOHANA, *Crowd behavior analysis: A survey*, in Recent Advances in Electronics and Communication Technology (ICRAECT), 2017 International Conference on, IEEE, 2017, pp. 169–178.
28. L. WANG, M. B. SHORT, AND A. L. BERTOZZI, *Efficient numerical methods for multiscale crowd dynamics with emotional contagion*, Mathematical Models and Methods in Applied Sciences, 27 (2017), pp. 205–230.
29. D. YU AND L. DENG, *Deep learning and its applications to signal and information processing*, IEEE Signal Processing Magazine, (2011).
30. C. ZHOU AND R. C. PAFFENROTH, *Anomaly detection with robust deep autoencoders*, in Proceedings of the 23rd ACM SIGKDD International Conference on Knowledge Discovery and Data Mining, KDD '17, New York, NY, USA, 2017, ACM, pp. 665–674.

The Matter of Shape: A computational Approach to Making in Architectural Heritage (Survey)

Mine Özkar

1 Introduction

One aspect of the support that computational methods offer to heritage-related disciplines is developing means to understand the role of making in how shapes come about and to represent this knowledge and the material properties in design models. This involves incorporating shapes as things [1] and deciphering design knowledge from heritage beyond just the visual properties.

The issue of how shapes are represented in design computation is multifaceted. Our research looks at computational shape representations in art and architecture history in the particular case of geometric patterns. It links to computer vision studies in unbiased pattern detection [2] and seeking new representations of shapes for use in design software [3]. It also links to devising computational methods to better understand and articulate from a pedagogical point of view the setting up of relations between shapes and their parts in design [4]. The term shape is used here in reference to shape grammars [5] which is a formalism for representing visual thinking in design. A visual design grammar can be a full or partial set of visual rules, depending on the purpose. The significance of visual rules is that they allow ambiguities which conventional symbolic computing discards. At another level, the theory offers a philosophical worldview. In this philosophy, practical, temporary, and contextual definitions trump those that are set once and for all.

That a whole is more than the sum of its parts recently resonated in the post-disaster insufficiency of a digital exact replica of the Cathedral of Notre Dame, however elaborate the gathered data. The late Andrew Tallon had produced an impressively large data point cloud for the cathedral to be used in recreating a precise 3D image of it [6]. Contrary to the make believe view of the world that

M. Özkar (✉)
Istanbul Technical University, Faculty of Architecture, Istanbul, Turkey

© The Author(s) and the Association for Women in Mathematics 2020
B. Acu et al. (eds.), *Advances in Mathematical Sciences*, Association for
Women in Mathematics Series 21, https://doi.org/10.1007/978-3-030-42687-3_22

is now ubiquitous with AI, this data alone is never enough to recreate the building physically. The multiplicitous relations between parts and the processes that bring these parts about need to be there as well. How an object of cultural heritage is realized is an important part of the data required for its representation.

Our research focuses on ornaments from thirteenth century Anatolian architecture (see Fig. 1). These patterns, even though they display distinct features based on period and location, are generally known as Islamic and span a time period between the ninth and eighteenth centuries, and a geography stretching from the western parts of Central Asia to the Iberian peninsula. There has been much mathematical interest in these patterns and they have been explored through many pattern books. The more recent generative models and programs continue this exploration of the geometry. There is a fascination with the variety that comes as a result of an underlying structure constructed from a system of interlocking circles and polygons. This variety is seemingly suitable for parametric modeling. However pattern variety depends on the visual and material redefining of parts and wholes rather than the changing values of a given set of parameters. Construction of the design is a moment for creating variation. At a first glance, patterns may look similar, but their underlying design systems may be very different altogether (see Fig. 2). With the premise of not only replicating or restoring these, but also understanding and passing down the design knowledge involved, our earlier work is on the geometric construction using visual rules [7].

2 Materiality of Geometric Patterns

Construction in architecture is used in multiple meanings: one is the construction of the geometry of the design, and another is the physical construction of the design with materials. Specifically in the case of the patterns, the first is achieved by interlocking circles and polygons merely using a compass and a ruler. Yet this underlying system is not just the instrument of design but also of its application to the material.

The geometric construction is for the most part very well documented and articulated in literature. Manuscripts from as early as the tenth century [8, 9] are known to have offered instructional drawings to the specialized audience of "scholar-practitioners, artisan-designers, builder-architects" [10]. The very same construction bridges geometry to the means of production. It primarily offers a convenient way to recreate the underlying system, merely using a compass and a ruler, onto a blank canvas such as a stone wall. However, literature and evidence, past and present, does not provide much information further this point. It is stipulated that the underlying structural geometry was inscribed on the stone using a sharp tool [11, 12]. In contemporary applications of traditional crafts, we see master craftsmen working with cardboard templates to chisel forms onto the stone, showing that the required skill was merely to copy the geometric construction. Studies on

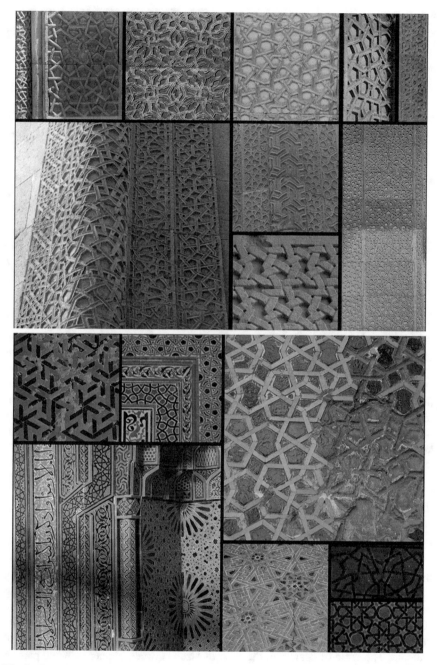

Fig. 1 Geometric patterns from thirteenth century Anatolian architecture, in stone, and in tile-mosaic

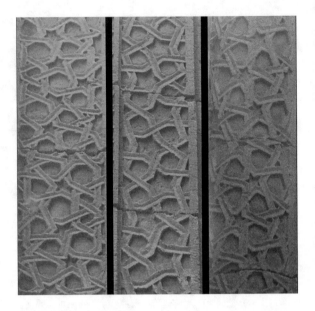

Fig. 2 Three very similar looking patterns from Kayseri Huand Hatun Complex employ very different underlying geometric systems

the traditional stone relief carving and tile mosaics of thirteenth century Anatolia neglect how the tools and techniques applied on the material surface factor into and determine the final shape. Representing the material properties and the construction knowledge behind these designs is a crucial issue waiting to be addressed in the field of architectural heritage preservation as it more and more involves digital processes and interfaces. Understanding what the specification of material aspects of shapes are for any given corpus is one of the first steps in deciphering past design knowledge from this perspective.

2.1 Specifications of Carved Material Applications

Capturing which properties should be incorporated to visual rules for these patterns has been the focal point of our research. One of the attributes to be included in the specifications is concerning the surface on which the design is applied. In a previous study [13], we mapped designs on different surfaces by means of drawing following two methods: one with a sheet template to transfer the lines, and the second with the compass (Fig. 3). Paper templates that wrap around developable curved surfaces, such as the outside of a cylinder, are useful when applying the patterns. The issue of scale, i.e. the ratio of sizes between elements of the pattern and the surface, is not trivial. When applying the design on the spherical surface, it was not possible to use the template, only the compass method worked. The size of the circles guiding the design were well within the perceived boundaries of the surface from the central viewpoint and it was possible to apply the pattern using a

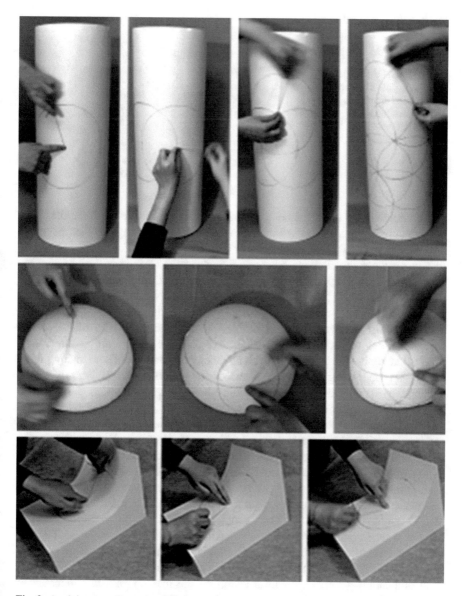

Fig. 3 Applying two-dimensional designs on curved surfaces [13]

rope compass. The exception in the third concave case, i.e. the curved squinch, was a real distortion due to varying geodesic distances. In the future, it will be important to apply these methods with variables of radius-surface size ratio and symmetry group of the pattern and to extend them to other materials such as brick, ceramic tiling, and wood.

Fig. 4 Left: A design drawn by the author based on an illustration from the Anonymous Compendium [10]. Right: Illustration drawn by the author based on a photograph of a brick pattern from Iran [15]

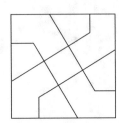

An alternative to using the compass and ruler is the tessellation of cut tiles [14]. Cutting uniform tiles and organizing them flat, similar to putting together the pieces of a puzzle, works for applications other than carving. Even for designs that seem to rely on these tessellations however, it is not very clear how designs that are often outlined in single line drawings are transformed into more complex relief designs of new parts. See Fig. 4 where the image on the left is representative of a common template presented to the artisans in historic documents that illustrate how to construct polygons using simple tools of geometry. The image on the right embodies this motif. The small square design elements are at the centers of tessellated pinwheels. Nevertheless, anyone, who attempts at constructing these patterns on paper using just a compass and ruler, will find out that giving a thickness, anticipating the off-axis distance that yields the right kind of pattern and fitting all pieces tightly together needs to be a calculated act.

For the patterns, we prepared a database that includes information on materials and techniques. In the documentation we have done, we determined shape rules for about 80 designs. Because there were many rules for each, we took the computation in intervals and reduced the number of steps. This still allowed us to comparatively analyze the two dimensional designs. For the three-dimensional features of the basic design element, we used cross-sections to introduce other rules and labels. We were able to see in comparison that exact geometric designs are differently made [16]. See Fig. 5 for a comparison of different making rules for the same design. The rule set on the left indicates a deeper and sculptural cut whereas the rule set on the right shows a shallower cut that results in a slightly etched surface.

Documenting patterns with information on the making and its parameters is unique. Additional to shape rules, we have identified rules to visualize the process of making, i.e. how these are physically produced. These rules distinguish the cross-section that the tools and craftsman's hands manipulate in parallel to the general visual design. Specifying carving tools parameters and material parameters, we reproduced on a CNC router old and new designs with the aim to study the parameters. In stone carving, the cross-section, the angle and size of the tool tip, the number of applications of the tool tip for one single line, the tool path sequencing, the speed of the chiseling were all relevant to shape emergence.

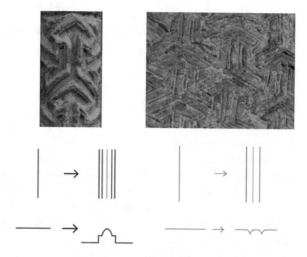

Fig. 5 The same geometric design is applied on stone in two different instances. Select rules show the detail of the difference in carving

2.2 Specifications of Assembled Material Applications

In addition to the technique of stone carving, we focused on another historical technique of pattern making by assembling pieces. In thirteenth century Anatolia, a new material, namely the tile mosaics, brought about new approaches to pattern making. Often, we see two instances of the same design, in two different materials, in one building. In tiles, the process requires pre-cutting the parts and assembling them. We studied the specifications of the process in terms of the predetermined shapes and assembly patterns, as well as sub-parts for hierarchic ordering of assembly.

Here the methods in which specifications are added to shape rules extend to the techniques of capturing and modeling the physical information from built heritage. For this type of pattern application, we started work on the giant dome of Konya Karatay Madrasah as part of a study on the material computation of tile mosaics [17]. We used photogrammetry to digitally capture the features of the surface with the further objective of putting the elaborate point cloud to use, to understand and represent how the pattern on the inside of the dome is constructed. The Karatay Madrasah dome interior consists of myriads of mosaic tiles assembled into a continuous geometric pattern that spans a wide band around the semi-sphere. In order to corroborate a renown historian's thesis on how it is assembled [18], our ongoing research aims to identify in the point cloud the larger parts and the smaller bits in the hierarchical ordering of the assembly and to capture the geometric structure that guides the construction.

We reverted to the similar but simpler case of mud-brick wall assemblies from the same period and geography for which we were able to develop two dimensional normal mapping based on the three dimensional photogrammetry data to make visible some of the alignments and figure outlines in the pixel-like mosaic assembly while running feature recognition algorithms to successfully detect the majority of the bricks [19].

3 Concluding Remarks

In order to overcome the problems in shape computation with not enough visual or material specifications, we need to augment the shape rules. In our research, we have so far identified specifications for the given corpus of stone ornaments and partly the tile mosaic patterns from thirteenth century Anatolian architecture. For the stone carving, material specifications are connected with the surface section, for a reference in curvature and size ratio, as well as the tool tip cross-section, angle, and size, the depth of subtraction, the number of applications of the tool tip for one single line, the tool path sequencing, and the speed of the chiseling. We show these in shape rules with labels to specify dimensions, parameters and their value ranges. Reference lines of the underlying geometric system also augment the rules. For the tile-mosaic and brick, the specifications extend to the hierarchy of parts and sub-parts in assembly, and the local and global geometries that form them. We continue seeking answers to how consequent representations of shapes in digital models may incorporate these material attributes, with the motivation to increase our understanding of past architectural and building traditions, and offering usable specifications and visual computation for the benefit of possible reconstructions.

Acknowledgements I would like to express my gratitude to TÜBİTAK BİDEB 2219 program for support during the academic year 2018–2019 I spent abroad at Carnegie Mellon University. Many thanks to my collaborators on the work cited here, particularly Sibel Tarı, Aslıhan Erkmen, Begüm Hamzaoğlu, Sibel Yasemin Özgan, and Demircan Taş. Unless noted otherwise, all visuals are from the database created as part of TÜBİTAK project 114K283 under the direction of the author in 2014–16.

References

1. Knight, T.: Shapes and other things. Nexus Netw. J. **17**, 963–980 (2015)
2. Adanova, V., Tari, S.: Beyond symmetry groups: A grouping study on Escher's Euclidean ornaments. Graphical Models. **83**, 15–27 (2016)
3. Keles, H.Y., Özkar, M., Tari, S.: Weighted Shapes for Embedding Perceived Wholes. Environment and Planning B: Planning and Design. **39**, 360–375 (2012)
4. Özkar, M.: Visual schemas: pragmatics of design learning in foundations studios. Nexus Netw. J. **13**, 113–130 (2011)
5. Stiny, G.: Shape: Talking about Seeing and Doing. The MIT Press, Cambridge, MA (2006)
6. Hertz, L.: Restoration of Notre Dame May Be Part of Professor Andrew Tallon's Legacy. Stories (2019) Available via Vassar University Website. https://stories.vassar.edu/2019/190417-notre-dame-andrew-tallon.html. Cited 1 July 2019

7. Özkar, M.: Repeating Circles, Changing Stars: Learning from the medieval art of visual computation. In Lee, N. (ed.) Digital Da Vinci: Computers in the Arts and Sciences, pp. 49–64. Springer (2014)
8. Özdural, A. Mathematicians, and "Conversazioni" with artisans. J. of the Soc. of Architectural Historians. **54**, 54–71 (1995)
9. Hogendijk, J. Mathematics and Geometric Ornament in the Medieval Islamic World. Newsletter of the European Mathematical Society. **86**, 37–43 (2012)
10. Necipoglu, G.: 1 — Ornamental Geometries: An Anonymous Persian Compendium at the Intersection of the Visual Arts and Mathematical Sciences. In: Necipoglu, G. (ed.) The Arts of Ornamental Geometry, pp. 11–78. Brill, Leiden, The Netherlands (2017)
11. Bakirer, Ö.: . The Story of Three Graffiti. Muqarnas. **16**, 42–69 (1999)
12. McClary, R.P.: Craftsmen in Medieval Anatolia: Methods and Mobility. In: Blessing, P., Gashgorian, R. (eds.) Architecture and Landscape in Medieval Anatolia, 1100–1500, pp. 27–58. Edinburgh University Press, Edinburgh (2017)
13. Hamzaoğlu, B., Özkar, M.: Geometric Patterns as Material Things: The Making of Seljuk Patterns on Curved Surfaces. In Bridges: Mathematics, Music, Art, Architecture, Education, Culture, pp. 331–336. Jyvaskylla, Finland (2016)
14. Lu, P.J. and Steinhardt, P.J. Decagonal and Quasicrystalline Tilings in Medieval Islamic Architecture. Science. **315**, 1106–1110 (2007)
15. Chorbachi, W.K. (1989). In the Tower of Babel: Beyond Symmetry in Islamic design. Computers and Mathematics with Applications. **7**, 751–789 (1989)
16. Hamzaoğlu, B., Özkar, M.: Design and Making: A Contemporary Look at Seljuk Geometric Patterns. In: (eds.) MSTAS 2016 — X. Computational Design in Architecture National Symposium, pp. 58–69. Istanbul Bilgi University, Istanbul, Turkey (2016)
17. Özgan, Y.S., Özkar, M.: A Reading of the Effects of Material and Craft Techniques on the Application of Geometric Patterns with a Focus on the Tile-Mosaic. In: Karakul, Ö., Dalkiran, A. (eds.) CRAFTARCH'18 International Art Craft Space Congress, Revitalizing Art Craft Space Relations: Proceedings Book, pp. 27–43. Eğitim Yayinevi, Konya, Turkey (2018)
18. Mülayim, S. Konya Karatay Medresesi'nin Ana Kubbe Geometrik Bezemesi. Sanat Tarihi Yilligi. **11**, 111–132 (1981)
19. Taş, D., Özkar, M.: Embedded Shape Matching in Photogrammetry Data for Modeling Making Knowledge. In: Çakici, N., Ezel, M., Öner, D., Baran, E. (eds.) MSTAS 2019 — Intersections Overlaps (XIII. Computational Design in Architecture National Symposium) pp. 313–326. Kocaeli University, Kocaeli, Turkey (2019)

Part VIII
Math Education

Being Research-Based and Research-Minded in Helping K-12 Mathematics Education (Survey)

Adem Ekmekci and Anne Papakonstantinou

1 Why Conduct Education Research?

Is it only the responsibility of colleges of education to do mathematics education research? Why is there such a disconnection between K-12 mathematics and higher education mathematics? What can mathematics faculty, hard-core mathematics research faculty do to build a bridge between the two? Are outreach activities that departments of mathematics undertake such as math circles, camps, or summer programs for students and teachers enough to establish a substantial bridge between K-12 and higher education mathematics? These questions are essential to seek the true meaning of mathematics and how it resonates and is understood and used in the community.

There is no doubt about why both federal and non-profit research organizations call for interdisciplinary work including the collaboration among departments of education and mathematics [6, 32]. Improving teacher education is between many facets of this collaboration and "teacher education must become a central focus of the entire institution, not just of schools or departments of education" [32]. Mathematics faculty can engage in mathematics education research with several goals in mind including: (a) to inform practice (critical especially in the context of very diverse and high-poverty urban schools and school districts); (b) to generate knowledge (for the sake of doing science); and (c) to develop proof of concept for future grant proposals benefiting both higher education and K-12 education. Besides, research products will help faculty advancement in several ways including contribution to their tenure requirements.

A. Ekmekci (✉) · A. Papakonstantinou
Rice University, Houston, TX, USA
e-mail: ae16@rice.edu; apapa@rice.edu

© The Author(s) and the Association for Women in Mathematics 2020
B. Acu et al. (eds.), *Advances in Mathematical Sciences*, Association for
Women in Mathematics Series 21, https://doi.org/10.1007/978-3-030-42687-3_23

There are many efforts to bridge the gap between research conducted by higher education institutions and the practice that takes place in public schools [5, 37]. However, mathematics faculty in higher education still have limited connection with elementary and secondary education practice. Through strong connections and collaborations between the research world and K-12 world, many issues facing K-12 education can be acted on and resolved [5, 32]. Mathematics faculty at the higher education can help identify issues and improve public education through studying these.

2 Where to Start? Important Theories That can Guide New Research Projects

The main areas of research that RUSMP conducts with robust theoretical frameworks to inform and improve pre-college mathematics education are as follows: professional development of teachers, teacher quality, teachers' motivational beliefs and technological pedagogical content knowledge, and student cognition and motivation. We provide a brief overview of some of these sound theories for higher education mathematics faculty so that they can build their research ideas and efforts on these theories as well or get inspired by them to embark on new research projects.

2.1 Professional Development

Core features of effective professional development are: (a) rigorous content focus, (b) active learning, (c) collaboration, (d) models of effective teaching practices, (e) coaching and expert support, (f) frequent feedback and reflection, and (g) long-term duration [14, 15]. RUSMP offers teacher professional development (PD) based on these core features and research indicates that these PD programs help teachers improve their content knowledge, pedagogical content knowledge, technological pedagogical content knowledge, self-efficacy beliefs, epistemic beliefs, beliefs about reform-based teaching practices, and beliefs about equity and diversity issues (e.g., [9, 11, 16, 18]).

2.2 Pedagogical Content Knowledge

Among the theories utilized in RUSMP research and professional development of teachers are the two major theories of mathematical knowledge for teaching (MKT; [21]) and technological pedagogical content knowledge (TPACK; [30]).

Both theories and related research have been primary built on Shulman's [36] work about pedagogical content knowledge (PCK).

The MKT framework is one of the most promising theories addressing the enduring question of what kind of knowledge is needed to teach mathematics effectively, and has also promoted studies of the effects of MKT on student learning and achievement [22]. Notably, recent studies at the elementary school level have found a significant positive association between MKT and student performance and between MKT and mathematical quality of instruction [21].

It should be noted that MKT is different than the pure mathematical knowledge (subject matter-knowledge) mathematicians or other professionals such as engineers use to perform their jobs. MKT transcends the pure content knowledge and includes knowledge about students' ideas, knowledge, and conceptual understanding of material. The MKT model is comprised of subject-matter knowledge and pedagogical content knowledge (PCK) in mathematics as depicted in Fig. 1.

More specifically, the MKT model describes six components. The first three are subdomains of "pure" content, or subject-matter knowledge [3]. The first, common content knowledge (CCK), is defined as general knowledge of mathematics that most educated people including teachers acquire. The second one is specialized content knowledge (SCK), which is mathematical knowledge that is unique to, and essential for, teaching mathematics. The third one, horizon content knowledge, is the knowledge about the next level of mathematics. Being familiar about what comes in the analysis topics after the multivariate calculus is an example of this. The last three

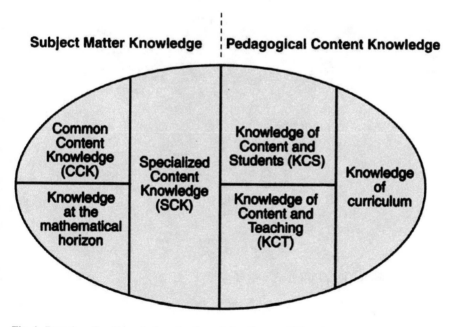

Fig. 1 Domains of mathematical content knowledge (Source: Hill et al.[21, p. 377])

components are subdomains of pedagogical content knowledge (PCK)—knowledge that combines content knowledge with student knowledge (KCS), knowledge that allows for the combination of content knowledge with teaching knowledge (KCT), and knowledge of content and curriculum and standards.

Since the arrival of accessible technology, many scholars have realized the necessity to include technology as part of the existing PCK that evolved into TPACK [2, 30, 33]. Being technology savvy does not directly translate into effective use of technology for teaching and learning [30]. Today, researchers and teacher educators unanimously agree upon on the importance of integrating technology into teacher preparation and professional development programs with several expected outcomes. Examples include effective utilization of technology for teaching particular topics; knowledge of students' understanding, thinking, and learning with technology in a particular subject; and knowledge of curriculum materials that integrate technology with learning in the subject area (see [33]). As a distinct form of knowledge, TPACK has been considered as complex, multi-faceted, integrative, and/or transformative [2, 7, 30]. Cox and Graham [12] define TPACK as "a teacher's knowledge of how to coordinate the use of subject-specific activities or topic-specific activities with topic-specific representations using emerging technologies to facilitate student learning" (p. 64) (Fig. 2).

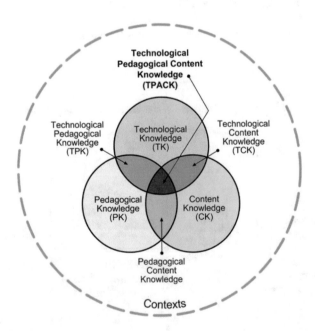

Fig. 2 TPACK, [24] image by http://tpack.org

2.3   Teacher Quality

Another important area that RUSMP's research and professional development focus on is the teacher quality concept. The framework for teacher quality [19] provides the most comprehensive framework to date based on a review and synthesis of the recent research regarding the impact teachers have on student achievement-related outcomes. The framework is comprised of three strands that are distinct but interrelated: inputs, processes, and outcomes. Inputs focus on two different but related ways of looking at teacher quality: teacher qualifications and teacher characteristics. Used as proxies for teacher quality, teacher qualifications include teachers' degrees, coursework, and grades in higher education as well as teacher preparation routes, certification types, years of experience, and continuing education such as internships, induction, coaching support, and professional development [13, 19, 23, 31]. The framework also conceptualizes teacher quality as encompassing soft attributes (teacher characteristics) such as subjective judgements, organization skills, critical thinking skills, and attitudes and beliefs (e.g., self-efficacy, epistemic beliefs, and beliefs about teaching and learning [31, 34]). The processes strand of the teacher quality framework focuses on factors related to teacher practices—i.e., what teachers actually enact in the classroom including instructional and classroom management practices. Processes also include instructional practices such as the emphasis placed on particular topics, questioning strategies, teacher interactions with students and with other colleagues outside of the classroom, school contextual interactions, and planning [19] (Fig. 3).

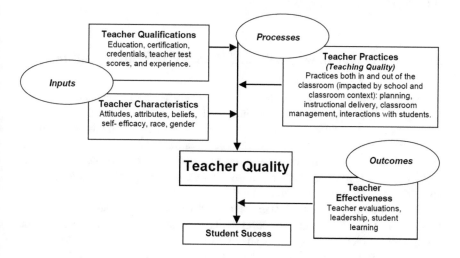

Fig. 3 Graphic representation of a framework for teacher quality. Adapted from Goe [19]

2.4   Social Cognitive Career Theory

RUSMP research also includes students' contemplation about their future (both college and career related outcomes). As an extension and application of social cognitive theory [4] to career choice, Lent et al. [28] social cognitive career theory (SCCT) provides the theoretical grounds for this study. SCCT posits that one's career choice is influenced by the beliefs that the individual develops and refines through the complex interaction between the individual, environment, and behavior [28, 39]. According to SCCT, the most important factors influencing career decisions relate to student motivation (e.g., task value, self-efficacy, interest, outcome expectations; see [38]). These psychological variables are considered as the mediators that connect other personal and contextual factors to future career choice and decisions [27, 39]. Empirical research has shown that students with higher mathematics and/or science self-efficacy and outcome expectations for engaging in mathematics and science are more likely to persist and be successful in these areas (e.g., [1, 25]).

In addition to personal motivation, the SCCT framework recognizes several contextual influences (e.g., supports and barriers at school and at home) that mold individuals' career aspirations and choices [26, 29, 39]. Specifically, several groups, including parents, peers, and teachers have socializing influences on students' academic and career-related outcomes.

RUSMP's recent research efforts include integrating teacher quality as a contextual factor (the most important contextual factor for student outcomes [20]) into social cognitive career theory to have a more comprehensive look into the factors affecting student outcomes (see Fig. 4).

3   A Brief Overview of Selected RUSMP Research Findings

In a study of predictive value of teachers' school-work environments on their self-efficacy and intrinsic value for teaching, Corkin et al. [10] found that principals' autonomy support positively predicted teachers' self-efficacy and intrinsic value for teaching beyond years of teaching experience, mathematics background, and grade level taught. Moreover, the negative effects of school-work environments dominated by high-stakes testing on teachers' motivation for teaching were moderated by the level of autonomy support provided by the school principal.

Ekmekci et al. [18] investigated the predictive value of teachers' beliefs (e.g., self-efficacy) and MKT on their level of TPACK and discovered that standards-based mathematics teaching beliefs positively predict mathematics teachers' level of TPACK for all teachers. Moreover, having a college/graduate mathematics degree is more predictive of TPACK for K-5 and middle school teachers while MKT is more predictive of TPACK for high school teachers. In addition, elementary teachers' mathematics self-concept and pedagogical preparedness and middle school

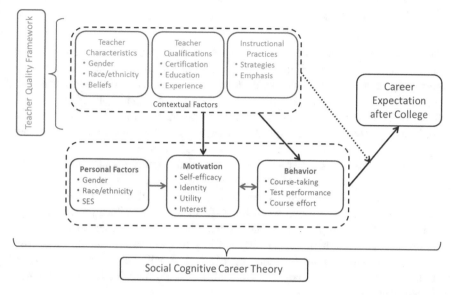

Fig. 4 The melding of teacher quality and social cognitive career theories to predict student outcomes [19, 28]

teachers' mathematics teaching interest are significantly related to their level of TPACK. In an earlier study, Ekmekci et al. [16] also found that teachers significantly improved their educational beliefs about mathematics after RUSMP's signature summer campus professional development program. Moreover, Papakonstantinou et al. [35] found that RUSMP summer PD institute positively impacted lead teachers' understanding of math concepts, equitable instructional strategies, and collaborative leadership skills.

In another study, Corkin et al. [9] found that the mathematics-teaching experience was positively associated with teachers' self-efficacy for teaching mathematics, and the number of mathematics college courses teachers had taken moderated their change in self-efficacy beliefs through professional development. Findings also indicated that epistemic beliefs about mathematics, which became more availing through professional development, were the strongest predictor of their mathematical knowledge for teaching.

In studies of teachers' impact on student outcomes, Ekmekci et al. [17] found that teachers' MKT and teaching experience had a significant effect on students' mathematics achievement. In addition, Corkin and Ekmekci [8] found that the degree to which teachers emphasized the development of deeper conceptual understanding of mathematics was a predictor of students' mathematics achievement, identity, and self-efficacy whereas the degree to which teachers emphasized the utility of mathematics predicted students' beliefs about the utility of mathematics.

4 Closing Remarks

RUSMP's research in K-12 education has several implications for education and community stakeholders. Findings provide school and district administrators with guidance on the ways mathematics teachers can be supported and developed professionally. Moreover, teachers can also benefit from the implications of these findings for their personal professional development and for setting work-related goals at a personal level. Lastly, the findings also have implications for teacher educators including faculty from both colleges of education and sciences or from both departments of teacher education and mathematics about how to design effective teacher education and professional development programs.

In closing, important theories that this paper summarized offer themselves as great resources to lay ground work for K-12 mathematics education research. In addition, it is our hope that RUSMP's featured research presented herein will set examples for mathematics faculty and inspire them to embark on productive research endeavors in mathematics education that will benefit K-12 education community and that will have a reciprocal impact on their institutions of higher education as well.

Acknowledgements The studies presented herein are based, in part, on different projects, each partially funded or supported by different sources including Code.org, Spencer Foundation, Texas Higher Education Coordinating Board, The University of Texas at Austin STEM Center, and National Science Foundation.

References

1. Andersen, L., & Ward, T. J. (2014). Expectancy-value models for the STEM persistence plans of ninth-grade, high-ability Students: A comparison between black, Hispanic, and white students. *Science Education, 98*(2), 216–242.
2. Angeli, C., & Valanides, N. (2009). Epistemological and methodological issues for the conceptualization, development, and assessment of ICT–TPCK: Advances in technological pedagogical content knowledge (TPCK). *Computers & Education, 52*(1), 154–168.
3. Ball, D. L., Thames, M. H., & Phelps, G. (2008). Content knowledge for teaching: What makes it special? *Journal for Teacher Education, 59*(5), 389–407.
4. Bandura, A. (1986). Social foundations of thought and action: A social cognitive theory. Englewood Cliffs, NJ: Prentice Hall.
5. Cacciatore, K. L., & Sevian, H. (2011). An urgent call for academic chemists to engage in precollege science education. *Journal of Chemical Education, 88*(3), 248–250.
6. Carr, K. (2002). Building bridges and crossing borders: Using service learning to overcome cultural barriers to collaboration between science and education departments. *School Science and Mathematics, 102*(6), 285–298.
7. Chai, C. S., Koh, J. H. L., & Tsai, C. C. (2013). A review of technological pedagogical content knowledge. *Journal of Educational Technology & Society, 16*(2), 31–51.
8. Corkin, D., & Ekmekci, A. (2019). The impact of mathematics teachers on student learning and motivation. In A. Redmond-Sanogo & J. Cribbs (Eds.), *Proceedings of the 46th Annual Meeting of the Research Council on Math Learning*, (pp. 34–41). Charlotte, NC: RCML.

9. Corkin, D., Ekmekci, A., & Papakonstantinou, A. (2015). Antecedents of teachers' educational beliefs about mathematics and mathematical knowledge for teaching among in-service teachers in high poverty urban schools. *Australian Journal of Teacher Education, 40*(9), 31–62. DOI: 10.14221/ajte.2015v40n9.3

10. Corkin, D., Ekmekci, A., & Parr, R. (2018). The effects of the school-work environment on mathematics teachers' motivation for teaching: A self-determination theoretical perspective. *Australian Journal of Teacher Education, 43*(6), 50–66. DOI: 10.14221/ajte.2018v43n6.4

11. Corkin, D., Ekmekci, A., White, C., & Fisher, A. (2016). Teachers' self-efficacy and knowledge for the integration of technology in mathematics instruction at urban schools. In K. V. Adolphson & T. M. Olson (Eds.), *Proceedings of the 43rd Annual Meeting of the Research Council on Mathematics Learning*, (pp. 101–108). Orlando, FL: RCML.

12. Cox, S., & Graham, C. R. (2009). Diagramming TPACK in practice: Using an elaborated model of the TPACK framework to analyze and depict teacher knowledge. *TechTrends: Linking Research & Practice to Improve Learning, 53*(5), 60–69.

13. Darling-Hammond, L. (2000). Teacher quality and student achievement. *Education policy Analysis Archives, 8*(1), 1–44.

14. Darling-Hammond, L., Hyler, M. E., & Gardner, M. (2017). *Effective teacher professional development*. Palo Alto, CA: Learning Policy Institute.

15. Desimone, L. M. (2009). Improving impact studies of teachers' professional development: Toward better conceptualizations and measures. *Educational Researcher, 38*(3), 181–199.

16. Ekmekci, A., Corkin, D., & Papakonstantinou, A. (2015a). The collective effects of teachers' educational beliefs and mathematical knowledge on students' mathematics achievement. In T. G. Bartell, K. N. Bieda, R. T. Putnam, K. Bradfield, & H. Dominguez, (Eds.), *Proceedings of the 37th PME-NA*, (pp. 884–887). East Lansing, MI: Michigan State Univ.

17. Ekmekci, A., Corkin, D., & Papakonstantinou, A. (2015b). The relationship between teacher related factors and mathematics teachers' educational beliefs about mathematics. In S. M. Che, & K. A. Adolphson (Eds.), *Proceedings of the 42nd Annual Meeting of the Research Council on Mathematics Learning*, (pp. 140–148). Las Vegas, NV.

18. Ekmekci, A., Papakonstantinou, A., Parr, R., & Shah, M. (2019). Knowledge, beliefs, and perceptions about the mathematics and mathematics teaching: How do they relate to teachers' technological pedagogical content knowledge? In M. L. Niess, H. Gillow-Wiles, & C. Angeli (Eds.), *Handbook of research on TPACK in the digital age*, (pp. 1–23). Hershey, PA: IGI Global. DOI: 10.4018/978-1-5225-7001-1

19. Goe, L. (2007). *The link between teacher quality and student outcomes: A research synthesis.* Washington, DC: National Comprehensive Center for Teacher Quality. Retrieved from http://eric.ed.gov/?id=ED521219

20. Hattie, J., Masters, D., & Birch, K. (2016). *Visible learning into action: International case studies of impact.* New York, NY: Routledge.

21. Hill, H. C., Ball, D. L., & Schilling, S. G. (2008). Unpacking pedagogical content knowledge: Conceptualizing and measuring teachers' topic-specific knowledge of students. *Journal for Research in Mathematics Education, 39*(4), 372–400.

22. Hill, H. C., Charalambous, C. Y., & Chin, M. J. (2018). Teacher characteristics and student learning in mathematics: A comprehensive assessment. *Educational Policy*, 1–32. DOI: 10.1177/0895904818755468 .

23. Ingersoll, R. (2007). *A comparative study of teacher preparation and qualifications in six nations* (CPRE RB-47). Philadelphia, PA: Consortium for Policy Research in Education. Retrieved from http://repository.upenn.edu/cpre_researchreports/47/

24. Koehler, M., & Mishra, P. (2009). What is technological pedagogical content knowledge (TPACK)? *Contemporary Issues in Technology and Teacher Education, 9*(1), 60–70.

25. Lee, S. W., & Min, S., Mamerow, G. P. (2015). Pygmalion in the classroom and the home: Expectation's role in the pipeline to STEMM. *Teachers College Record, 117*(9), 1–36.

26. Lent, R. W., & Brown, S. D. (1996). Social cognitive approach to career development: An overview. *The Career Development Quarterly, 44*(4), 310–321.

27. Lent, R. W., & Brown, S. D. (2006). On conceptualizing and assessing social cognitive constructs in career research: A measurement guide. *Journal of Career Assessment, 14*(1), 12–35.

28. Lent, R. W., Brown, S. D., & Hackett, G. (1994). Toward a unifying social cognitive theory of career and academic interest, choice, and performance. *Journal of Vocational Behavior, 45*(1), 79–122.

29. Maltese, A. V., & Tai, R. H. (2011). Pipeline persistence: Examining the association of educational experiences with earned degrees in STEM among US students. *Science Education, 95*(5), 877–907.

30. Mishra, P., & Koehler, M. J. (2006). Technological pedagogical content knowledge: A framework for teacher knowledge. *Teachers College Record, 108*(6), 1017–1054.

31. National Council on Teacher Quality (2004). *Increasing the odds: How good policies can yield better teachers*. Washington, DC: Author.

32. National Research Council. (2000). *Educating teachers of science, mathematics, and technology: New practices for the new millennium*. Washington, DC: The National Academies Press.

33. Niess, M. L. (2005). Preparing teachers to teach science and mathematics with technology: Developing a technology pedagogical content knowledge. *Teaching and Teacher Education, 21*(5), 509–523.

34. Pajares, M. F. (1992). Teachers' beliefs and educational research: Cleaning up a messy construct. *Review of Educational Research, 62*(3), 307–332.

35. Papakonstantinou, A., Ekmekci, A., & Parr, R. (2014). Mathematics teacher leadership: A Sustainable approach to improve mathematics education. In S. Oesterle, C. Nicol, P. Liljedahl, & D. Allan (Eds.), *Proceedings of the 38th Conference of the PME and the 36th Conference of the PME-NA*, (Vol. 6), (p. 379). Vancouver, Canada: PME.

36. Shulman, L. (1986). Those who understand: Knowledge growth in teaching. *Educational Researcher, 15*(2), 4–14.

37. Skerrett, A., & Sevian, H. (2010). Identity and biography as mediators of science and mathematics faculty's involvement in K-12 service. *Cultural Studies of Science Education, 5*(3), 743–766.

38. Wigfield, A. & Eccles, J. S. (2002). The development of competence beliefs, expectancies for success, and achievement values from childhood through adolescence. *Development of Achievement Motivation, 91*(120), 91–120.

39. Yu, S. L., Corkin, D. M., & Martin, J. P. (2016). STEM motivation and persistence among underrepresented minority students: A social cognitive perspective. In J. T. DeCuir-Gunby & P. A. Schutz (Eds.), *Race and ethnicity in the study of motivation in education* (pp. 67–81). New York, NY: Taylor & Francis.

The Rice University School Mathematics Project: Supporting Excellence in K-16 Mathematics Since 1987 (Survey)

Anne Papakonstantinou and Adem Ekmekci

1 Introduction

This chapter describes the wide-ranging contributions of the Rice University School Mathematics Project (RUSMP) from its inception to the present. The efforts of RUSMP can serve as a model for outreach efforts at other institutions that wish to create bridges between pre-college and university mathematics communities.

Since 1987, over 10,000 teachers and teacher leaders from over 100 districts and private and charter schools and over 12,000 K-12 students have benefited from RUSMP programs. RUSMP has evolved into an important regional STEM center and continues to grow and impact the K-16 educational community. Through its wide-ranging work, RUSMP is valued as a vital partner in meeting the educational needs of the greater Houston community and beyond.

2 Mission and Goals

RUSMP's mission is to create a better understanding of the nature, beauty, and importance of mathematics and to promote effective teaching of mathematics. The mission has expanded to include supporting science, technology, engineering, and the arts as they relate to mathematics.

RUSMP's major goal is to increase the content and pedagogical knowledge of K-12 STEM teachers and support them in implementing more effective programs. In order to achieve this goal, RUSMP

A. Papakonstantinou (✉) · A. Ekmekci
Rice University, Houston, TX, USA
e-mail: apapa@rice.edu; ae16@rice.edu

© The Author(s) and the Association for Women in Mathematics 2020
B. Acu et al. (eds.), *Advances in Mathematical Sciences*, Association for
Women in Mathematics Series 21, https://doi.org/10.1007/978-3-030-42687-3_24

- increases the STEM knowledge of Houston-area teachers,
- promotes and models effective teaching and assessment of mathematics as it relates to the other STEM disciplines and the arts with active student involvement in the learning process,
- encourages the appropriate use of instructional tools in the teaching of STEM disciplines,
- provides strategies to actively engage struggling students, English language learners, and underrepresented minorities in learning the STEM disciplines so that all students have the best possible instruction in these fields,
- provides a forum for communication and collaboration between and among teachers, university faculty, and the community,
- develops teachers' leadership capacity,
- supports instructional leaders in designing effective programs in STEM disciplines and better supporting their teachers,
- creates innovative curricular materials for STEM disciplines,
- assists developers of STEM programs and curricula,
- conducts research on important aspects of STEM education, and
- promotes research-based teaching and learning of STEM disciplines.

3   Programs for Educators

RUSMP has developed an extensive array of programs, courses, and interventions available to teachers, teacher leaders, and administrators. These include long-term, intensive summer programs; after-school academic-year courses; personalized professional development for schools which may include workshops along with classroom support; seminars for teachers and leaders; and opportunities for networking across schools and districts. RUSMP continues to create and share resources that support mathematics instruction which can be accessed from RUSMP's award-winning web site (http://rusmp.rice.edu). RUSMP programs focus on teachers' competence in content-knowledge and pedagogical skills, while integrating and promoting the need for teachers to care about all their students, especially students from populations traditionally underrepresented in STEM. RUSMP programs provide a model for teaching STEM from a problem-solving approach. Improving teachers' understanding of the concepts developed in RUSMP programs increases student understanding of the concepts. Students with a sound understanding of these concepts are more likely to pursue the study of more advanced STEM courses.

RUSMP hosts annual Fall and Spring Networking Conferences for the RUSMP network of teachers, administrators, and others from the educational community. Distinguished educators and scientists share their current research and interests, and RUSMP alumni and members of the RUSMP instructional team demonstrate exemplary mathematics lessons, new resources, and successful teaching strategies.

RUSMP expanded its efforts to include computer science. As a Code.org Regional Partner, RUSMP is the local hub for computer science (CS) professional learning for K-12 teachers to expand pre-college students' access to CS. RUSMP is dedicated to expanding access to CS and, in particular, to increasing the participation of women and underrepresented ethnic minority students in CS.

In addition, to further develop educational leaders, RUSMP is leading an NSF Robert Noyce Teaching Fellowship grant which is developing 15 exceptional secondary mathematics teachers in the Houston Independent School District into National Science Foundation Robert Noyce Master Teaching Fellows. The program is creating school-based leaders in mathematics deeply grounded in sound mathematical content and research-based pedagogical, leadership, adult education, and mathematics advocacy skills. Their leadership roles include work with preservice teachers from Rice University, teaching and mentoring inservice teachers in the greater Houston area through RUSMP professional development programs, and teaching in and managing student math camps and programs during the school year and summer through RUSMP. This work is supported by the National Science Foundation under Grant No. 1556006.

4 Opportunities for Students and Support for Parents

RUSMP offers a wide variety of camps and programs for students during both the academic year and the summer. These activity-filled camps for students engage students in fun, hands-on mathematics investigations and games with connections to science, technology, engineering, and the visual arts typically not found in classrooms. Students leave the camps more energized and enthusiastic about mathematics and the other STEAM (Science, Technology, Engineering, Arts, and Mathematics) disciplines.

RUSMP also offers summer programs for students who want to enrich their mathematics backgrounds, prepare for their next math courses, and explore mathematics topics that are not typically emphasized during the school year.

RUSMP is embarking on new STEM programs. Its newest program focusing on the oil and gas industry provided students with valuable insight into the oil and gas industry to motivate them to pursue further STEM courses in school and to consider careers in the oil and gas industry.

In addition, RUSMP supports parents by providing information about school and district mathematics programs, opportunities to engage with the school community through mathematics, and resources for supporting mathematics instruction at home including resources on the award-winning RUSMP web site (http://rusmp.rice.edu).

5 Support to Schools, School Districts, and the Community

RUSMP's excellent reputation as a change agent and a trusted partner has resulted in RUSMP receiving numerous requests from schools and school districts for support. RUSMP evaluates the effectiveness of mathematics programs, makes recommendations for improvement, and provides targeted professional development, school-based support, coaching, and mentoring to improve student achievement. RUSMP also conducts mathematics curriculum audits for school districts and private schools and collaborates to revise their curricula so that their curricula are vertically aligned and support current state and national standards. In addition, school-based programs for teachers and students can be developed and supported by RUSMP. These programs provide imbedded professional development for teachers while providing students engaging, activity-based instruction.

6 Support for Rice University and Other Universities

RUSMP provides support to departments and faculty across Rice University as well as support to other universities. RUSMP's multi-faceted university-level support includes:

- assisting with preparation of grant proposals, in particular NSF CAREER grants,
- collaborating with faculty in joint grant submissions,
- collaborating in research initiatives,
- serving as external evaluators for faculty programs and projects,
- conceptualizing, planning, recruiting for, promoting, and assisting with faculty and department programs, in particular for K-12 broader impact activities,
- providing pedagogical feedback to graduate students and post-docs,
- providing teaching resources for university courses,
- assisting graduate students and post-docs in writing teaching statements and job searches,
- assisting university students in the development of resumes and cover letters and in job searches,
- providing outreach opportunities for university students,
- providing practicum experiences and guidance in education for university students,
- writing letters of recommendation and letters of support for university students and faculty,
- serving on thesis and dissertation committees,
- participating in tenure and promotion reviews, and
- conducting mathematics tours of the Rice campus.

From its inception, RUSMP has naturally had a close working relationship with the Rice Mathematics Department. RUSMP currently collaborates with the Mathematics Department to offer the Rice University Math Circle which meets on the Rice University campus during the school year. Each session includes a lecture on an easy-to-understand math topic followed by activities related to the lecture. The goal is to share why mathematicians find math so interesting and how it is a part of the everyday lives of people.

7 Research and Evaluation Efforts

RUSMP research and evaluation efforts provide evidence-based insights and strategies to improve mathematics instruction, student learning, student achievement, and the professional development of K-12 mathematics teachers, teacher leaders, and administrators. RUSMP contributes to the growing body of research on teaching and learning and professional development in K-12 mathematics education. RUSMP's research team evaluates the impact of RUSMP on teachers, teacher leaders, students, administrators, and schools. RUSMP's research and evaluation clarify the role RUSMP plays in educational reform at classroom, campus, district, state, and national levels. RUSMP also conducts research and program evaluation for other organizations.

8 Conclusion

RUSMP can serve as an exemplar for faculty who wish to deepen ties with the pre-college mathematics community while providing meaningful educational and pedagogical support to their colleagues at the university level. RUSMP Directors are more than willing to serve as mentors and guides for those who wish to take the first steps toward creating similar meaningful outreach endeavors.

Looking to the future, in addition to its work with its current collaborators, RUSMP will continue develop educational programs and conduct research and will seek to build new relationships to positively impact K-16 education.

9 Selected RUSMP Publications

9.1 The Early Years

Austin, J. D. (1989). Technology and transactions in mathematics education. In T.J. Cooney (Ed.), American perspectives in the Sixth International Congress on Mathematical Education (pp. 25–26). Reston, VA: National Council of Teachers of Mathematics.

Wells, R. O. (1989). A critique of ICME-6. In T.J. Cooney (Ed.), American perspectives in the Sixth International Congress on Mathematical Education (p. 56). Reston, VA: National Council of Teachers of Mathematics.

Austin, J. D., Herbert, E., & Wells, R. O. (1990). Master teachers as teacher role models. Mathematicians and Education Reform, 1, 189–196.

Papakonstantinou, A., Berger, S., Wells, R. O., & Austin, J. D. (1996, November/December). The Marshall Plan: Rice University Mathematics Affiliates Program. Schools in the Middle: Theory into Practice, 39–46.

Papakonstantinou, A., & Casey, J. (1998). Professional growth opportunities for grades 7–12 mathematics teachers: An urban perspective. In Proceedings from the Moving On Mathematically in Ohio Conference, Ohio University: Athens, OH.

Schweingruber, H. A. (1999). The Rice University School Mathematics Project. The Mathematics Teacher, 92(7), 644.

Nease, A. (1999). Do motives matter: An examination of reasons for attending training and their influence on training effectiveness (Unpublished doctoral dissertation). Rice University, Houston, TX.

Killion, J. (1999). The Rice University School Mathematics Project. In What works in the middle: Results-based staff development (pp. 94–97). Oxford, OH: National Staff Development Council.

Eaves, E. (2000). Progress report and next steps: North District - Houston Independent School District. In F. Curcio (Ed.), Proceedings of Diversity, Equity, and Standards: An Urban Agenda in Mathematics Education (pp. 97–98). New York University, New York, NY.

Killion, J. (2002). Rice University School Mathematics Project Summer Campus Program. In What works in the elementary school: Results-based staff development (pp. 120–123). Oxford, OH: National Staff Development Council.

Killion, J. (2002). Algebra Initiative. In What works in the high school: Results-based staff development (pp. 80–83). Oxford, OH: National Staff Development Council.

Killion, J. (2002). The Rice University School Mathematics Project Summer Campus Program. In What works in the high school: Results-based staff development (pp. 84–90). Oxford, OH: National Staff Development Council.

9.2 The Middle Years

Cannon, R. Parr, R., & Webb, A. (2003). Advanced mathematics educational support: Support, recommendations, and resources for facilitating collaboration, between higher education mathematics faculty and Texas public high schools. Austin, TX: The University of Texas, Charles A. Dana Center.

National Council of Supervisors of Mathematics. (2004). NSF announces third-year Math and Science Partnership grants and eases institutional cost-sharing requirements. NCSM Newsletter, 35(2), 28.

Parr, R., Papakonstantinou, A., Schweingruber, H. A., & Cruz, P. (2004). Professional development to support the NCTM Standards: Lessons from the Rice University School Mathematics Project's Summer Campus Program. National Council of Supervisors of Mathematics Journal of Mathematics Education Leadership, 7(1), 3–12.

Ward, R. (2005). Impact of mentoring on teacher efficacy. Academic Exchange Quarterly 9(4), 148–154.

Ofiesh, N., Rohas, C., & Ward, R. (2006). Applying principles of universal design to the assessment of student learning. Journal of Postsecondary Education and Disability, 19(2), 173–181.

Anhalt, C., Ward, R., & Vinson, K. (2006). Teacher candidates' growth in designing mathematical tasks as exhibited in their lesson planning. Teacher Educator, 41(3), 172–186.

Science Daily (2007, July 18). Student results show benefits of math and science partnerships.

Hill, A., McCoy, A., Papakonstantinou, A., Parr, R., & Sack, J. (2007). Strengthening mathematics teachers' pedagogical content knowledge through collaborative investigations in combinatorics. In T. Lamberg & L. Wiest (Eds.), Proceedings of the 29th annual meeting of the North American Chapter of the International Group for the Psychology of Mathematics Education (pp. 887–889). Stateline (Lake Tahoe), NV: University of Nevada.

Sack, J. J. (2008). Commonplace intersections within a high school mathematics leadership institute. Journal of Teacher Education, 59(2), 189–199.

Sack, J. J., & Vazquez, I. (2008). Three-dimensional visualization: Children's non-conventional verbal representations. In O. Figueras, J.L. Cortina, S. Alatorre, T. Rojano, & A. Sepulveda (Eds.), Proceedings of the Joint Meeting of PME 32 and PME-NA XXX. Mexico: Cinvestav-UMSNH, 4, 217–224.

Sack, J., & Kamau, N. (2009). The impact of the lead teacher professional learning community within the Rice University Mathematics Leadership Institute. The Journal of Mathematics and Science: Collaborative Explorations, 11, 141–162.

Bergin, K., & Hamos, J. (2009). Math and Science Partnership (MSP) Program: A research and development effort.

Martinez, D., Desiderio, M., & Papakonstantinou, A. (2010). Teaching: A job or a profession? The perceptions of educators. The Educational Forum, 74(4), 289–296.

Parr, R., & Kamau, N. (2013). Teachers' beliefs regarding reform standards, equity and self-efficacy for teaching. In M.V., Martinez & A.C. Superfine (Eds.) Proceedings of the Thirty-fifth Annual Meeting of the North American Chapter of the International Group for the Psychology of Mathematics Education. (p. 952). Chicago, IL: University of Illinois-Chicago.

Cruz, P., Kamau, N., Papakonstantinou, A., Parr, R., Troutman, S., Ward, R., & White, C. (2013). Effective professional development: Defining the vital role of the master teacher. National Council of Supervisors of Mathematics (NCSM) Journal of Mathematics Leadership, 14(2), 48–60.

Copur-Gencturk, Y., & Lubienski, S. T. (2013). Measuring mathematics knowledge for teaching: A longitudinal study using two measures. Journal of Mathematics Teacher Education, 16(3), 211–236.

9.3   Recent Years

Ekmekci, A., Corkin, D., & Papakonstantinou, A. (2015). The relationship between teacher related factors and mathematics teachers' educational beliefs about mathematics. In S.M. Che & K.A. Adolphson (Eds.), Proceedings of the 42nd Annual Meeting of the Research Council on Mathematics Learning, (pp. 140–148). Las Vegas, NV.

Parr, R. (2015).Investigating the effects of sustained professional development on campus and district mathematics programs: An analysis of the Rice University Mathematics Leadership Institute. (Unpublished doctoral dissertation). University of Houston, Houston, TX.

Troutman, S. (2015). Examining the evolution of a university summer campus program for mathematics teachers. (Unpublished doctoral dissertation). University of Houston, Houston, TX.

Ekmekci, A., Corkin, D., & Papakonstantinou, A. (2015). The collective effects of teachers' educational beliefs and mathematical knowledge on students' mathematics achievement. In T. G. Bartell, K.N. Bieda, R.T. Putnam, K. Bradfield, & H. Dominguez (Eds.), Proceedings of the 37th annual meeting of the North American Chapter of the International Group for the Psychology of Mathematics Education, (pp. 884–887). East Lansing, MI: Michigan State University.

Yu, S.L., Corkin, D. M., & Trenor, J. M. (2016). STEM motivation and persistence among underrepresented minority students: A social cognitive perspective. In J. DeCuir-Gunby & P. A. Schutz (Eds.), Race and ethnicity in the study of motivation in education (pp. 67–81). NY: Routledge.

Copur-Gencturk, Y., & Papakonstantinou, A. (2015). Sustainable changes in teacher practices: a longitudinal analysis of the classroom practices of high school mathematics teachers. Journal of Mathematics Teacher Education, 1–20.

Corkin, D., Ekmekci, A., & Papakonstantinou, A. (2015). Antecedents of teachers' educational beliefs about mathematics and mathematical knowledge for teaching among in-service teachers in high poverty urban schools. The Australian Journal of Teacher Education, 40(9), 31–62.

Corkin, D., Ekmekci, A. & Fan, W. (2016). The significance of teachers' mathematical knowledge for teaching and their math background on students' math achievement. Research Brief for the Houston Independent School District, 4(6), 1–6. Houston, TX: Houston Education Research Consortium, Rice Kinder Institute for Urban Research.

Corkin, D. M., Ekmekci, A., White, C., & Fisher, A. (2016). Teachers' self-efficacy and knowledge for the integration of technology in mathematics instruction at urban schools. In K. V. Adolphson & T. M. Olson (Eds.), Proceedings of the 43rd Annual Meeting of the Research Council on Mathematics Learning (pp. 101–108). Orlando, FL.

Glynn, J. (2017, August). Opening doors: How selective colleges and universities are expanding access for high-achieving, low-income students. Landsdowne VA: Jack Kent Cooke Foundation.

Corkin, D., Ekmekci, A., & Coleman, S. (2017). Barriers to implementation of constructivist teaching in a high-poverty urban school district. In T. A. Olson & L. Venenciano (Eds.), Proceedings of the 44th Annual Meeting of the Research Council on Mathematics Learning (pp. 57–64). Fort Worth, TX.

Ekmekci, A., Parr, R., & Fisher, A. (2018). Results from a computer science teaching collaborative: Serving teachers with different needs through variety of pathways. In E. Langran & J. Borup (Eds.), Proceedings of Society for Information Technology & Teacher Education International Conference (ISBN 978-1-939797-32-2 , (pp. 2025–2030). Washington, DC: Association for the Advancement of Computing in Education. Available at https://www.learntechlib.org/primary/p/182806/

Corkin, D., Coleman, S., & Ekmekci, A. (2018). Implementation fidelity: Teachers' perceptions of barriers to constructivist teaching learned through professional development. The Urban Review: Issues and Ideas in Public Education, 1–34. https://doi.org/10.1007/s11256-018-0485-6

Ekmekci, A., Papakonstantinou, A., Parr, R., & Shah, M. (2019). Knowledge, beliefs, and perceptions about the mathematics and mathematics teaching: How do they relate to teachers' technological pedagogical content knowledge? In M. L. Niess, C. Angeli, & H. Gillow-Wiles (Eds.), Handbook of research on TPACK in the digital age, (pp. 1–23). Hershey PA: IGI Global. https://doi.org/10.4018/978-1-5225-7001-1.ch001

Corkin, D., & Ekmekci, A.(2019). The impact of mathematics teachers on student learning and motivation. In A. Redmond-Sanogo & J. Cribbs (Eds.), Proceedings of the 46th Annual Meeting of the Research Council on Mathematics Learning, (pp. 34–41). Charlotte, NC: Research Council on Mathematics Learning.

Printed in the United States
by Baker & Taylor Publisher Services